Forest Pathology and Plant Health

Forest Pathology and Plant Health

Edited by Kelly Davidson

SYRAWOOD
PUBLISHING HOUSE
New York

Published by Syrawood Publishing House,
750 Third Avenue, 9th Floor,
New York, NY 10017, USA
www.syrawoodpublishinghouse.com

Forest Pathology and Plant Health
Edited by Kelly Davidson

International Standard Book Number: 978-1-64740-434-5 (Hardback)

Cataloging-in-publication Data

Forest pathology and plant health / edited by Kelly Davidson.
p. cm.
Includes bibliographical references and index.
ISBN 978-1-64740-434-5
1. Forest health. 2. Trees--Diseases and pests. 3. Forest plants--Diseases and pests. I. Davidson, Kelly.
SB765 .F67 2023
634.9--dc23

TABLE OF CONTENTS

PREFACE

Healthy plants and trees are critical for sustainable and lucrative crop production as well as for ensuring the supply of fiber, food and fuel for a country. Ensuring plant and tree health involves protecting them from plant pathogens and pests using scientific and regulatory frameworks. Forest pathology refers to the study of diseases of woody plants across tree nurseries, natural forests, urban environments and planted landscapes. It is a branch of forestry that serves the needs of the environment and the general public by using scientific principles for the control and prevention of tree diseases. It is important to maintain effective plant health for increasing sustainability, productivity, resilience and profitability of agrifood systems. This book aims to shed light on some of the unexplored aspects of forest pathology and plant health, and the recent researches in this area of study. It will serve as a valuable source of reference for graduate and postgraduate students.

Various studies have approached the subject by analyzing it with a single perspective, but the present book provides diverse methodologies and techniques to address this field. This book contains theories and applications needed for understanding the subject from different perspectives. The aim is to keep the readers informed about the progresses in the field; therefore, the contributions were carefully examined to compile novel researches by specialists from across the globe.

Indeed, the job of the editor is the most crucial and challenging in compiling all chapters into a single book. In the end, I would extend my sincere thanks to the chapter authors for their profound work. I am also thankful for the support provided by my family and colleagues during the compilation of this book.

Editor

1

Scientometrics of Forest Health and Tree Diseases

Marco Pautasso [1,2,*]

Academic Editors: Jan Stenlid, Jonas Oliva and Audrius Menkis

1 Animal and Plant Health Unit, European Food Safety Authority (EFSA), 43126 Parma, Italy; marpauta@gmail.com

2 Forest Pathology & Dendrology, Institute of Integrative Biology (IBZ), ETH Zurich, 8092 Zurich, Switzerland

* The positions and opinions presented in this article are those of the author alone and are not intended to represent the views or scientific works of EFSA.

Abstract: Maintaining forest health is a worldwide challenge due to emerging tree diseases, shifts in climate conditions and other global change stressors. Research on forest health is thus accumulating rapidly, but there has been little use of scientometric approaches in forest pathology and dendrology. Scientometrics is the quantitative study of trends in the scientific literature. As with all tools, scientometrics needs to be used carefully (e.g., by checking findings in multiple databases) and its results must be interpreted with caution. In this overview, we provide some examples of studies of patterns in the scientific literature related to forest health and tree pathogens. Whilst research on ash dieback has increased rapidly over the last years, papers mentioning the Waldsterben have become rare in the literature. As with human health and diseases, but in contrast to plant health and diseases, there are consistently more publications mentioning "tree health" than "tree disease," possibly a consequence of the often holistic nature of forest pathology. Scientometric tools can help balance research attention towards understudied emerging risks to forest trees, as well as identify temporal trends in public interest in forests and their health.

Keywords: bibliometrics; data science; disturbance ecology; forest ecosystems; *Hymenoscyphus fraxineus*; long-term monitoring; publication growth; research assessment; sustainable development; tree diseases

---

1. Introduction

Forest health is an important aim of ecosystem management [1–3]. Forests deliver essential ecosystem services from boreal to tropical latitudes [4–6]. Without healthy forests, humanity and many other species would be in trouble [7–9]. There is thus concern about the impacts of global change on forest health, and much research is focusing on how to predict and manage such impacts [10–13].

But what is a healthy forest [14–16]? Researchers are increasingly realizing that a certain amount of disease plays a key role in the functioning of forest ecosystems, e.g., by diversifying uniform stands, recycling nutrients and facilitating adaptation to changing environmental conditions [17–19]. At the same time, exotic tree diseases can jeopardize the existence of native tree species, as shown by Dutch Elm Disease, chestnut blight, ash dieback and other emerging tree diseases [20–23]. Outbreaks of exotic tree pathogens can lead to a sudden spike in research on particular pathosystems (Figure 1), potentially leading to a relatively reduced research attention on the many other risks to tree health. Our understanding of forest health is also developing due to the availability of new tools to study the microbiome of forest trees and its role in maintaining forest health despite human activities and other biotic disturbances [24–26]. Just as importantly, there is increasing recognition of the role that various stakeholders play in shaping how societies react to new challenges to forest health [27–29].

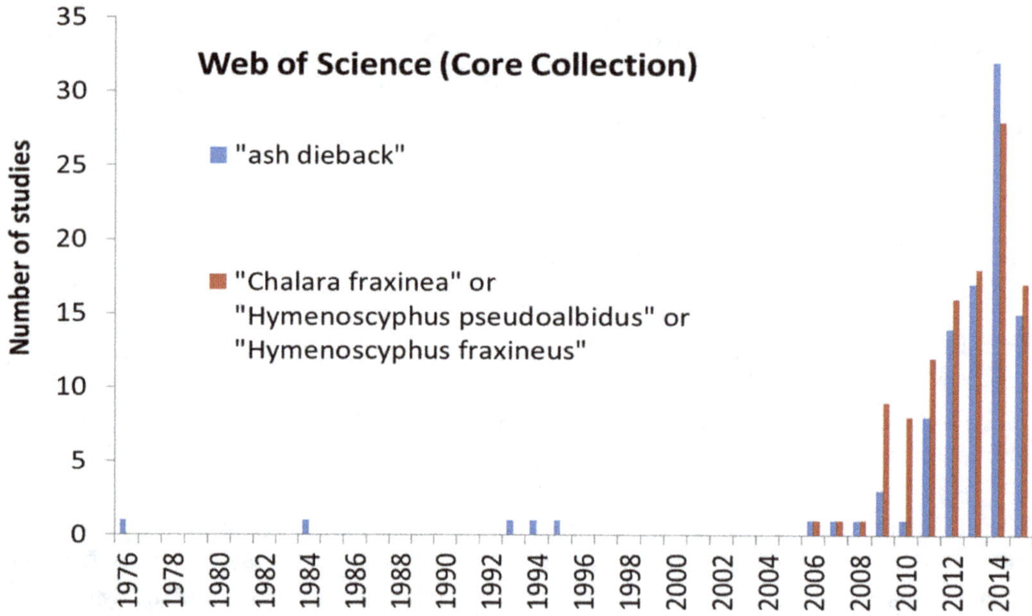

Figure 1. Temporal trend in Web of Science (core collection, since 1975), as of September 2015, of the number of studies retrieved with the keywords "ash dieback" (total = 98) and the names of its causal agent ("*Chalara fraxinea*" OR "*Hymenoscyphus pseudoalbidus*" OR "*Hymenoscyphus fraxineus*") (total = 111), an ascomycete shown in 2006 to be responsible for the dieback of *Fraxinus excelsior* in Europe [30–32]. Note that many relevant studies are not included in this database. Even so, the sudden increase in research shown by the figure is likely to reflect what happened in reality.

Given the importance of forest health, over the past decades, an increasing number of studies have dealt with this topic each year. For example, of the nearly 1700 studies retrieved in Scopus by searching for "forest health" (as of September 2015), about half appeared since 2007 and a third since 2010. Similarly, of the 162 items retrieved in Web of Science searching for "tree disease*" (core collection, 1991–2015, as of September 2015), about half were published since 2008 and a third since 2011. This is not a special feature of the literature related to forest health, as the number of scientific papers published each year has been increasing through time for most topics [33–35].

This increase in publication growth rates can have unintended consequences: although we have at our disposal much more information on forest health than, e.g., at the time of the Waldsterben (the 1980s), thus in principle leading to more pondered and hopefully evidence-based forest management decisions, it is now more difficult to keep updated with the literature, which becomes outdated more rapidly. A solution to the problems due to this growing number of studies is to try and summarize the gist of papers by writing reviews of the literature. However, literature reviews can also contribute to information overload. In addition, it is possible to study patterns in the literature using scientometric approaches [36–38].

Incidentally, the Waldsterben is a clear exception to the rule that scientific attention to most topics tends to increase through time (Figure 2), thus providing a potential example of a paradigm shift. Whilst this topic rapidly became the object of much research in the 1980s [39], the improved forest health conditions in the following decade, likely due to reduced air pollution and rarer acid rain events, both in North America and in Europe, resulted in a decreased use of this keyword. However, research on the links between forest health and air/soil pollution continued after the 1980s [40–42]. The forest damages due to air pollution currently observed in China [43–45] do not seem to be leading to a resurgence in the literature of the use of the term "Waldsterben".

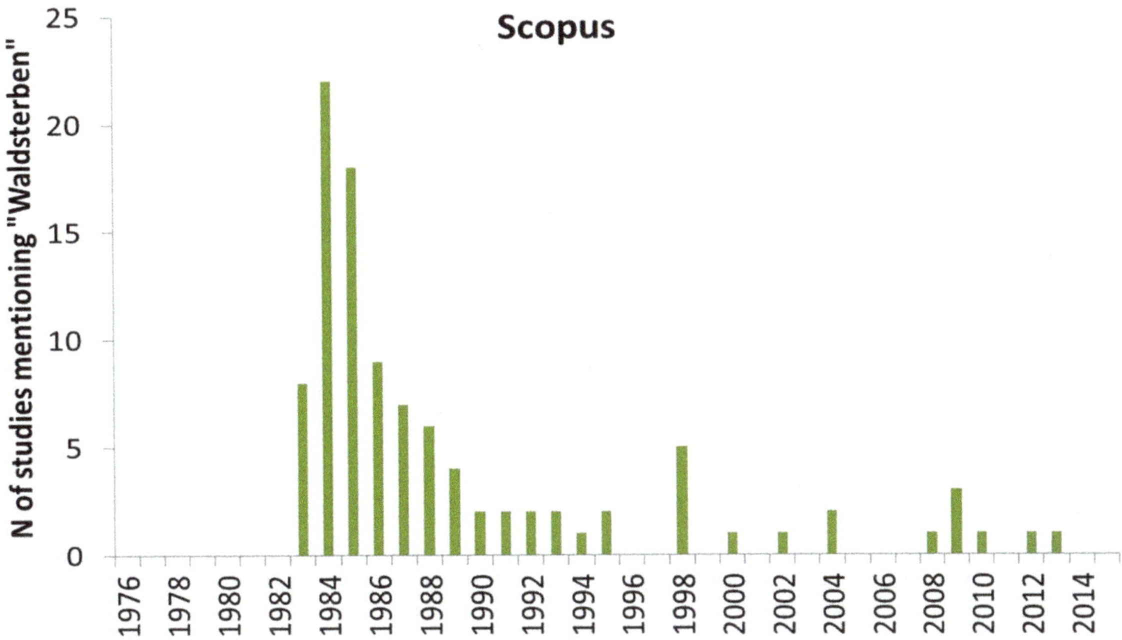

Figure 2. Number of studies mentioning "Waldsterben" (German for "forest mortality") in Scopus, as of September 2015. Total number = 101.

2. What is Scientometrics?

Scientometrics is the quantitative study of patterns in the scientific literature [46–48]. This approach can be applied, e.g.,:

(i) to examine temporal trends in the number of publications on a certain topic [49–51],
(ii) to investigate the spatial distribution of the research attention given to a certain topic [52–54],
(iii) to study the relative number of papers dealing with various research areas [55–57],
(iv) to analyze the reception of papers (or items therein, e.g., anecdotes) by posterity [58–60],
(v) to assess the productivity of researchers and its likely quality [61–63],
(vi) to compare the research performance and diversity of institutions and countries [64–66],
(vii) to document patterns and determinants of the length of peer review processes [67–69], and
(viii) to study interdisciplinary collaboration activities [70–72].

As with all scientific endeavours, scientometrics needs to be carried out carefully and its findings must be interpreted with caution [73–75]. For example, not all research papers on forests, forestry and forest health are published in forest-, forestry- or forest health-related journals, respectively [76,77]. Moreover, scientometric patterns can differ among different databases, e.g., Web of Science and Google Scholar, as the latter includes more journals and also tends to cover the full text of articles available online [78].

Even though more and more attention is given to scientometric issues such as the impact factor of journals and the h-index of researchers, often these metrics have limited predictive, comparative and explanatory power [79–83]. However, despite the increasing popularity of scientometric approaches over the last years across the spectrum of scientific fields, there has been relatively little use of such tools in forest pathology. This brief overview aims to present some examples of scientometric studies related to forest health and tree pathogens.

3. Scientometrics of Forest Health and Tree Pathogens

3.1. Scientometrics of Forest Health

There is evidence that, over the last two decades, the proportion of forest-related research studies mentioning "forest health" has been fairly stable, despite a steady increase in the number of both forest-related studies and those mentioning "forest health" [26]. The same pattern applies to the proportion of forest-related studies mentioning "tree diseases." The stable proportion of forest-related studies mentioning forest health (or tree diseases) over the last two decades is remarkable, given the emergence of many new threats to forests, some of which have attracted considerable research interest (e.g., Sudden Oak Death [84], the Emerald Ash Borer [85] and European ash dieback [32]).

Interestingly, fewer publications are retrieved when searching for "tree disease" compared to "tree health" (as well as for "forest disease" compared to "forest health"). This pattern tends to be consistent regardless of the publication year, so that the literature on forest health and diseases resembles that on human health and diseases (where publications mentioning "human health" tend to be more frequent than those mentioning "human disease") and differs from the literature on plant health and diseases, where the opposite is the case (Figure 3). This pattern might be a consequence of the traditional focus on specific crop diseases in plant pathology, whereas the science of forest pathology often has to deal with abiotic stressors, complex diseases and many factors other than diseases contributing to tree and forest health.

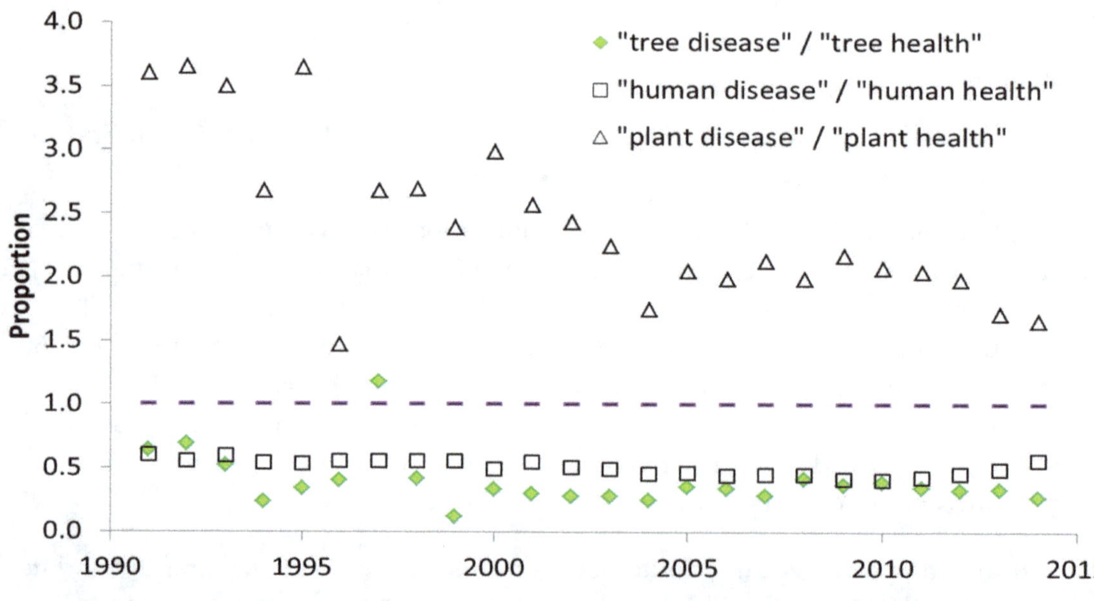

Figure 3. Number of publications retrieved in Google Scholar (1991–2014, as of September 2015) searching for the keyword "tree disease" relative to the number of publications retrieved using the keyword "tree health", and for human/plant disease in comparison with human/plant health, respectively. Modified from [86].

Tree mortality is an important component of forest health. In a sample of 67 studies modeling tree mortality published between 1997 and 2006, about 40% of the studies were based on data retrieved from studies shorter than five years [87]. This pattern might be inevitable due to the short-term nature of many current research projects, but basing models on short-term data is problematic when trying to predict the long-term effects of climate changes on forest health.

Scientometric studies can deliver insights about long-term trends in the literature, thus ultimately helping improve our long-term management of forests. For example, although fungi are essential organisms in forest ecosystems, fungi are relatively rarely mentioned in the scientific literature on forests (only about 3%–5% of the papers mentioning forests also mention fungi) [88]. Nevertheless, there is evidence for a slow increase in the proportion of forest-related papers mentioning fungi over the last two decades from Web of Science [88]. Although the proportion of forest-related papers mentioning insects is similar to that mentioning fungi, there has been no such increase through time for forest-related papers mentioning insects over the last two decades. These findings call for increased funding, training and hiring of mycologists, entomologists and forest pathologists at Universities and research institutions.

3.2. Scientometrics of Tree Pathogens

Scientific attention given to exotic tree pathogens appears to be skewed towards a few famous ones, with many tree fungal diseases receiving little attention. About ~20% of the exotic tree pathogens reported in Switzerland account for ~90% of the publications on such pathogens indexed in Web of Science [89]. This skewed attention is confirmed when searching within a more comprehensive database (Google Scholar) (Figure 4). Although there is a positive correlation between the numbers of findings on a certain tree pathogen in the two databases, the correlation is not perfect, thus confirming the importance of using multiple databases when searching the literature (Figure 4).

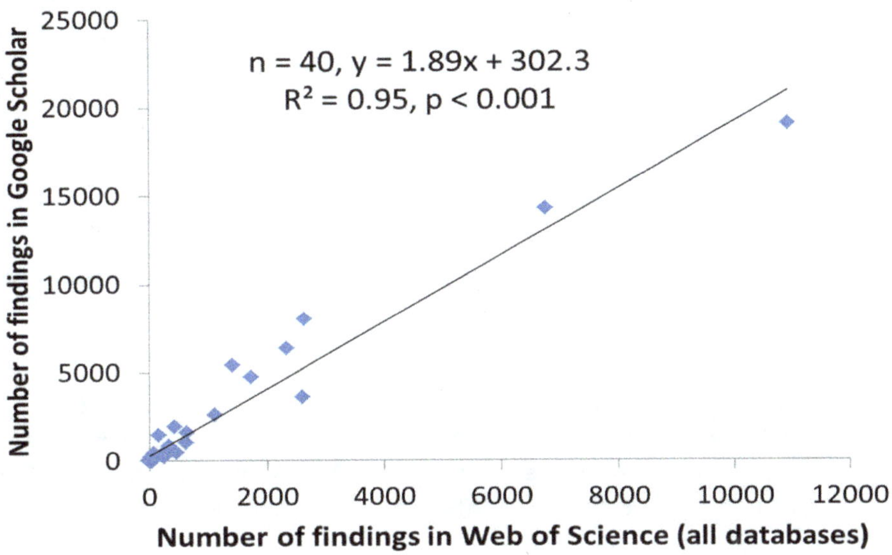

Figure 4. Correlation between number of findings in Google Scholar and Web of Science on the exotic tree pathogens reported in Switzerland up to October 2011 (as in [89], data kindly provided by Ottmar Holdenrieder, with addition of *Hymenoscyphus fraxineus*, the causal agent of ash dieback; searching for just this current name [90] will generally not retrieve publications only using the previous name *Hymenoscyphus pseudoalbidus* or the name of its anamorph, *Chalara fraxinea*).

One prominent example of an exotic tree pathogen that has received increased attention from the scientific community is *Phytophthora ramorum* (the oomycete responsible for Sudden Oak Death in the USA and Sudden Larch Death in the British Isles), now the object of just as many publications (Figure 5) as, e.g., *Cryphonectria parasitica*, the long studied causal agent of chestnut blight both in North America and Europe [91]. Figure 5 shows that few early publications on a topic tend to accumulate citations by the many late publications on the same topic, even if those early publications to some extent inevitably become outdated, as they could not consider all the knowledge made available in later publications. Note that not all articles retrieved by searching for "*Phytophthora ramorum*" are really on *Phytophthora ramorum*, as some only mention it as an aside.

4. Conclusions

Scientometric studies can illustrate the temporal development in the research focus towards particular issues. The drivers for such attention include overall interest in such topics, the number of researchers working on them, administrative incentives and funding availability. Scientometrics needs to be used carefully and interpreted with caution. Indeed, "not everything that can be counted counts and not everything that counts can be counted" [92]. Most scientometric indexes are (sub)field specific, thus making comparisons of publications, researchers, journals and institutions among (sub)fields of little value [93–95]. For example, whilst about 27% of the papers published in Australasian Plant Pathology between 2001 and 2007 were not cited as of 2011 [96], the proportion of not cited papers (as of 2014) was 40% for entomological papers published between 2005 and 2009 [97] and only about 7% for papers published in 2004 (as of 2010) in eight mathematical journals [98].

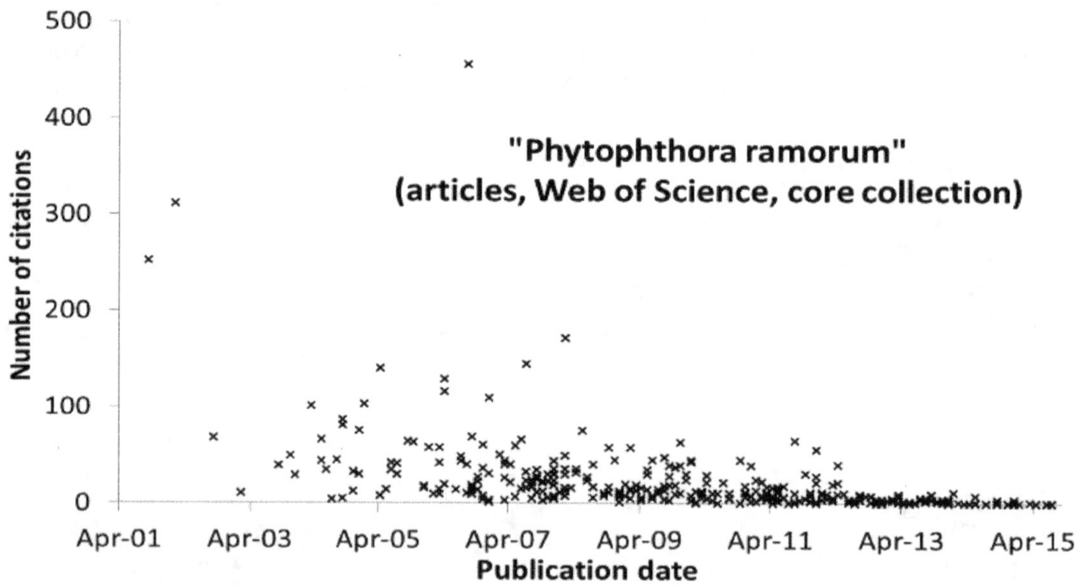

Figure 5. Number of citations received by the 353 articles (April 2001–September 2015) retrieved with the keyword "*Phytophthora ramorum*" in Web of Science (core collection, as of September 2015, excluding reviews, book reviews, meeting abstracts, editorials and news items). Note that the Web of Science database also lists some review papers as articles (but the pattern shown is essentially the same even if further refining the list of articles considered).

Just as when searching the literature for writing a literature review or a research paper, there is the need to use multiple databases and to be creative with the keywords used [99]. It is worth repeating that different databases have their strengths and weaknesses; for example, Web of Science focuses on journals with an impact factor, thus excluding much of the grey literature. Conversely, Google Scholar includes literature in languages other than English (as long as this is available on the internet), but does not exclude papers that have not been peer reviewed [100]. A clear documentation of the search strategy is needed, so as to enable repetition of the same search at a later stage.

It is important to keep in mind that the number of publications or of citations is a rough and insufficient indicator of the scientific importance of a topic or of the achievements within individual articles and of researchers [101–103]. But studying trends in the scientific literature, when combined with other lines of evidence, can help document gaps in knowledge and research priorities [104]. Of course, this approach needs to be complemented with the actual reading and pondering of the retrieved publications, as shown by an assessment of research intensity and its perception by experts on ecosystem services in mountain regions [105].

Approaches related to scientometrics can be used to monitor the attention of the media to forest and tree health [106], as well as public interest in various tree diseases, e.g., ash dieback [107], and, more generally, in forests (Figure 6). Figure 6 suggests that whilst in India there has been no clear trend or seasonality of public interest in forests over the last years, in Japan there appears to have been a decline through time, whereas in the UK there has been marked seasonality, with public interest peaking during the summer. A similar method can be used to study bibliometric patterns. For example, the proportion of published books mentioning "forest pathology" appears to have declined in recent decades for books in American and British English, French, German and Italian [108]. This finding is of concern, given the importance of books in inspiring students to study a certain topic, although this influence of books on career and lifestyle choices might now have been largely replaced by the internet and social media [109]. There has partly been a counter-trend for books mentioning "forest health" in the same languages [108].

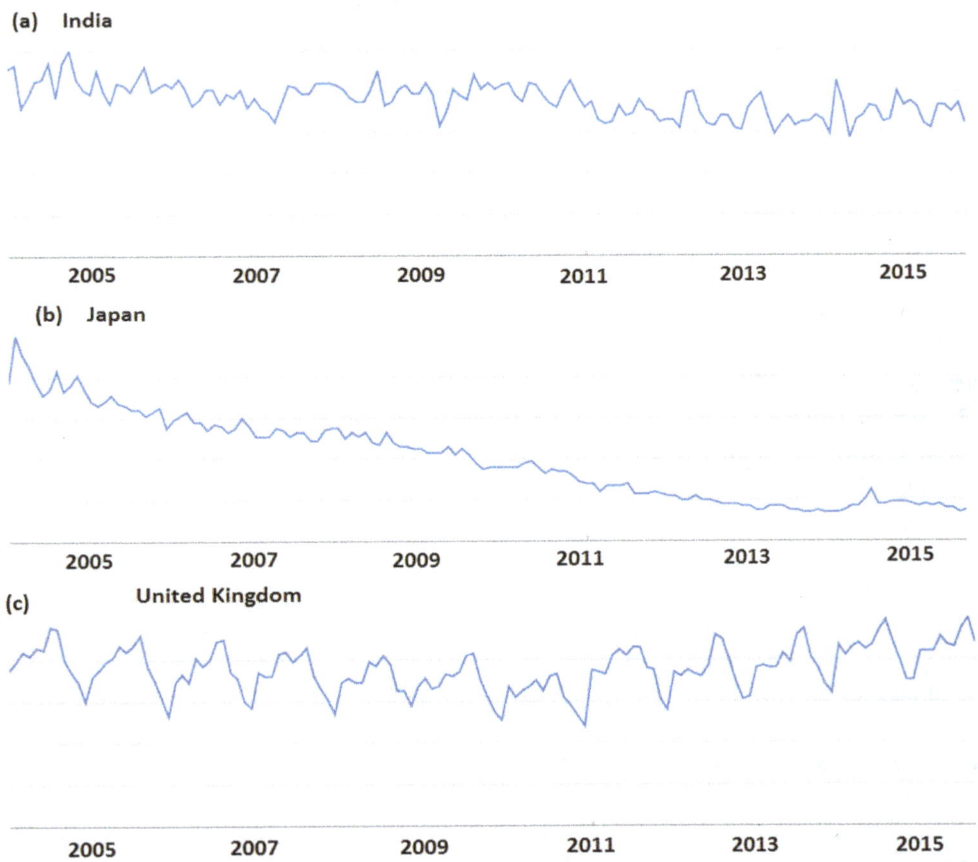

Figure 6. Temporal trend (2004–September 2015), as of September 2015, in Google searches by the general public for "forest" in (**a**) India; (**b**) Japan; and (**c**) the UK.

The absolute number of published papers per year is increasing for most scientific topics, including various tree fungal pathogens (although not all of them). Documenting an increase in the absolute number of papers per year for a certain topic is thus not particularly informative, although it can raise awareness of the amount of literature already available and the frequent lack of enduring value of many papers [97,110]. More interesting than documenting an absolute increase in publications is to study whether there is a temporal trend in the proportion of studies on a certain topic relative to the number of studies retrievable in a related area [111–113]. Given the skewed focus in research attention and funding to a few tree fungal pathogens, scientometric studies can help rebalance scientific priorities to a larger suite of potentially emerging threats to tree health under changing environmental conditions.

Acknowledgments: Many thanks to T. Döring, O. Holdenrieder, M. Jeger, C. Pautasso, S. Prospero and T. Sieber for insights and discussions, to C. Almqvist, J. Boberg, A. Menkis, J. Oliva and J. Stenlid for the excellent organization of a very insightful IUFRO meeting, and to T. Matoni and anonymous reviewers for helpful comments on a previous draft.

References

1. Holdenrieder, O. Der Forstschutz—Objekte, Probleme, Strategien. *Schweiz. Z. Forstwes.* **1991**, *142*, 795–807.
2. DellaSala, D.A.; Olson, D.M.; Barth, S.E.; Crane, S.L.; Primm, S.A. Forest health: Moving beyond rhetoric to restore healthy landscapes in the inland Northwest. *Wildl. Soc. Bull.* **1995**, *23*, 346–356.
3. Edmonds, R.L.; Agee, J.K.; Gara, R.I. *Forest Health and Protection*; McGraw-Hill: NewYork, NY, USA, 2000.
4. Gauthier, S.; Bernier, P.; Kuuluvainen, T.; Shvidenko, A.Z.; Schepaschenko, D.G. Boreal forest health and global change. *Science* **2015**, *349*, 819–822. [CrossRef] [PubMed]
5. Lewis, S.L.; Edwards, D.P.; Galbraith, D. Increasing human dominance of tropical forests. *Science* **2015**, *349*, 827–832. [CrossRef] [PubMed]
6. Millar, C.I.; Stephenson, N.L. Temperate forest health in an era of emerging megadisturbance. *Science* **2015**, *349*, 823–826. [CrossRef] [PubMed]
7. Kadoya, T.; Takenaka, A.; Ishihama, F.; Fujita, T.; Ogawa, M.; Katsuyama, T.; *et al*. Crisis of Japanese vascular flora shown by quantifying extinction risks for 1618 taxa. *PLoS ONE* **2014**, *9*, e98954. [CrossRef] [PubMed]
8. Ghazoul, J. *Forests: A Very Short Introduction*; Oxford University Press: Oxford, UK, 2015.
9. Le Roux, D.S.; Ikin, K.; Lindenmayer, D.B.; Manning, A.D.; Gibbons, P. Single large or several small? Applying biogeographic principles to tree-level conservation and biodiversity offsets. *Biol. Conserv.* **2015**, *191*, 558–566. [CrossRef]
10. Sturrock, R.N.; Frankel, S.J.; Brown, A.V.; Hennon, P.E.; Kliejunas, J.T.; Lewis, K.J.; Worrall, J.J.; Woods, A.J. Climate change and forest diseases. *Plant Pathol.* **2011**, *60*, 133–149. [CrossRef]
11. Pautasso, M. Forest ecosystems and global change: The case study of Insubria. *Annali Bot.* **2013**, *3*, 1–29.
12. Trumbore, S.; Brando, P.; Hartmann, H. Forest health and global change. *Science* **2015**, *349*, 814–818. [CrossRef] [PubMed]
13. Wingfield, M.J.; Brockerhoff, E.G.; Wingfield, B.D.; Slippers, B. Planted forest health: The need for a global strategy. *Science* **2015**, *349*, 832–836. [CrossRef] [PubMed]
14. Pautasso, M.; Dehnen-Schmutz, K.; Holdenrieder, O.; Pietravalle, S.; Salama, N.; Jeger, M.J.; Lange, E.; Hehl-Lange, S. Plant health and global change–some implications for landscape management. *Biol. Rev.* **2010**, *85*, 729–755. [CrossRef] [PubMed]
15. Döring, T.F.; Pautasso, M.; Finckh, M.R.; Wolfe, M.S. Concepts of plant health—Reviewing and challenging the foundations of plant protection. *Plant Pathol.* **2012**, *61*, 1–15. [CrossRef]
16. Holdenrieder, O.; Pautasso, M. Wie viel Krankheit braucht der Wald? *Bündner Wald* **2014**, *2014*, 5–10.
17. Hansen, E.M. Disease and diversity in forest ecosystems. *Australas. Plant Pathol.* **1999**, *28*, 313–319. [CrossRef]
18. Pautasso, M.; Holdenrieder, O.; Stenlid, J. Susceptibility to fungal pathogens of forests differing in tree diversity. In *Forest Diversity and Function*; Scherer-Lorenzen, M., Körner, C., Schulze, E.-D., Eds.; Springer: Berlin, Germany, 2005; pp. 263–289.
19. Ostry, M.E.; Laflamme, G. Fungi and diseases—Natural components of healthy forests. *Botany* **2009**, *87*, 22–25. [CrossRef]
20. Boyd, I.L.; Freer-Smith, P.H.; Gilligan, C.A.; Godfray, H.C.J. The consequence of tree pests and diseases for ecosystem services. *Science* **2013**, *342*, 1235773. [CrossRef] [PubMed]
21. Pautasso, M.; Aas, G.; Queloz, V.; Holdenrieder, O. European ash (*Fraxinus excelsior*) dieback—A conservation biology challenge. *Biol. Conserv.* **2013**, *158*, 37–49. [CrossRef]
22. Britton, K.O.; Liebhold, A.M. One world, many pathogens! *New Phytol.* **2013**, *197*, 9–10. [CrossRef] [PubMed]
23. Sieber, T.N. Neomyzeten—Eine anhaltende Bedrohung für den Schweizer Wald. *Schweiz. Z. Forstwes.* **2014**, *165*, 173–182. [CrossRef]
24. Sieber, T.N. Endophytic fungi in forest trees: Are they mutualists? *Fungal Biol. Rev.* **2007**, *21*, 75–89. [CrossRef]
25. Crann, S.E.; Fairley, C.; Badulescu, D.; Mohn, W.W.; O'Doherty, K.C. Soils, microbes, and forest health: A qualitative analysis of social and institutional factors affecting genomic technology adoption. *Technol. Soc.* **2015**, *43*, 1–9. [CrossRef]

26. Pautasso, M.; Schlegel, M.; Holdenrieder, O. Forest health in a changing world. *Microb. Ecol.* **2015**, *69*, 826–842. [CrossRef] [PubMed]
27. Stenlid, J.; Oliva, J.; Boberg, J.B.; Hopkins, A.J. Emerging diseases in European forest ecosystems and responses in society. *Forests* **2011**, *2*, 486–504. [CrossRef]
28. Sulak, A.; Huntsinger, L. Perceptions of forest health among stakeholders in an adaptive management project in the Sierra Nevada of California. *J. For.* **2012**, *110*, 312–317. [CrossRef]
29. Marzano, M.; Dandy, N.; Bayliss, H.R.; Porth, E.; Potter, C. Part of the solution? Stakeholder awareness, information and engagement in tree health issues. *Biol. Invasions* **2015**, *17*, 1961–1977. [CrossRef]
30. Kowalski, T. *Chalara fraxinea* sp. nov. associated with dieback of ash (*Fraxinus excelsior*) in Poland. *For. Pathol.* **2006**, *36*, 264–270. [CrossRef]
31. Queloz, V.; Grünig, C.R.; Berndt, R.; Kowalski, T.; Sieber, T.N.; Holdenrieder, O. Cryptic speciation in *Hymenoscyphus albidus*. *For. Pathol.* **2011**, *41*, 133–142. [CrossRef]
32. Gross, A.; Holdenrieder, O.; Pautasso, M.; Queloz, V.; Sieber, T.N. *Hymenoscyphus pseudoalbidus*, the causal agent of European ash dieback. *Mol. Plant Pathol.* **2014**, *15*, 5–21. [CrossRef] [PubMed]
33. Dobbertin, M.K.; Nobis, M.P. Exploring research issues in selected forest journals 1979–2008. *Ann. For. Sci.* **2010**, *67*, 800. [CrossRef]
34. Pautasso, M. Publication growth in biological sub-fields: Patterns, predictability and sustainability. *Sustainability* **2012**, *4*, 3234–3247. [CrossRef]
35. Bornmann, L.; Mutz, R. Growth rates of modern science: A bibliometric analysis based on the number of publications and cited references. *J. Assoc. Inf. Sci. Technol.* **2015**, *66*, 2215–2222. [CrossRef]
36. Armbruster, W.S. Floral specialization and angiosperm diversity: Phenotypic divergence, fitness trade-offs and realized pollination accuracy. *AoB Plants* **2014**, *6*, plu003. [CrossRef] [PubMed]
37. Habermann, G. Plant physiology in South America: Diagnostics and perspectives. *Theor. Exp. Plant Physiol.* **2015**, *27*, 87–89. [CrossRef]
38. Ladle, R.J.; Malhado, A.C.M.; Correia, R.A.; Guedes dos Santos, J.; Santos, A.M.C. Research trends in biogeography. *J. Biogeogr.* **2015**, *42*, 2270–2276. [CrossRef]
39. Zimmermann, W. *Zur Politischen Karriere des Themas Waldsterben*; ETH Zürich: Zürich, Switzerland, 1991.
40. Paoletti, E.; Schaub, M.; Matyssek, R.; Wieser, G.; Augustaitis, A.; Bastrup-Birk, A.M.; Bytnerowicz, A.; Günthardt-Goerg, M.S.; Müller-Starck, G.; Serengil, Y. Advances of air pollution science: From forest decline to multiple-stress effects on forest ecosystem services. *Environ. Poll.* **2010**, *158*, 1986–1989. [CrossRef] [PubMed]
41. Matyssek, R.; Kozovits, A.R.; Wieser, G.; Augustaitiene, I.; Augustaitis, A. Biological reactions of forests to climate change and air pollution. *Eur. J. For. Res.* **2014**, *133*, 671–673. [CrossRef]
42. Weigel, H.J.; Bergmann, E.; Bender, J. Plant-mediated ecosystem effects of tropospheric ozone. In *Progress in Botany*; Lüttge, U., Beyschlag, W., Eds.; Springer: Berlin, Germany, 2015; pp. 395–438.
43. Wang, Y.; Solberg, S.; Yu, P.; Myking, T.; Vogt, R.D.; Du, S. Assessments of tree crown condition of two Masson pine forests in the acid rain region in south China. *For. Ecol. Manag.* **2007**, *242*, 530–540. [CrossRef]
44. Feng, Z.; Sun, J.; Wan, W.; Hu, E.; Calatayud, V. Evidence of widespread ozone-induced visible injury on plants in Beijing, China. *Environ. Poll.* **2014**, *193*, 296–301. [CrossRef] [PubMed]
45. Wan, W.; Manning, W.J.; Wang, X.; Zhang, H.; Sun, X.; Zhang, Q. Ozone and ozone injury on plants in and around Beijing, China. *Environ. Poll.* **2014**, *191*, 215–222. [CrossRef] [PubMed]
46. Garfield, E. Is citation analysis a legitimate evaluation tool? *Scientometrics* **1979**, *1*, 359–375. [CrossRef]
47. Ellegaard, O.; Wallin, J.A. The bibliometric analysis of scholarly production: How great is the impact? *Scientometrics* **2015**, *105*, 1809–1831. [CrossRef] [PubMed]
48. Stork, H.; Astrin, J.J. Trends in biodiversity research—A bibliometric assessment. *Open J. Ecol.* **2014**, *4*, 354. [CrossRef]
49. Dietrich, M.R.; Ankeny, R.A.; Chen, P.M. Publication trends in model organism research. *Genetics* **2014**, *198*, 787–794. [CrossRef] [PubMed]
50. Pautasso, M. The jump in network ecology research between 1990 and 1991 is a Web of Science artefact. *Ecol. Model.* **2014**, *286*, 11–12. [CrossRef]
51. Flage, R.; Aven, T. Emerging risk—Conceptual definition and a relation to black swan type of events. *Reliab. Eng. Syst. Saf.* **2015**, *144*, 61–67. [CrossRef]

52. Pautasso, M. Geographical genetics and the conservation of forest trees. *Perspect. Plant Ecol. Syst. Evol.* **2009**, *11*, 157–189. [CrossRef]
53. Song, Y.; Zhao, T. A bibliometric analysis of global forest ecology research during 2002–2011. *SpringerPlus* **2013**, *2*, 1–9. [CrossRef] [PubMed]
54. Bullock, R.; Lawler, J. Community forestry research in Canada: A bibliometric perspective. *For. Policy Econ.* **2015**, *59*, 47–55. [CrossRef]
55. Lonsdale, D.; Pautasso, M.; Holdenrieder, O. Wood-decaying fungi in the forest: Conservation needs and management options. *Eur. J. For. Res.* **2008**, *127*, 1–22. [CrossRef]
56. Matić, R.; Stamenković, S.; Popović, Z.; Stefanović, M.; Vidaković, V.; Smiljanić, M.; Bojović, S. Tree responses, tolerance and acclimation to stress: Does current research depend on the cultivation status of studied species? *Scientometrics* **2015**, *105*, 1209–1222. [CrossRef]
57. McKenzie, A.J.; Robertson, P.A. Which species are we researching and why? A case study of the ecology of British breeding birds. *PLoS ONE* **2015**, *10*, e0131004. [CrossRef] [PubMed]
58. Steele, T.W.; Stier, J.C. The impact of interdisciplinary research in the environmental sciences: A forestry case study. *J. Am. Soc. Inf. Sci.* **2000**, *51*, 476–484. [CrossRef]
59. Sarringhaus, L.A.; McGrew, W.C.; Marchant, L.F. Misuse of anecdotes in primatology: Lessons from citation analysis. *Am. J. Primatol.* **2005**, *65*, 283–288. [CrossRef] [PubMed]
60. Steiner, F.M.; Pautasso, M.; Zettel, H.; Moder, K.; Arthofer, W.; Schlick-Steiner, B.C. A falsification of the citation impediment in the taxonomic literature. *Syst. Biol.* **2015**, *64*, 860–868. [CrossRef] [PubMed]
61. Pautasso, M. Focusing on publication quality would benefit all researchers. *Trends Ecol. Evol.* **2013**, *28*, 318–320. [CrossRef] [PubMed]
62. Brew, A.; Boud, D.; Namgung, S.U.; Lucas, L.; Crawford, K. Research productivity and academics' conceptions of research. *High. Educ.* **2016**. in press. [CrossRef]
63. Kwiek, M. Academic generations and academic work: Patterns of attitudes, behaviors and research productivity of Polish academics after 1989. *Stud. High. Educ.* **2015**, *40*, 1354–1376. [CrossRef]
64. Huisman, J.; Lepori, B.; Seeber, M.; Frølich, N.; Scordato, L. Measuring institutional diversity across higher education systems. *Res. Eval.* **2015**, *24*, 369–379. [CrossRef]
65. Tchetchik, A.; Grinstein, A.; Manes, E.; Shapira, D.; Durst, R. From research to practice: Which research strategy contributes more to clinical excellence? Comparing high-volume *versus* high-quality biomedical research. *PLoS ONE* **2015**, *10*, e0129259. [CrossRef] [PubMed]
66. Wooding, S.; van Leeuwen, T.N.; Parks, S.; Kapur, S.; Grant, J. UK doubles its "world-leading" research in life sciences and medicine in six years: Testing the claim? *PLoS ONE* **2015**, *10*, e0132990. [CrossRef] [PubMed]
67. Pautasso, M.; Schäfer, H. Peer review delay and selectivity in ecology journals. *Scientometrics* **2010**, *84*, 307–315. [CrossRef]
68. Nguyen, V.M.; Haddaway, N.R.; Gutowsky, L.F.; Wilson, A.D.; Gallagher, A.J.; Donaldson, M.R.; Hammerschlag, N.; Cooke, S.J. How long is too long in contemporary peer review? Perspectives from authors publishing in conservation biology journals. *PLoS ONE* **2015**, *10*, e0132557. [CrossRef] [PubMed]
69. Pautasso, M.; Jeger, M. How long does it take to deliver a peer review report? A decade of data from the European Journal of Plant Pathology. *Eur. J. Plant Pathol.* **2016**, *144*, 231–233. [CrossRef]
70. Hickey, G.M.; Nitschke, C.R. Crossing disciplinary boundaries in forest research: An international challenge. *For. Chron.* **2005**, *81*, 321–323. [CrossRef]
71. Kiss, I.Z.; Broom, M.; Craze, P.G.; Rafols, I. Can epidemic models describe the diffusion of topics across disciplines? *J. Informetr.* **2010**, *4*, 74–82. [CrossRef]
72. Lužar, B.; Levnajić, Z.; Povh, J.; Perc, M. Community structure and the evolution of interdisciplinarity in Slovenia's scientific collaboration network. *PLoS ONE* **2014**, *9*, e94429. [CrossRef] [PubMed]
73. Künzli, N. Trashing epidemiology and public health with bibliometry? In defence of science in Germany (and elsewhere). *Int. J. Public Health* **2015**, *60*, 877–878. [CrossRef] [PubMed]
74. Génova, G.; Astudillo, H.; Fraga, A. The scientometric bubble considered harmful. *Sci. Eng. Ethics* **2016**. in press. [CrossRef]
75. Van Wesel, M. Evaluation by citation: Trends in publication behavior, evaluation criteria, and the strive for high impact publications. *Sci. Eng. Ethics* **2016**. in press. [CrossRef] [PubMed]
76. Vanclay, J.K. Ranking forestry journals using the h-index. *J. Informetr.* **2008**, *2*, 326–334. [CrossRef]

77. Bojović, S.; Matić, R.; Popović, Z.; Smiljanić, M.; Stefanović, M.; Vidaković, V. An overview of forestry journals in the period 2006–2010 as basis for ascertaining research trends. *Scientometrics* **2014**, *98*, 1331–1346. [CrossRef]
78. Pautasso, M.; Jeger, M.J. Network epidemiology and plant trade networks. *AoB Plants* **2014**, *6*, plu007. [CrossRef] [PubMed]
79. Pautasso, M. Fungal under-representation is (indeed) diminishing in the life sciences. *Fungal Ecol.* **2013**, *6*, 460–463. [CrossRef]
80. De Sutter, A.; van Driel, M.; Maier, M.; de Maeseneer, J. The new impact factor has arrived. Who cares? *Eur. J. Gen. Pract.* **2015**, *21*, 153–154. [CrossRef] [PubMed]
81. Dilger, A.; Lütkenhöner, L.; Müller, H. Scholars' physical appearance, research performance, and feelings of happiness. *Scientometrics* **2015**, *104*, 555–573. [CrossRef]
82. Fox, C.W.; Burns, C.S. The relationship between manuscript title structure and success: Editorial decisions and citation performance for an ecological journal. *Ecol. Evol.* **2015**, *5*, 970–980. [CrossRef] [PubMed]
83. Von Bartheld, C.S.; Houmanfar, R.; Candido, A. Prediction of junior faculty success in biomedical research: Comparison of metrics and effects of mentoring programs. *PeerJ* **2015**, *3*, e1262. [CrossRef] [PubMed]
84. Frankel, S.J. Sudden oak death and *Phytophthora ramorum* in the USA: A management challenge. *Australas. Plant Pathol.* **2008**, *37*, 19–25. [CrossRef]
85. Herms, D.A.; McCullough, D.G. Emerald ash borer invasion of North America: History, biology, ecology, impacts, and management. *Annu. Rev. Entomol.* **2014**, *59*, 13–30. [CrossRef] [PubMed]
86. Pautasso, M.; Jeger, M.J. Impacts of climate change on plant diseases: New scenarios for the future. In *The Influence* of Global Environmental Change on Infectious Disease Dynamics: Workshop Summary; Choffnes, E.R., Mack, A., Eds.; National Academy of Sciences: Washington, DC, USA, 2014; pp. 359–374.
87. Jeger, M.J.; Pautasso, M. Plant disease and global change—The importance of long-term data sets. *New Phytol.* **2008**, *177*, 8–11. [CrossRef] [PubMed]
88. Pautasso, M. Fungal under-representation is (slowly) diminishing in the life sciences. *Fungal Ecol.* **2013**, *6*, 129–135. [CrossRef]
89. Pautasso, M. Responding to diseases caused by exotic tree pathogens. In *Infectious Forest Diseases*; Gonthier, P., Nicolotti, G., Eds.; CABI: Wallingford, UK, 2013; pp. 592–612.
90. Baral, H.O.; Queloz, V.; Hosoya, T. *Hymenoscyphus fraxineus*, the correct scientific name for the fungus causing ash dieback in Europe. *IMA Fungus* **2014**, *5*, 79–80. [CrossRef] [PubMed]
91. Pautasso, M. *Phytophthora ramorum*—A pathogen linking network epidemiology, landscape pathology and conservation biogeography. *CAB Rev.* **2013**, *8*, 024. [CrossRef]
92. Cameron, W.B. *Informal Sociology: A Casual Introduction to Sociological Thinking*; Random House: New York, NY, USA, 1963.
93. Vanclay, J.K. An evaluation of the Australian Research Council's journal ranking. *J. Informetr.* **2011**, *5*, 265–274. [CrossRef]
94. Vanclay, J.K. Publication patterns of award-winning forest scientists and implications for the Australian ERA journal ranking. *J. Informetr.* **2012**, *6*, 19–26. [CrossRef]
95. Vanclay, J.K. What was wrong with Australia's journal ranking? *J. Informetr.* **2012**, *6*, 53–54. [CrossRef]
96. Calver, M.C.; O'Brien, P.A.; Lilith, M. Australasian plant Pathology: An analysis of authorship and citations in the 21st century. *Australas. Plant Pathol.* **2012**, *41*, 179–187. [CrossRef]
97. Chouvenc, T.; Su, N.Y. How do entomologists consume and produce their science? *Am. Entomol.* **2015**, *61*, 252–257. [CrossRef]
98. Mauleón, E.; Bordons, M. Authors and editors in mathematics journals: A gender perspective. *Int. J. Gend. Sci. Technol.* **2012**, *4*, 267–293.
99. Pautasso, M. Ten simple rules for writing a literature review. *PLoS Comput. Biol.* **2013**, *9*, e1003149. [CrossRef] [PubMed]
100. Archambault, É.; Campbell, D.; Gingras, Y.; Larivière, V. Comparing bibliometric statistics obtained from the Web of Science and Scopus. *J. Am. Soc. Inf. Sci. Technol.* **2009**, *60*, 1320–1326. [CrossRef]
101. Ayling, M. Journal citation metrics in forestry research. *For. Chron.* **2015**, *91*, 343–346. [CrossRef]
102. Belter, C.W. Bibliometric indicators: Opportunities and limits. *J. Med. Libr. Assoc.* **2015**, *103*, 219–221. [CrossRef] [PubMed]

103. Moustafa, K. Aberration of the citation. *Account. Res. Policies Qual. Assur.* **2016**. in press. [CrossRef] [PubMed]
104. Pasgaard, M.; Dalsgaard, B.; Maruyama, P.K.; Sandel, B.; Strange, N. Geographical imbalances and divides in the scientific production of climate change knowledge. *Glob. Environ. Chang.* **2015**, *35*, 279–288. [CrossRef]
105. Haida, C.; Rüdisser, J.; Tappeiner, U. Ecosystem services in mountain regions: Experts' perceptions and research intensity. *Reg. Environ. Chang.* **2016**. in press. [CrossRef]
106. Alomar, O.; Batlle, A.; Brunetti, J.M.; García, R.; Gil, R.; Granollers, A.; Jiménez, S.; Laviña, A.; Linge, J.P.; Pautasso, M.; *et al.* Development and testing of the media monitoring tool MedISys for early identification and reporting of existing and emerging plant health threats. *EPPO Bull.* **2015**, *45*, 288–293. [CrossRef]
107. Pautasso, M.; Petter, F.; Rortais, A.; Roy, A.S. Emerging risks to plant health: A European perspective. *CAB Rev.* **2015**, *10*, 21. [CrossRef]
108. Pautasso, M. Long-term trends in books of terms related to forest pathology. *EPPO Bull.* **2015**, *45*, 323–327. [CrossRef]
109. Lazzeretti, L.; Sartori, A.; Innocenti, N. Museums and social media: The case of the Museum of Natural History of Florence. *Int. Rev. Public Nonprofit Mark.* **2015**, *12*, 267–283. [CrossRef]
110. Gregoris, N.; Shorvon, S. What is the enduring value of research publications in clinical epilepsy? An assessment of papers published in 1981, 1991, and 2001. *Epilepsy Behav.* **2013**, *28*, 522–529. [CrossRef] [PubMed]
111. Pautasso, M.; Pautasso, C. Peer reviewing interdisciplinary papers. *Eur. Rev.* **2010**, *18*, 227–237. [CrossRef]
112. Pautasso, M. Worsening file-drawer problem in the abstracts of natural, medical and social science databases. *Scientometrics* **2010**, *85*, 193–202. [CrossRef]
113. Döring, T.F.; Vieweger, A.; Pautasso, M.; Vaarst, M.; Finckh, M.R.; Wolfe, M.S. Resilience as a universal criterion of health. *J. Sci. Food Agric.* **2015**, *95*, 455–465. [CrossRef] [PubMed]

2

New R-Based Methodology to Optimize the Identification of Root Endophytes against *Heterobasidion parviporum*

Linda Rigerte, Kathrin Blumenstein and Eeva Terhonen *

Forest Pathology Research Group, Büsgen-Institute, Department of Forest Botany and Tree Physiology, Faculty of Forest Sciences and Forest Ecology, University of Göttingen, Büsgenweg 2, 37077 Göttingen, Germany; lindarigerte@gmail.com (L.R.); kathrin.blumenstein@uni-goettingen.de (K.B.)

* Correspondence: terhonen@uni-goettingen.de

Abstract: Many root fungal endophytes inhabiting forest trees have potential impact on the health and disease progression of certain tree species. Hence, the screening of root endophytes for their biocontrol abilities is relevant for their potential to protect their hosts against invaders. The aim of this research is to screen for the potential inhibitory effects of selected conifer root endophytes during interaction, in vitro, with the root rot pathogen, *Heterobasidion parviporum*. Here, we introduce a guideline that facilitates the use of root fungal endophytes as biocontrol agents. We isolated fungal root endophytes from eight different conifers. These root fungal endophytes were evaluated for their antagonism against the root rot pathogen, *H. parviporum*, by means of paired-culture antagonism assays. We determined the antagonism of the isolated root fungal endophytes to elucidate potential biocontrol applications. For the analysis, a software package in R was developed. Endophyte candidates with antagonistic potential were identified.

Keywords: fungal endophytes; inhibitory activity; biocontrol; forest pathogens; conifers

---

1. Introduction

Endophytes are organisms that, at some stage in their lifecycle, colonize a portion of a plant's tissue internally without causing any visible disease symptoms [1–4]. The increasing interest in plant endophyte communities is derived from their apparent potential to positively influence stress tolerance in trees, further providing new sources to be exploited in tree health protection [5]. To understand the function of root endophytes in forest trees, inoculation studies on host trees have been performed. The observed effects vary from beneficial interactions [6–10], to neutral [11], and sometimes even to pathogenic [8,12]. The nature of root endophyte–host interactions seems to depend on the strain of endophyte and the experimental environment [12–15]. In the context of forest pathology, the most beneficial aspect of endophytes is the production of unique secondary metabolites, which may limit pathogen growth [6–8,16,17]. Tellenbach and Sieber [6] isolated compounds from *Phialocephala europaea*, of which, sclerin and sclerotinin A were demonstrated to significantly reduce the growth of *Phytophtora citricola sensu lato*. Similarly, Terhonen et al. [8] found that metabolites extracted from liquid cultures of root endophytes, *Phialocephala sphareoides* and *Cryptosporiopsis* sp., inhibit the growth of several plant pathogens (*Heterobasidion parviporum*, *Phytophtora pini*, and *Botrytis cinerea*). Fungal endophytes can also decrease the infections of pathogens in their host roots, which can be observed in a study conducted by Tellenbach and Sieber [6], who showed that some strains of *Phialocephala subalbina* could reduce disease severity caused by the two oomycete root rot pathogens, *Elongisporangium undulatum* and *Phytophtora plurivora*, in Norway spruce seedlings. Terhonen et al. [8] demonstrated that the root endophyte, *P.*

sphareoides, was able to prevent the infections of Norway spruce seedling roots by the pathogen, *H. parviporum*, in vitro. Indeed, fungal endophytes might play an important role in ecosystem dynamics, affecting not just the fitness of the specific host plant, but also the overall structure and well-being of tree communities [5]. This is supported by the fact that entire assemblages of endophytes have adapted to specialized habitats along with their hosts; for example, aquatic fungi that inhabit submerged plant roots [18].

The dominating group of root endophytes in tree roots are the 'dark septate endophytes' (DSEs) [15,19–27]. DSEs feature a very characteristic darkly pigmented ("melanized") hyphal growth inside roots [9,16]. The *Phialocephala fortinii* s.l.–*Acephala applanata* species complex (PAC) is the dominant group of DSEs and they are among the best-characterized DSEs [15,25,28–32]. Members of the PAC species complex are cryptic and they are very widespread and abundant in roots of conifers and ericaceous plants [15,19,20,23–25,27]. Generally, DSEs do not seem to have host specificity; however, there are some species that seem to prefer certain host taxa and some of the PAC species preferentially occur on the Pinaceae family [15,19–21,24,33,34].

Heterobasidion annosum sensu lato (s.l.) is the main cause of the root rot of conifers in Europe [35–38]. The *H. annosum* s.l. complex in Europe consists of three native species: *H. annosum sensu stricto*, *H. parviporum*, and *H. abietinum* [35,39,40]. Of these, *H. parviporum* is the major pathogen for the economically important conifer, Norway spruce (*Picea abies*). *H. parviporum* infects the tree through deposition of airborne basidiospores on freshly cut stumps, where infection can easily initiate [41–43]. As the infection progresses, wood material is rapidly colonized by *H. parviporum*. After the infection is established, *Heterobasidion* species can remain viable in these stumps for decades [44], resulting in an inoculum source for new tree generations [43–45] by spreading through root contacts to neighbouring trees [45,46]. In addition, as *H. parviporum* can remain undetected in the host tree for many years, the risk of *H. parviporum* infecting neighbouring healthy trees via root contacts increases [43–46]. As such, in search of control measures against *H. parviporum*, one interesting direction to pursue is the study of root endophytes. The possible inhibitory effect of root endophytes on root pathogens, such as *H. parviporum*, can be harnessed for potential biocontrol applications. These kinds of fungi could be used in infected sites to protect the roots of newly planted seedlings. Terhonen et al. [19] demonstrated that 17% of root endophytes isolated from a spruce-dominated site are capable of inhibiting the growth of *H. parviporum*. Particularly, the root endophyte, *P. sphareoides*, not only inhibits the necrotrophic *H. parviporum*, but also promotes host growth and health [8]. However, the ecological and potential functional role of these abundant endophytic fungi in the host-endophyte interaction remains largely unexplored.

In this study, the root fungal endophytes of eight conifer tree species from the Forest Botanical Garden of the University of Göttingen (Germany) were isolated and tested for their antagonistic reaction against *H. parviporum*. The identities and phylogeny of the isolated root endophytes were established on the basis of their internal transcribed spacer (ITS) region sequences. The antagonism of the endophytes to *H. parviporum* was investigated by dual culture assays on agar media. This work aims to determine the antagonism of the isolated root endophytes and elucidate potential biocontrol applications within the contextual constraints described in the sections above. It also aims to evaluate the potential inhibitory effect of a subset of the isolated endophytes on the pathogen, *H. parviporum*. To analyse this, a software package in R was developed.

2. Materials and Methods

Fine root samples were collected in early January 2018 from mature conifer trees with no visible disease symptoms in the Forest Botanical Garden, Göttingen, Germany (51°33′26.55″ N/9°57′40.72″ E). A simple random sampling pattern was applied for tree selection to minimize sampling bias. Fine roots from conifer hosts exhibiting no visible signs of mycorrhizae were collected, placed in Falcon tubes, and stored at −20 °C. Eight different conifer species were sampled. Overall, only one tree was sampled each for *Picea pungens*, *Picea sitzenchis*, and *Pinus peuce*, two trees each for *Picea omorika*, *Picea glauca*, and *Pinus jeffreyi*, and three trees each for *P. abies* and *Pinus sylvestris* (Table 1).

Table 1. List of sampled host conifers, their common names, native ranges, and sample abbreviations used in the data.

Host Species	Common Name	Native to	Abbreviation
Picea abies	Norway spruce	Northern, Central, and Eastern Europe	Abie
Picea glauca	White/Canadian spruce	Alaska through central Canada to Newfoundland	Glau
Picea omorika	Serbian spruce	Endemic to Drina river valley, Serbia	Omori
Picea pungens	Blue/Colorado spruce	Rocky Mountains, USA	Pung
Picea sitchensis	Sitka spruce	West coast Canada, down to California	Sitch
Pinus jeffreyi	Jeffrey pine	Oregon-California, USA	Jeff
Pinus peuce	Macedonian/Balkan pine	Mountains of Balkan region	Peuc
Pinus sylvestris	Scots pine	Eurasia	Sylv

The collected fine roots were processed in 24 h. First, they were washed under running tap water to remove soil particles and cut into small pieces. Root segments were disinfected in 99% ethanol (C_2H_5OH) for 1 min, 2% sodium hypochlorite (NaOCl) for 3 min, followed by 70% ethanol for 1 min; the roots were then rinsed in autoclaved ultrapure water H_2O for 3 to 5 times. The surface sterilized roots were placed on 1.5% malt agar plates (MEA) and Modified Melin–Norkrans medium (MMN) plates [47]. The MMN plates were composed as follows: 0.066 g $CaCl_2X2H_2O$ (calcium chloride dihydrate), 0.025 g NaCl (sodium chloride), 0.150 g $MgSO_4X7H_2O$ (magnesium sulfate heptahydrate), 0.500 g $NH_4(H_2PO_4)$ (ammonium dihydrogen phosphate), 0.500 g KH_2PO_4 (potassium dihydrogen phosphate), 0.5 mL $FeCl_3X6H_2O$ (iron(III) chloride hexahydrate), 0.200 g $C_6H_{12}O_6$ (glucose), 15.000 g agar, and 0.5 mL thiamine-HCl dissolved in 1000 mL distilled water (pH of 5.5–5.8). Altogether, three short root tip segments from each sample tube were placed in MEA and MMN plates. The plates were incubated in darkness at 19 °C with 75% relative humidity. The incubation period necessary for the growth of various root endophytes differed between plates, ranging from 14 days to one month. When the growth of a root endophyte was observed, a small piece of the hyphae was transferred as a pure culture. Sixty-five (65) root endophytes were isolated, and grouped together based on their morphological features to 19 different groups; 1 to 3 members of each group were then chosen for molecular identification (28 samples).

DNA of fungal endophytes was extracted from 150 mg of the homogenized mycelium sample using the "innuPREP Plant DNA Kit" (Analytik Jena AG, Jena, Germany), according to the manufacturer's instructions. Taq DNA polymerase (Qiagen, Hilden, Germany) was used for PCR amplification of ITS regions with the primer pair, ITS1-F and ITS4 [48]. Briefly, the PCR protocol was as follows: 1X CoralLoad PCR Buffer, 200 μM dNTP, 0.5 μM primer 1, 0.5 μM primer 2, 100 ng template DNA, and 0.2 U/μL DNA polymerase; the reaction was adjusted to 25 μL with autoclaved MQ H2O. Thermocycler parameters were as follows: Initial denaturation at 95 °C for 3 min, followed by 15 cycles of: denaturation at 95 °C for 30 s, annealing at 55 °C for 1 min, extension at 72 °C for 1 min; 15 cycles of: denaturation at 95 °C for 30 s, annealing at 63 °C for 1 min, extension at 72 °C for 1 min; final extension at 72 °C for 10 min and held at 4 °C indefinitely. The quality of the obtained PCR products was checked on 1% agarose gel (with GelRed nucleic acid stain) under UV illumination. Sequencing was performed with ITS4 primer at Microsynth SEQLAB (Göttingen, Germany). For 24 samples, the full coverage of the ITS1-5.8S-ITS2 region was achieved.

The ITS1 and ITS2 regions of all sequenced samples were extracted using the ITSx program [49]. New sequences consisting of ITS1 to ITS2 regions were constructed—concatenation was performed using the sequence concatenation tool available in MEGA X [50]. Sequence identity was established through a search for homology using the National Center for Biotechnology Information's (NCBI) Basic Local Search Alignment Tool (BLAST) [51]. The top three matches for each sequence were selected based on query coverage (QC) and identity (Mi%) with regional proximity acting as a tiebreaker in case of conflicts. Of these, only the match that had the best coverage and identity was selected as the homolog establishing positive identification of the sample. The homologous sequences were downloaded as an unaligned FASTA file. The concatenated ITS1 to ITS2 regions and their BLAST matches were aligned using MAFFT [52]; alignment parameters were determined automatically by the program (–auto command). The final dataset consisted of 41 nucleotide sequences (24 samples, 16 references, and 1 outgroup sequence) with 635 positions (inclusive of gaps). The Kimura 2-parameter model [53] with gamma distributed rates (5 rate categories; G = 1.915) and a proportion of invariant sites ($I = 0.032$) was selected to model the evolution of the sequences on the basis of its lowest negative log likelihood score (−lnL = −4987.465) in comparison to the evaluated fits for all other models available in MEGA X (evaluated using MEGA X's internal model selection tool). The initial tree(s) for the ML phylogeny was/were obtained automatically by applying Neighbor-Joining (NJ) and BioNJ algorithms to a matrix of pairwise distances estimated using the maximum composite likelihood (MCL) approach, and then selecting the topology/topologies with superior log likelihood value(s). Molecular phylogeny was established by means of maximum likelihood (ML) analysis with 1000 bootstrap replicates in MEGA X [50,54]. Final adjustments to the generated ML tree were performed in TreeGraph2 [55] and Microsoft PowerPoint. The ITS1 sequences used in BLAST searches were deposited to GenBank with the following accession numbers, MK589294-MK589317.

The antagonism of the isolated root endophyte against the pathogen, *H. parviporum*, was investigated by means of a paired-growth assay. The ability of a root endophyte to antagonize the pathogen was determined based on the inhibition level over a given period of time. This was achieved by assessing and measuring the concurrent growth of both the endophyte and the pathogen simultaneously on a shared 1.5% MEA nutrient media surface (Figure 1).

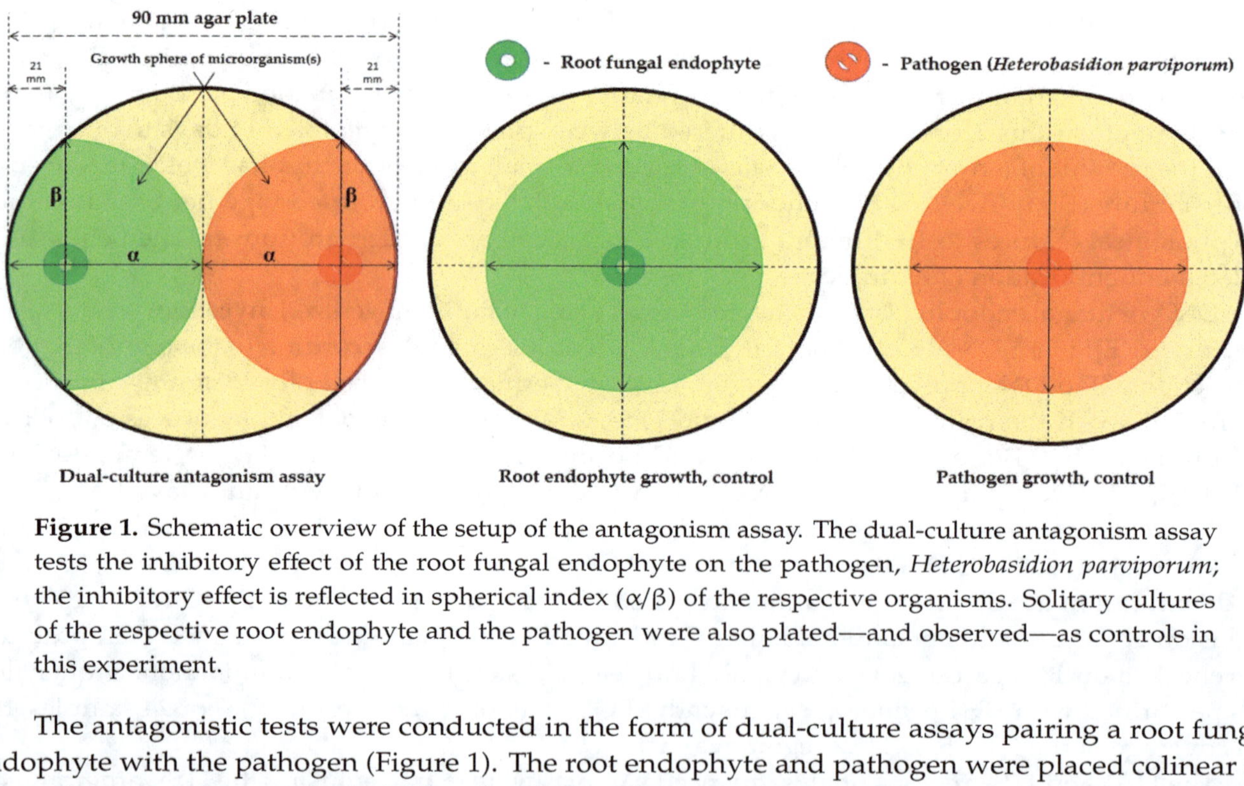

Figure 1. Schematic overview of the setup of the antagonism assay. The dual-culture antagonism assay tests the inhibitory effect of the root fungal endophyte on the pathogen, *Heterobasidion parviporum*; the inhibitory effect is reflected in spherical index (α/β) of the respective organisms. Solitary cultures of the respective root endophyte and the pathogen were also plated—and observed—as controls in this experiment.

The antagonistic tests were conducted in the form of dual-culture assays pairing a root fungal endophyte with the pathogen (Figure 1). The root endophyte and pathogen were placed colinear on the surface of malt agar plates (pH = 6) within a distance of 2.1 cm from the perimeter of the petri dish. Such pairings were generated in triplicates for statistical rigor. For each pair of endophyte and pathogen, a control plate containing a single solitary endophyte (spotted identically on the plate as to the paired setup) was also prepared as the endophyte growth control; each endophyte received three controls. Control plates to monitor the pathogen's growth were also prepared in an identical manner (12 replicates in total). All these plates were then incubated under the same conditions to stimulate the growth of the resident microorganisms (darkness, humidity 75%, 19 °C). Growth measurements were taken 3, 7, and 10 days after inoculation. Measurements were performed for both the root endophytes as well as the pathogen in the antagonism test plate (Figure 2). All isolated endophytes participated in the antagonism assay(s). The measured parameters were the major (α) and minor (β) axes of progression along the colinear axis joining the original spotting locations of the endophyte and/or pathogen. In theory, α measures the progression of the antagonists towards one another (and conversely indicates a measure of the antagonistic effect) while β measures the orthogonal progression (i.e., to α) of the antagonists over the available surface area (an indirect measure of competitive spread). The ratio, α/β, is the spherical index, which is considered a measure of the antagonism exhibited by the endophyte and pathogen against one another [56,57]. The reasoning is that an organism growing in isolation on a circular surface would grow uninhibited radially outwards from the site of inoculation—this corresponds to a spherical index of 1 (since $\alpha = \beta$). Therefore, should an inhibitory source be present, radial progression in that particular direction would be negated or minimized, and the spherical index would no longer equal 1 (since $\alpha \neq \beta$). In this regard, a spherical index > 1 (for the endophyte) indicates that the endophyte is inhibiting the pathogen, and an index < 1 indicates it has minimal or no inhibitory effect. Conversely, a spherical index < 1 for the pathogen indicates that it is being (successfully) antagonized by the endophyte, while a spherical index > 1 indicates the endophyte has no effect on the pathogen's growth or existence. This data included observations under the aforementioned variables for unique endophyte samples ($n = 65$) at every point of measurement (3, 7, and 10 days, respectively) for a total of 195 measurements. This resulted in a large raw dataset to be screened for the identification of successful fungal antagonists to *H. parviporum*. Thus, a filtering process was necessary, and data filtering pipeline was implemented in R to mitigate this challenge; the filtering logic as well as the pipeline are briefly described below.

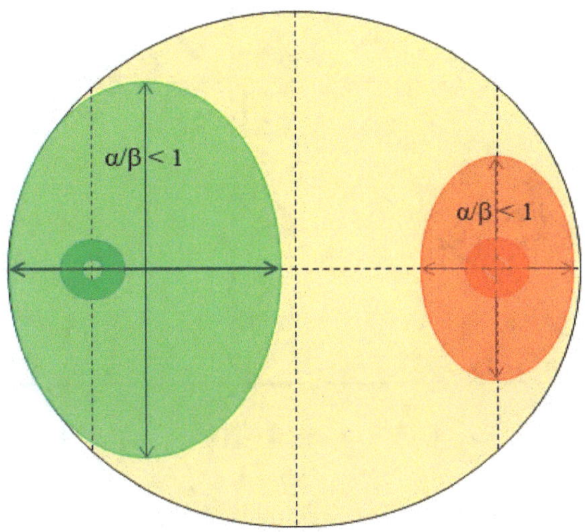

Figure 2. Example schematic antagonism assay plate where both the endophyte antagonist (green) as well as the pathogen (red) have a spherical index < 1.

The explanation provided earlier in this section concerning the spherical index of the root endophyte or the pathogen alone, as a determinant of the fungus' antagonistic effectiveness against the pathogen, is a naïve interpretation of this number. The spherical index is a ratio resulting from measurements specific to the organism only (and not its surroundings or context), which cannot provide reliable information on the effects of biotic or abiotic factors upon the test organism (nor vice versa) even with replication and statistical testing. For example, consider the case presented in Figure 2: If successful antagonism were indicated only by the spherical index of the endophyte, then this would be an unsuccessful case; conversely, if it were indicated by the spherical index of the pathogen only, then this would be considered a successful case. In truth, it is neither, as neither the antagonist nor the pathogen have grown considerably towards one another. The problem lies in the fact that even under the best-optimized control conditions, biological organisms do not grow evenly nor predictably. In this context, the only way to distinguish a case of an actual, successful antagonism from a spurious one is by considering the spherical ratios of both the antagonist endophyte as well as the pathogen. It can be stated that true antagonism in the experimental conditions described in this section only occurs when the spherical indices indicate that the growth of the endophyte antagonist along the colinear axis (joining the locations of the inoculates of the antagonist and the pathogen) is greater than its growth orthogonal to the colinear axis and also the growth of the pathogen along the colinear axis is less than its growth orthogonal to this axis.

This is illustrated in Figure 3 using simulated data (mimicking actual observations). Each data point presented in Figure 3 corresponds to a spherical index pair (X-axis—spherical index of root endophyte, Y-axis—spherical index of pathogen). In Figure 3, the logic described in the paragraphs above corresponds to the unshaded quadrant occupied by the green coloured data points in Figure 3 (pathogen, $\alpha/\beta < 1$ and endophyte, $\alpha/\beta > 1$; quadrant 4). The data points of every other quadrant can be considered spurious in the context of them being indicative of antagonism from the endophyte to the pathogen.

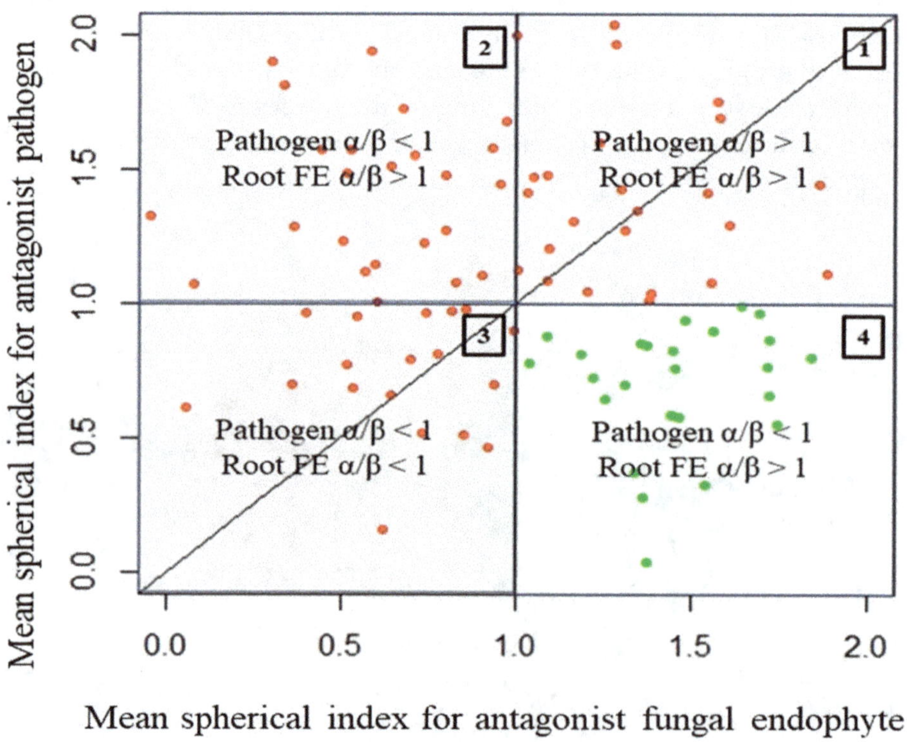

Figure 3. A graph of the mean spherical indices of a set of root fungal endophyte (FE) antagonists plotted against their single test pathogen. This is a simulated dataset generated from a random distribution (n = 100, mean = 1, standard deviation = 0.5) that mimics empirical data observed in this experiment. In the graph above, the data points coloured green are non-spurious instances of successful antagonistic interaction while the red data points indicate spurious ones.

The reasoning is as follows:

1. Quadrant described by pathogen, $\alpha/\beta > 1$, and endophyte, $\alpha/\beta > 1$ (quadrant 1): There is clearly no antagonism occurring in these samples as the growth of neither organism along the colinear axis appears to be suppressed or prohibited.

2. Quadrant described by pathogen, $\alpha/\beta > 1$, and endophyte, $\alpha/\beta < 1$ (quadrant 2): The root endophytes described by their datapoints in this quadrant likely have no effect on the growth of

the pathogen (or may conversely even be getting suppressed by the pathogen themselves) as its progression along the colinear axis is higher in comparison to its orthogonal spread and also higher in comparison to the progression of the antagonist along the colinear axis itself.

3. Quadrant described by pathogen, $\alpha/\beta < 1$, and endophyte, $\alpha/\beta < 1$ (quadrant 3): There is likely no interaction at all between the root endophytes and the pathogen in the cases described by this quadrant as both spherical indices are less than 1; this likely indicates purely natural growth with no antagonistic effects taking place.

Thus, for analysis of this dataset, a short filtering pipeline was constructed in R to select only those samples that met the above-mentioned criteria for consideration as instances of true antagonism (pathogen $\alpha/\beta < 1$ and endophyte $\alpha/\beta > 1$). The code for the pipeline as well as the statistics are presented in the appendix (Supplementary Files 1 and 2). The means and standard deviations of the pathogen spherical indices and the endophyte spherical indices ('PRabMean', 'PRabSD', 'FRabMean', and 'FRabSD' of the 3 replicates each) were first calculated and fed into this pipeline to filter out observations that did not match the selection criteria (pathogen $\alpha/\beta < 1$ and endophyte $\alpha/\beta > 1$). The original raw dataset was then subdivided into subsets consisting of raw observations (i.e., individual replicate-level observations) from day 3, day 7, and day 10, respectively (Supplementary File 2), consisting of only the selected observations, which were then used for comparative statistics. Comparative statistics involved the 12 pathogen control replicates and the triplicate measurements of the pathogen's spherical index from the antagonism assays—since the data had already been filtered for spurious measurements, this comparison would be strongly indicative of the presence of actual antagonist endophytes in the testing regime. Comparisons were achieved by means of an unpaired t-test accounting for inequality in the variances of the sample populations (as sample sizes were unequal). The confidence interval for the test statistic was set to 95%. The means of the spherical indices of different groupings of the test data were then tested against the controls. The groupings were: All root endophytes common to a specific tree host, and only the triplicates of individual endophytes.

The null hypothesis was formulated as follows: The difference between the means of the spherical indices of pathogens under antagonism (S1) and the means of the spherical indices of pathogen controls (S2) is zero (i.e., S1–S2 = 0). The alternative hypothesis was formulated as follows: The difference between the means of the spherical indices of pathogens under antagonism (S1) and the means of the spherical indices of pathogen controls (S2) is less than zero (i.e., S1–S2 < 0). The rationale for assuming this alternative hypothesis was that the spherical index of the pathogen under antagonism would be less than one (<1) while the controls, having grown in the absence of any biotic/abiotic pressure(s), should have a spherical index $\approx$1. Thus, if the effect (i.e., antagonism from that specific endophyte) is real, then the means would also record the same behaviour, and the subtraction of the means of controls from the pathogen replicates involved in the test would then be a negative number (i.e., less than 0). The results of the statistical analysis were then evaluated in the context of the other findings obtained in this study.

3. Results

3.1. Identification and Diversity of Root Endophytes

A total of 65-root endophytes were obtained from eight coniferous hosts and grouped to 19 different groups based on their morphological differences (Tables 2 and 3). Only *P. pungens* did not yield any isolates.

Table 2. The isolated fungal root endophytes' frequency by tree host species.

Identification	Best Match Accession no.	Order	Class	*Picea*				*Pinus*			Sum
				abies	*glauca*	*omorika*	*sitchensis*	*jeffreyi*	*peuce*	*sylvestris*	
Pseudogymnoascus sp.	MH857254	Incertae sedis	Ascomycetes						1		1
Cladosporium sp.	MH863129	Capnodiales	Dothideomycetes					2			2
Pyrenochaeta sp.	KT269928	Pleosporales	Dothideomycetes	2			4				6
Pyrenochaeta sp.	LT821390	Pleosporales	Dothideomycetes		1			7			8
Penicillium sp.	MG821367	Eurotiales	Eurotiomycetes			1					1
Penicillium sp.	MH865458	Eurotiales	Eurotiomycetes			1					1
PAC [b]	AY606286	Helotiales	Leotiomycetes			1					1
Fungal sp.	KY322665	NA	NA			2					2
Pezizales sp.	MH859398	Pezizales	Pezizomycetes							1	1
Clonostachys sp.	KY977560	Hypocreales	Sordariomycetes	3							3
Dactylonectria sp.	MH865183	Hypocreales	Sordariomycetes		8				1		9
Fusarium sp.	MG252283	Hypocreales	Sordariomycetes			2		6			8
Fusarium sp.	MG704912	Hypocreales	Sordariomycetes	2							2
Ilyonectria sp.	MH865177	Hypocreales	Sordariomycetes	1		2					3
Trichoderma sp.	MH930456	Hypocreales	Sordariomycetes							5	5
Unknown		NA	NA				6				6
Unknown		NA	NA					2			2
Unknown		NA	NA					1			1
Unknown		NA	NA	3							3
			TOTAL	11	9	9	10	18	2	6	**65**

[b] PAC: *Phialocephala fortinii* s.l.—*Acephala applanata* species complex.

Table 3. Sequenced root endophytes and their best matches in the National Center for Biotechnology Information (NCBI) database based on concatenated internal transcribed spacer (ITS1-ITS2) sequences. The samples are grouped according to host tree species, and the GenBank accession numbers, sequence similarity (maximum identity/query coverage), as well as region of the homologous sequences used to establish molecular identity are shown.

Sample ID	Host	Best Match	Accession	Mi/Qc [a]	Region	Our Definition
Abie23	*Picea abies*	*Clonostachys candelabrum*	KY977560	100/100	Croatia	*Clonostachys* sp.
Abie26	*Picea abies*	*Fusarium tricinctum*	MG704912	100/099	S. Korea	*Fusarium* sp.
Abie27	*Picea abies*	*Pyrenochaeta sp.*	KT269928	099/100	Greece	*Pyrenochaeta* sp.
Glau13	*Picea glauca*	*Dactylonectria torresensis*	MH865183	098/100	Portugal	*Dactylonectria* sp.
Glau17	*Picea glauca*	*Dactylonectria torresensis*	MH865183	099/100	Portugal	*Dactylonectria* sp.
Glau18	*Picea glauca*	*Pyrenochaeta inflorescentiae*	LT821390	099/100	Germany	*Pyrenochaeta* sp.
Glau21	*Picea glauca*	*Dactylonectria torresensis*	MH865183	100/100	Portugal	*Dactylonectria* sp.
Jeff49	*Pinus jeffreyi*	*Fusarium solani*	MG252283	099/100	China	*Fusarium* sp.
Jeff51	*Pinus jeffreyi*	*Pyrenochaeta inflorescentiae*	LT821390	099/100	Germany	*Pyrenochaeta* sp.
Jeff56	*Pinus jeffreyi*	*Cladosporium subinflatum*	MH863129	099/100	Slovenia	*Cladosporium* sp.
Jeff59	*Pinus jeffreyi*	*Pyrenochaeta inflorescentiae*	LT821390	099/100	Germany	*Pyrenochaeta* sp.
Jeff60	*Pinus jeffreyi*	*Pyrenochaeta inflorescentiae*	LT821390	099/100	Germany	*Pyrenochaeta* sp.
Jeff61	*Pinus jeffreyi*	*Pyrenochaeta inflorescentiae*	LT821390	099/100	Germany	*Pyrenochaeta* sp.
Jeff65	*Pinus jeffreyi*	*Pyrenochaeta inflorescentiae*	LT821390	099/100	Germany	*Pyrenochaeta* sp.
Omori01	*Picea omorika*	*Penicillium glandicola*	MG821367	099/100	Italy	*Penicillium* sp.
Omori02	*Picea omorika*	*Penicillium granulatum*	MH865458	099/100	USA	*Penicillium* sp.
Omori03	*Picea omorika*	Fungal sp. KK15	KY322665	099/100	Montenegro	NA
Omori07	*Picea omorika*	*Phialocephala fortinii*	AY606286	099/100	Sweden	PAC [b]
Omori08	*Picea omorika*	*Ilyonectria lusitanica*	MH865177	099/100	Portugal	*Ilyonectria* sp.
Peuc68	*Pinus peuce*	*Dactylonectria torresensis*	MH865183	100/100	Portugal	*Dactylonectria* sp.
Peuc69	*Pinus peuce*	*Pseudogymnoascus pannorum*	MH857254	100/100	Germany	*Pseudogymnoascus* sp.
Sylv36	*Pinus sylvestris*	*Trichoderma koningii*	MH930456	099/100	Spain	*Trichoderma* sp.
Sylv37	*Pinus sylvestris*	*Trichoderma koningii*	MH930456	099/100	Spain	*Trichoderma* sp.
Sylv38	*Pinus sylvestris*	*Ascobolus lineolatus*	MH859398	097/099	Tanzania	Pezizales sp.

[a] Mi/Qc: Match identity (%)/Query coverace (%); [b] PAC: *Phialocephala fortinii s.l.* – *Acephala applanata* species complex.

The Helotiales sample (Omori07) was identified to be a member of the *Phialocephala fortinii* s.l.—*Acephala applanata* species complex (PAC). The sample, Peuc69, was positively identified as a *Pseudogymnoascus* sp. (100% query coverage and 100% concatenated ITS sequence similarity), but the precise placement of this taxon within the class, Ascomycetes, remains unresolved (Incertae sedis). Although the sample, Sylv38, only had relatively low query coverage and sequence similarity (99% and 97%, respectively, Table 3), potentially being identified as Ascobolus sp., the ML tree firmly established it as belonging to the Pezizales clade (Figure 4). The placements of the root endophytes from the various host trees within the phylogenetic space appears to be freely intermixed, with no readily apparent preferential patterns visible.

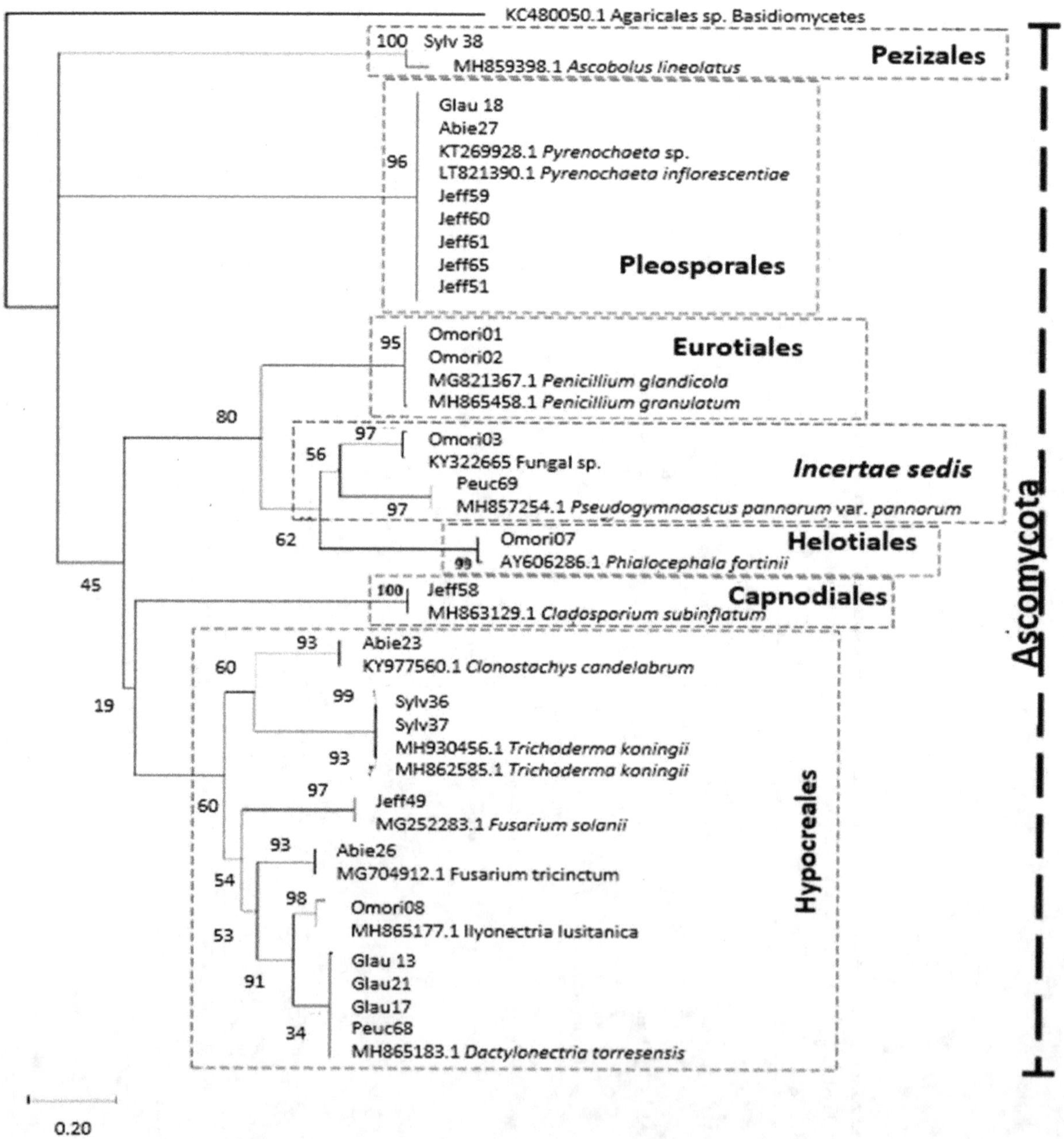

Figure 4. Molecular phylogenetic analysis by maximum likelihood (ML) based on concatenated ITS sequences. The percentage of replicate trees in which the associated taxa clustered together in the bootstrap test are shown next to the branches. Also indicated are the orders to which the samples belong to (dashed rectangles w/descriptions), the phylum (dashed vertical bar), and the outgroup taxon (the Basidiomycete KC480050.1 Agaricales sp. isolated from *P. abies* by Terhonen et al. [19].

The phylogeny was well-resolved up to the species level (Figure 4), with the nodes identifying species relatives featuring strong bootstrap support. Resolution was poorer at the higher taxonomic levels, as the node support was weaker, but yielded 13 distinct clades in total (i.e., lower bootstrap values, Figure 4). Root fungal endophytes belonged to the phylum, Ascomycota, and to the classes, Ascomycetes, Dothideomycetes, Eurotiomycetes, Leotiomycetes, Pezizomycetes, and Sordariomycetes (Table 2). A majority of the identified taxa fell under the orders, Hypocreales and Pleosporales (Table 2, Figure 4). Two taxa were identified as belonging to Eurotiales. One taxa in the orders, Capnodiales, Helotiales, and Pezizales, was also identified.

3.2. Antagonisms Assay

The antagonism assay identified 26 of the 65 root endophytes as being statistically successful antagonists against the pathogen, *H. parviporum* (Table 4). None of the assayed samples displayed statistically significant antagonism on day 3. Half the successful samples (13/26) displayed statistically significant antagonism on day 7, and 19/26 samples displayed statistically significant antagonism on day 10. Only six samples showed statistically significant antagonism on both days 7 and 10. These samples were *Dactylonectria* sp. (Glauc15), *Pyrenochaeta* sp. (Glauc18 and Sitch48), Fungal sp. (Omori03), Fusarium sp. (Jeff50), and unknown (Sitch46). *Pinus jeffreyi* was the host with the largest number of successful antagonist root endophytes (9 samples), followed by *Pinus sitchensis* (6 samples), *Picea omori* (5 samples), *Picea glauca* (4 samples), and *Picea abies* (2 samples) (Table 4).

Table 4. Root endophytes identified as successful antagonists against the pathogen, *H. parviporum*, in the antagonism assay. These samples were identified on the basis of the significance of a t-test (p-value = 0.05) comparing the mean spherical index of their pathogen partners (n = 3 trials) with the mean spherical index of the pathogen controls (n = 12).

Sample ID	Host	Identification	Sampling Time (days) [a]		
			3	7	10
Abie25	*Picea abies*	*Fusarium* sp.	N	Y	N
Abie29	*Picea abies*	Unknown	N	N	Y
Glauc15	*Picea glauca*	*Dactylonectria* sp.	N	Y	Y
Glauc19	*Picea glauca*	*Dactylonectria* sp.	N	N	Y
Glauc13	*Picea glauca*	*Dactylonectria* sp.	N	Y	N
Glauc18	*Picea glauca*	*Pyrenochaeta* sp.	N	Y	Y
Omori01	*Picea omorika*	*Penicillium* sp.	N	Y	N
Omori03	*Picea omorika*	Fungal sp.	N	Y	Y
Omori07	*Picea omorika*	PAC [b]	N	N	Y
Omori08	*Picea omorika*	*Ilyonectria* sp.	N	Y	N
Omori06	*Picea omorika*	Fungal sp.	N	Y	N
Jeff50	*Pinus jeffreyi*	*Fusarium* sp.	N	Y	Y
Jeff55	*Pinus jeffreyi*	*Fusarium* sp.	N	N	Y
Jeff58	*Pinus jeffreyi*	*Fusarium* sp.	N	N	Y
Jeff62	*Pinus jeffreyi*	*Pyrenochaeta* sp.	N	N	Y
Jeff63	*Pinus jeffreyi*	*Pyrenochaeta* sp.	N	N	Y
Jeff64	*Pinus jeffreyi*	Unknown	N	N	Y
Jeff51	*Pinus jeffreyi*	*Pyrenochaeta* sp.	N	Y	N
Jeff56	*Pinus jeffreyi*	*Cladosporium* sp.	N	Y	N
Jeff65	*Pinus jeffreyi*	*Pyrenochaeta* sp.	N	N	Y
Sitch40	*Pinus sitchensis*	*Pyrenochaeta* sp.	N	N	Y
Sitch41	*Pinus sitchensis*	*Pyrenochaeta* sp.	N	N	Y
Sitch42	*Pinus sitchensis*	Unknown	N	N	Y
Sitch46	*Pinus sitchensis*	Unknown	N	Y	Y
Sitch47	*Pinus sitchensis*	*Pyrenochaeta* sp.	N	N	Y
Sitch48	*Pinus sitchensis*	*Pyrenochaeta* sp.	N	Y	Y

[a] 'Y' is statistically significant, 'N' is insignificant (for p-value = 0.05). [b] PAC: *Phialocephala fortinii* s.l.—*Acephala applanata* species complex.

Assuming that smaller values of α/β (i.e., $\alpha/\beta \rightarrow 0$) for the pathogen indicate more successful root endophyte antagonists (Figures 3 and 5), the endophyte, *Cladosporium* sp. (Jeff56), from the host,

Pinus jeffreyi, is the most successful antagonist in the sample set 7 days after the start of the experiment (pathogen spherical index 0.74 ± 0.13, endophyte spherical index 1.04 ± 0.03). Similarly, all averages of the spherical indices of all antagonists' root endophytes with statistically significant effects against the pathogen, *H. parviporum*, were identified with the R-script (Supplementary Files 1 and 2). To study the temporal stability of the antagonism of samples that yielded statistically significant results on both days 7 and 10, the ratios of their mean spherical indices (i.e., MEAN (pathogen spherical index)/MEAN (endophyte spherical index) for every sample) from both time points of measurement were plotted and visualized (Figure 5).

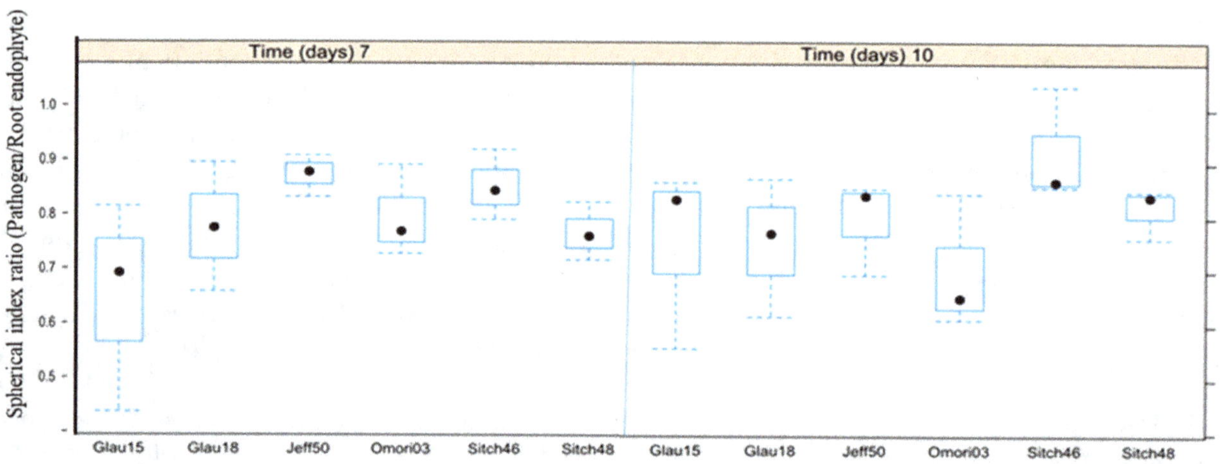

Figure 5. Ratios of the mean spherical indices of the pathogen over its antagonistic partner from day 7 and day 10 measurements of the antagonism assay visualized side-by-side in box-and-whisker plots (solid black circles—medians). Samples shown here are only those which displayed statistically significant antagonism on both days of measurement.

4. Discussion

Fungal root endophytes could not be isolated from *P. pungens*. It is unlikely that this conifer species hosts no endophytes. Hence, it is plausible to suggest the presence of uncultivable root endophytes in this particular species, by which the methodology used in this study is unable to recover them [58,59]. This experimental constraint is consolidated by the fact that no relevant information on the abundances and diversity of this conifer's root fungal endophyte community is currently available in the literature. The overall community composition observed in this study greatly differs from other studies of conifer roots [19,20,24,60,61]. This study encountered only a single PAC species (from *P. omorika*, sample Omori07). This contrasts with previous studies that assume that those DSEs—especially members of the PAC—are predominant and abundant among coniferous hosts [15,19–21,24,31,33,34,62]. This result is of special relevance since no PAC species were recovered from well-studied hosts, such as *P. abies* and *P. sylvestris*. Taking into account that the samples in this study were collected from a non-forest environment, a botanical garden, one explanation could be that the environment influences the composition of endophytic assemblage [63,64]. However, the methodological approach used for the assessment of endophytes' antagonism against *H. parviporum* is methodologically robust. This enables the identification of endophytes that represent potential candidates for future investigations in the context of host-enhancement effects and biocontrol applications.

Species in the genus, *Cladosporium*, are common endophytes also found on conifer hosts [65,66]. The *Cladosporium* sp. identified in this study (sample Jeff56) also appears to impart a benefit to its host (*P. jeffreyi*) as it was the most successful antagonist against *H. parviporum* in this study (Table 4; Figure 5). The fungal endophyte, *Clonostachys* sp., was identified in the roots of *Picea glauca*. While there are no known instances of this specific root endophyte–host pairing, *Clonostachys* itself is a well-known root endophyte genus with some species imparting potential benefits to the host plant [67,68]. However, in this study, it was not a successful antagonist against the root rot pathogen (Table 4). Interestingly, several occurrences of the Cylindrocarpon-like pathogenic fungus, *Dactylonectria* sp. (best match *Dactylonectria torresensis*), were found in this study (in *P. glauca* and *P. peuce*, respectively), in which the sample, Glauc13, was one of the successful antagonists of *H. parviporum*. This fungus has been reported as the cause of root rot and decline in host health in strawberry, raspberry, olive trees, and in apple orchards [69–71]. Although species from the genus, *Dactylonectria*, have been encountered in European old-growth forests [72], no study has reported the presence of this genus in *Picea* or *Pinus* before. *Fusarium* spp. were recovered from multiple hosts (Table 2). *Fusarium* is a well-documented and ubiquitous root endophyte-containing genus [73–75] of which some species are pathogenic upon their hosts [76]. The results obtained in this study demonstrated that some *Fusarium* strains proved to be successful in the antagonism assay (Table 4). The Ilyonectria was also encountered in this study (in *P. omorika*). This is a widely-occurring genus of endophytic fungi [77–80] and at least one member has also been found in *P. glauca* [81]. Thus, this study is also likely the first time a member of this genus has been recorded in *P. omorika*. Ilyonectria spp. are also among frequently occurring non-DSE root endophytes. Whether the presence of these fungi in the hosts from this study is pathogenic or beneficial has yet to be defined. The results from this study indicates that the Ilyonectria (isolate Omori08) is a good antagonist of *H. parviporum*, suggesting that it could compete for niche space with other fungi [82].

Penicillium is a genus of well-known root endophytes that include non-DSE species found in grasses [83], orchids [67], and woody plants [84]. *Penicillium* spp. are also encountered in the roots of conifer hosts [85]. One *Penicillium* spp. (sample Omori01) proved to be a successful antagonist against *H. parviporum* (Table 4). This sample's closest *Penicillium* relative is *P. glandicola*, which was only recently revealed as an endophyte in the cactus, *Opunita ficus-indica* [86]. *Pseudogymnoascus* sp. in this study was the only sample isolated from *P. peuce*. *Pseudogymnococcus* spp. are well-known root endophytes recovered from a variety of hosts, including Ericaceae, *Pinus*, and *Picea* [87–89]. *Pyrenochaeta* is a genus of fungi that contains several DSEs in grasses [83], shrubs [90], and woody species [42,62]. Although a *Pyrenochaeta* species was reported as an endophyte evoking inhibition of growth in the ash dieback pathogen, *Hymenoscyphus fraxineus* [91]—indicating that it can be a beneficial endophyte to its host—it has never before been reported as a conifer endophyte. *Pyrenochaeta* spp. have only been reported in conifers in the role of fungal decomposers of dead wood [92]. In this study, not only *Pyrenochaeta* spp. have been found in the roots of multiple host species (*P. abies*, *P. glauca*, *P. sitchensis*, and *Pinus jeffreyi*, Table 2), but we also observed they can be successful antagonists against *H. parviporum* (Table 4). The *Pyrenochaeta* sp. (Glauc18 and Sitch48) exhibited statistically significant antagonism on both day 7 and day 10 of measurement (Table 4). Thus, the genus, *Pyrenochaeta*, may hold interesting possibilities for the biocontrol of pathogenicity in conifer hosts. *Trichoderma* is a genus of well-known endophytes found in orchids [93], herbaceous plants [94], woody species [95], and conifer hosts [96]. *Trichoderma* sp. were found in this study from *P. sylvestris*. None of the isolates, however, were an effective antagonist against *H. parviporum*, indicating that *Trichoderma* in the roots does not play a defensive role against this pathogen. Pezizales sp. observed in this study are also a known root endophyte of conifer hosts as at least one member of this order has been encountered in a conifer previously [62].

Molecular phylogenetic analysis by maximum likelihood (ML) based on concatenated ITS sequences (Figure 4) suggests that the unidentified fungal sp. (Omori03) is likely a close relative of Pseudogymnoascus, indicating that this fungus might belong to the Ascomycetes class. Unfortunately, the node support for the branches from which these relationships were inferred is <90%, so a final conclusion in the absence of other corroborating data cannot be made. This isolate, fungal sp. (Omori03), was one of the strongest antagonists against *H. parviporum* as it displayed statistically significant antagonism on both day 7 and day 10 of measurement (Table 4). The only PAC species was isolated from the host, *P. omorika*. The screened literature does not contain any records of the PAC occurring in

P. omorika, making our result the first confirmation of the species complex' presence in the roots of this conifer species. This PAC species was also found to be a strong antagonist against *H. parviporum*. With regards to PAC species as antagonist candidates, the PAC species identified in this study (Omori07) did offer above-average antagonism to *H. parviporum*, but only on day 10 of measurement (Table 4).

In theory, assuming strength by variety and numbers, *P. jeffreyi* with nine successful root endophytic antagonists would be the best protected tree species against *H. parviporum*. Hence, we can assume a possible correlation between hosting a greater number of antagonist endophytes and level of defensive capabilities against root pathogens. This, however, remains a hypothesis and requires validation for further confirmation, especially as *P. jeffreyi* is not the main host tree of *H. parviporum* [35]).

Root fungal endophytes are important co-inhabitants of the rhizosphere of coniferous host species. This study has identified a variety of DSE and non-DSE root endophytes from coniferous hosts, including undiscovered root fungal endophyte diversity (e.g., in *P. omorika*). The antagonism assays performed indicate that potential antagonist root endophytes against *H. parviporum* can be found in a wide variety of coniferous hosts, and that some of these antagonists may also be suitable for long-term biocontrol of the pathogen due to their temporally stable and persistent antagonism. The results of this study also indicate that *Cladosporium* sp. and *Pyrenochaeta* sp. are very strong antagonists to *H. parviporum* under laboratory conditions in different ways. *Cladosporium* sp. did not exhibit temporally persistent antagonism, and while *Pyrenochaeta* sp. exhibited temporally persistent antagonism, it was not the strongest antagonist. Results established in this paper and the surveyed literature demonstrate that the endophyte–host interactions are complex and affect a wide variety of interactions [5,97], and that even pathogenic fungi (e.g., *Dactylonectria*) can act as successful antagonists against other pathogens (e.g., *H. parviporum*), offering protection to the colonized hosts in the process. Schlegel et al. [98] demonstrated that there is probably no clear-cut distinction between a pathogenic endophyte and a mutualistic one, with the endophyte likely switching roles seamlessly as environmental and host-associated factors vary [99]. The factors affecting the antagonist–pathogen relationship also need to be fully elucidated. In addition, the mechanisms mediating these interactions are also not fully understood nor generalized [62]. It was only recently demonstrated experimentally that root endophytes safeguard hosts from root pathogens by quickly occupying the newly growing roots [8]. In this context, antagonistic endophytes as biocontrol measures offer an enticing alternative to chemical methods in forestry [5]. Successful use of root fungal endophytes as biocontrol agents would thus require a thorough understanding of the endophyte–host relationship, the endophyte–pathogen relationship, the host–pathogen relationship, and the influence of environmental factors on these relationships.

Only six of the 65 root endophyte samples displayed statistically significant antagonism on both time points, 7 and 10 days, of the assay. This is indicative of a temporal aspect to the antagonism under this study's experimental conditions—namely, that in a purely 1 vs. 1 scenario with limited space and nutrient availability, a root endophyte might be initially effective against a pathogen only for its antagonistic ability to falter later where the pathogen is potentially the better competitor/survivor under a more resource-constrained situation. With this in mind, selection of endophytes for potential biocontrol applications should then take into account the temporal aspects of the antagonism offered by the fungus. The temporal aspects primarily relate to the instantaneous stability of the antagonism (i.e., there must be minimal variation in the measurable antagonistic effects against the pathogen over time) and to the persistence of the antagonistic effect (i.e., the antagonistic effect must not taper off with time). In the experimental context, this implies that promising candidates from initial, short-term antagonism screenings should then be subjected to extended screening (preferably over multiple years) under lab and field conditions.

5. Conclusions

Root endophytes are effective antagonists against *H. parviporum*, suggesting that they may be of benefit to their host tree(s), should these endophytes form either mutualistic or commensalistic relationships with the host tree. On the other hand, two types of pathogenic fungi were also recovered as root endophytes, with some occurrences of them actually being good antagonists against *H. parviporum* (Table 4). Further, it is also possible that the sampled trees host entirely different root endophytes compared to their forest counterparts, as these were old, mature trees situated in a botanical garden. Nonetheless, these results imply that there is no simplistic picture of the root endophyte–host tree relationship, and that these interactions need to be investigated in detail with a more all-encompassing range of methodologies on a case-by-case basis. The mechanistic implications of the presence of the discovered root endophytes upon their respective hosts are beyond the scope of this study. In conclusion, this study identified the root fungal endophytes of several conifer hosts, which demonstrated successful antagonism against the white rot pathogen, *H. parviporum*. The data, conclusions, and inferences generated by this study will serve as the basis for executing future investigations into their potential for acting as biocontrol agents against root pathogens.

Author Contributions: L.R. collected the samples, performed the laboratory experiments and analyzed the data. L.R. wrote the R-script and the first draft. E.T. designed the experimental protocol, conceived the experiment, and contributed to the final draft. K.B. advised and contributed on the experimental protocols, and contributed to the writing of the manuscript.

Acknowledgments: We would like to thank Julia Quintana and Susanna Keriö for valuable comments on the manuscript. We are obliged to Xenia Yeoh Hao-Yi for kindly improving the manuscript as a native English speaker.

References

1. Carroll, G. Fungal Endophytes in Stems and Leaves: From Latent Pathogen to Mutualistic Symbiont. *Ecology* **1988**, *69*, 2–9. [CrossRef]
2. Saikkonen, K.; Faeth, S.H.; Helander, M.; Sullivan, T.J. Fungal endophytes. A continuum of interactions with host plants. *Annu. Rev. Ecol. Syst.* **1998**, *29*, 319–343.
3. Clay, K. Fungi and the food of the gods. *Nature* **2004**, *427*, 401–402. [CrossRef]
4. Petrini, O. Fungal Endophytes of Tree Leaves. In *Microbial Ecology of Leaves*; Andrews, J.H., Hirano, S.S., Eds.; Springer: New York, NY, USA, 1991; pp. 179–197.
5. Terhonen, E.; Blumenstein, K.; Kovalchuk, A.; Asiegbu, F.O. Forest Tree Microbiomes and associated Fungal Endophytes: Functional Roles and Impact on Forest Health. *Forests* **2019**, *10*, 42. [CrossRef]
6. Tellenbach, C.; Sieber, T.N. Do colonization by dark septate endophytes and elevated temperature affect pathogenicity of oomycetes? *FEMS Microbiol. Ecol.* **2012**, *82*, 157–168. [CrossRef] [PubMed]
7. Martín, J.A.; Witzell, J.; Blumenstein, K.; Rozpedowska, E.; Helander, M.; Sieber, T.N.; Gil, L. Resistance to Dutch Elm Disease Reduces Presence of Xylem Endophytic Fungi in Elms (Ulmus spp.). *PLoS ONE* **2013**, *8*, e56987.
8. Terhonen, E.; Sipari, N.; Asiegbu, F.O. Inhibition of phytopathogens by fungal root endophytes of Norway spruce. *Biol. Control* **2016**, *99*, 53–63. [CrossRef]
9. Jumpponen, A.; Trappe, J.M. Performance of *Pinus contorta* inoculated with two strains of root endophytic fungus, *Phialocephala fortinii*: Effects of synthesis system and glucose concentration. *Can. J. Bot.* **1998**, *76*, 1205–1213.
10. Newsham, K.K. A meta-analysis of plant responses to dark septate root endophytes. *New Phytol.* **2011**, *190*, 783–793. [CrossRef]
11. Vohník, M.; Lukančič, S.; Bahor, E.; Regvar, M.; Vosátka, M.; Vodnik, D. Inoculation of Rhododendron cv. Belle-Heller with two strains of *Phialocephala fortinii* in two different substrates. *Folia Geobot.* **2003**, *38*, 191. [CrossRef]
12. Tellenbach, C.; Grünig, C.R.; Sieber, T.N. Negative effects on survival and performance of Norway spruce seedlings colonized by dark septate root endophytes are primarily isolate-dependent. *Environ. Microbiol.* **2011**, *13*, 2508–2517. [CrossRef]

13. Fernando, A.A.; Currah, R.S. A comparative study of the effects of the root endophytes *Leptodontidium orchidicola* and *Phialocephala fortinii* (Fungi Imperfecti) on the growth of some subalpine plants in culture. *Can. J. Bot.* **1996**, *74*, 1071–1078. [CrossRef]
14. Jumpponen, A.; Mattson, K.G.; Trappe, J.M. Mycorrhizal functioning of *Phialocephala fortinii* with *Pinus contorta* on glacier forefront soil: Interactions with soil nitrogen and organic matter. *Mycorrhiza* **1998**, *7*, 261–265. [CrossRef]
15. Grünig, C.R.; Queloz, V.; Sieber, T.N.; Holdenrieder, O. Dark septate endophytes (DSE) of the *Phialocephala fortinii* s.l.—*Acephala applanata* species complex in tree roots: Classification, population biology, and ecology. *Botany* **2008**, *86*, 1355–1369. [CrossRef]
16. Mandyam, K.; Jumpponen, A. Seeking the elusive function of the root-colonising dark septate endophytic fungi. *Stud. Mycol.* **2005**, *53*, 173–189. [CrossRef]
17. Witzell, J.; Martín, J.A. Phenolic metabolites in the resistance of northern forest trees to pathogens—Past experiences and future prospects. *Can. J. For. Res.* **2008**, *38*, 2711–2727. [CrossRef]
18. Schulz, B.J.E.; Boyle, C.J.C.; Sieber, T.N. *Microbial Root Endophytes*; Soil Biology; Springer: Berlin/Heidelberg, Germany, 2006; ISBN 978-3-540-33525-2.
19. Terhonen, E.; Keriö, S.; Sun, H.; Asiegbu, F.O. Endophytic fungi of Norway spruce roots in boreal pristine mire, drained peatland and mineral soil and their inhibitory effect on *Heterobasidion parviporum* in vitro. *Fungal Ecol.* **2014**, *9*, 17–26. [CrossRef]
20. Kovalchuk, A.; Mukrimin, M.; Zeng, Z.; Raffaello, T.; Liu, M.; Kasanen, R.; Sun, H.; Asiegbu, F.O. Mycobiome analysis of asymptomatic and symptomatic Norway spruce trees naturally infected by the conifer pathogens Heterobasidion spp. *Environ. Microbiol. Rep.* **2018**, *10*, 532–541. [CrossRef]
21. Addy, H.D.; Piercey, M.M.; Currah, R.S. Microfungal endophytes in roots. *Can. J. Bot.* **2005**, *83*, 1–13. [CrossRef]
22. Jumpponen, A.; Trappe, J.M. Dark septate endophytes: A review of facultative biotrophic root-colonizing fungi. *New Phytol.* **1998**, *140*, 295–310. [CrossRef]
23. Queloz, V.; Sieber, T.N.; Holdenrieder, O.; McDonald, B.A.; Grünig, C.R. No biogeographical pattern for a root-associated fungal species complex. *Glob. Ecol. Biogeogr.* **2011**, *20*, 160–169. [CrossRef]
24. Sietiö, O.-M.; Tuomivirta, T.; Santalahti, M.; Kiheri, H.; Timonen, S.; Sun, H.; Fritze, H.; Heinonsalo, J. Ericoid plant species and *Pinus sylvestris* shape fungal communities in their roots and surrounding soil. *New Phytol.* **2018**, *218*, 738–751. [CrossRef]
25. Grünig, C.R.; Duò, A.; Sieber, T.N.; Holdenrieder, O. Assignment of species rank to six reproductively isolated cryptic species of the *Phialocephala fortinii* s.l.-*Acephala applanata* species complex. *Mycologia* **2008**, *100*, 47–67. [CrossRef]
26. Stoyke, G.; Currah, R.S. Endophytic fungi from the mycorrhizae of alpine ericoid plants. *Can. J. Bot.* **1991**, *69*, 347–352. [CrossRef]
27. Tejesvi, M.V.; Sauvola, T.; Pirttilä, A.M.; Ruotsalainen, A.L. Neighboring *Deschampsia flexuosa* and *Trientalis europaea* harbor contrasting root fungal endophytic communities. *Mycorrhiza* **2013**, *23*, 1–10. [CrossRef]
28. Grünig, C.R.; Sieber, T.N.; Rogers, S.O.; Holdenrieder, O. Spatial distribution of dark septate endophytes in a confined forest plot. *Mycol. Res.* **2002**, *106*, 832–840. [CrossRef]
29. Grünig, C.R.; McDonald, B.A.; Sieber, T.N.; Rogers, S.O.; Holdenrieder, O. Evidence for subdivision of the root-endophyte *Phialocephala fortinii* into cryptic species and recombination within species. *Fungal Genet. Biol.* **2004**, *41*, 676–687. [CrossRef]
30. Queloz, V.; Duò, A.; Grünig, C.R. Isolation and characterization of microsatellite markers for the tree-root endophytes *Phialocephala subalpina* and *Phialocephala fortinii* s.s. *Mol. Ecol. Resour.* **2008**, *8*, 1322–1325. [CrossRef]
31. Queloz, V.; Grunig, C.R.; Sieber, T.N.; Holdenrieder, O. Monitoring the spatial and temporal dynamics of a community of the tree-root endophyte *Phialocephala fortinii* s.l. *New Phytol.* **2005**, *168*, 651–660. [CrossRef]
32. Queloz, V.; Duò, A.; Sieber, T.N.; Grünig, C.R. Microsatellite size homoplasies and null alleles do not affect species diagnosis and population genetic analysis in a fungal species complex. *Mol. Ecol. Resour.* **2010**, *10*, 348–367. [CrossRef]
33. Sieber, T.N. Endophytic fungi in forest trees: Are they mutualists? *Fungal Biol. Rev.* **2007**, *21*, 75–89. [CrossRef]
34. Saikkonen, K. Forest structure and fungal endophytes. *Fungal Biol. Rev.* **2007**, *21*, 67–74. [CrossRef]

35. Garbelotto, M.; Gonthier, P. Biology, epidemiology, and control of Heterobasidion species worldwide. *Annu. Rev. Phytopathol.* **2013**, *51*, 39–59. [CrossRef]
36. Gonthier, P.; Garbelotto, M.M.; Nicolotti, G. Seasonal Patterns of Spore Deposition of Heterobasidion Species in Four Forests of the Western Alps. *Phytopathology* **2005**, *95*, 759–767. [CrossRef]
37. Lind, M.; Stenlid, J.; Olson, Å. Chapter Twelve—*Heterobasidion annosum* s.l. Genomics. In *Advances in Botanical Research*; Martin, F.M., Ed.; Academic Press: London, UK, 2014; Volume 70, pp. 371–396.
38. Asiegbu, F.O.; Adomas, A.; Stenlid, J. Conifer root and butt rot caused by *Heterobasidion annosum* (Fr.) Bref. s.l. *Mol. Plant Pathol.* **2005**, *6*, 395–409. [CrossRef]
39. Korhonen, K. Intersterility groups of *Heterobasidion annosum*. *Commun. Inst. For. Fenn.* **1978**, *94*, 1–25.
40. Capretti, P.; Korhonen, K.; Mugnai, L.; Romagnoli, C. An intersterility group of *Heterobasidion annosum* specialized to *Abies alba*. *Eur. J. For. Pathol.* **1990**, *20*, 231–240. [CrossRef]
41. Rishbeth, J. Dispersal of *Fomes annosus* Fr. and *Peniophora gigantea* (Fr.) massee. *Trans. Br. Mycol. Soc.* **1959**, *42*, 243–260. [CrossRef]
42. Isomäki, A.; Kallio, T. Consequences of injury caused by timber harvesting machines on the growth and decay of spruce (*Picea abies* (L.) Karst.). *Acta For. Fenn.* **1974**, *136*, 1–34. [CrossRef]
43. Redfern, D.B.; Stenlid, J. Spore dispersal and infection. In *Heterobasidion Annosum: Biology, Ecology, Impact and Control*; Woodward, S., Stenlid, J., Karjalainen, R., Hüttermann, A., Eds.; CAB International: Wallingford, UK, 1998; pp. 105–124.
44. Piri, T. The spreading of the S type of *Heterobasidion annosum* from Norway spruce stumps to the subsequent tree stand. *Eur. J. For. Pathol.* **1996**, *26*, 193–204. [CrossRef]
45. Oliva, J.; Zhao, A.; Zarei, S.; Sedlák, P.; Stenlid, J. Effect of temperature on the interaction between *Phlebiopsis gigantea* and the root-rot forest pathogen Heterobasidion spp. *For. Ecol. Manag.* **2015**, *340*, 22–30. [CrossRef]
46. Piri, T.; Hamberg, L. Persistence and infectivity of *Heterobasidion parviporum* in Norway spruce root residuals following stump harvesting. *For. Ecol. Manag.* **2015**, *353*, 49–58. [CrossRef]
47. Marx, D.H. The influence of ectotrophic mycorrhizal fungi on the resistance of pine roots to pathogenic infections. I. Antagonism of mycorrhizal fungo to root pathogenic fungi and soil bacteria. *Phytopathology* **1969**, *59*, 153–163.
48. Gardes, M.; Bruns, T.D. ITS primers with enhanced specificity for basidiomycetes—Application to the identification of mycorrhizae and rusts. *Mol. Ecol.* **1993**, *2*, 113–118. [CrossRef]
49. Bengtsson-Palme, J.; Ryberg, M.; Hartmann, M.; Branco, S.; Wang, Z.; Godhe, A.; Wit, P.D.; Sánchez-García, M.; Ebersberger, I.; de Sousa, F.; et al. Improved software detection and extraction of ITS1 and ITS2 from ribosomal ITS sequences of fungi and other eukaryotes for analysis of environmental sequencing data. *Methods Ecol. Evol.* **2013**, *4*, 914–919. [CrossRef]
50. Kumar, S.; Stecher, G.; Li, M.; Knyaz, C.; Tamura, K. MEGA X: Molecular evolutionary genetics analysis across computing platforms. *Mol. Biol. Evol.* **2018**, *35*, 1547–1549. [CrossRef]
51. Altschul, S.F.; Gish, W.; Miller, W.; Myers, E.W.; Lipman, D.J. Basic local alignment search tool. *J. Mol. Biol.* **1990**, *215*, 403–410. [CrossRef]
52. Katoh, K.; Standley, D.M. MAFFT Multiple Sequence Alignment Software Version 7: Improvements in Performance and Usability. *Mol. Biol. Evol.* **2013**, *30*, 772–780. [CrossRef]
53. Kimura, M. A simple method for estimating evolutionary rates of base substitutions through comparative studies of nucleotide sequences. *J. Mol. Evol.* **1980**, *16*, 111–120. [CrossRef]
54. Felsenstein, J. Confidence Limits on Phylogenies: An Approach Using the Bootstrap. *Evolution* **1985**, *39*, 783–791. [CrossRef]
55. Stöver, B.C.; Müller, K.F. TreeGraph 2: Combining and visualizing evidence from different phylogenetic analyses. *BMC Bioinform.* **2010**, *11*, 7. [CrossRef] [PubMed]
56. Blumenstein, K. Endophytic Fungi in Elms: Implications for the Integrated Management of Dutch Elm Disease. Ph.D. Thesis, Swedish University of Agricultural Sciences, Alnarp, Sweden, 2016.
57. Martín, J.A.; Macaya-Sanz, D.; Witzell, J.; Blumenstein, K.; Gil, L. Strong in vitro antagonism by elm xylem endophytes is not accompanied by temporally stable in planta protection against a vascular pathogen under field conditions. *Eur. J. Plant Pathol.* **2015**, *142*, 185–196. [CrossRef]
58. Sun, X.; Guo, L.-D. Endophytic fungal diversity: Review of traditional and molecular techniques. *Mycology* **2012**, *3*, 65–76.

59. Siddique, A.B.; Khokon, A.M.; Unterseher, M. What do we learn from cultures in the omics age? High-throughput sequencing and cultivation of leaf-inhabiting endophytes from beech (*Fagus sylvatica* L.) revealed complementary community composition but similar correlations with local habitat conditions. *MycoKeys* **2017**, *20*, 1–16. [CrossRef]
60. Lazarević, J.; Menkis, A. Fungi inhabiting fine roots of Pinus heldreichii in the Montenegrin montane forests. *Symbiosis* **2018**, *74*, 189–197. [CrossRef]
61. Ahlich, K.; Sieber, T.N. The profusion of dark septate endophytic fungi in non-ectomycorrhizal fine roots of forest trees and shrubs. *New Phytol.* **1996**, *132*, 259–270. [CrossRef]
62. Sieber, T.N.; Grünig, C.R. Fungal root endophytes. In *Plant Roots: The Hidden Half*, 4th ed.; Eshel, A., Beeckman, T., Eds.; CRC Press LLC: Boca Raton, FL, USA, 2013; pp. 38.1–38.49.
63. Grünig, C.R.; Duò, A.; Sieber, T.N. Population genetic analysis of *Phialocephala fortinii* s.l. and *Acephala applanata* in two undisturbed forests in Switzerland and evidence for new cryptic species. *Fungal Genet. Biol.* **2006**, *43*, 410–421. [CrossRef]
64. Stroheker, S.; Dubach, V.; Queloz, V.; Sieber, T.N. Resilience of *Phialocephala fortinii* s.l.—*Acephala applanata communities*—Effects of disturbance and strain introduction. *Fungal Ecol.* **2018**, *31*, 19–28. [CrossRef]
65. Bensch, K.; Braun, U.; Groenewald, J.Z.; Crous, P.W. The genus Cladosporium. *Stud. Mycol.* **2012**, *72*, 1–401. [CrossRef]
66. Paul, N.C.; Yu, S.H. Two species of endophytic Cladosporium in Pine trees in Korea. *Mycobiology* **2008**, *36*, 211–216. [CrossRef]
67. Yuan, Z.; Chen, Y.; Yang, Y. Diverse non-mycorrhizal fungal endophytes inhabiting an epiphytic, medicinal orchid (*Dendrobium nobile*): Estimation and characterization. *World J. Microbiol. Biotechnol.* **2009**, *25*, 295. [CrossRef]
68. Sutton, J.C.; Liu, W.; Ma, J.; Brown, W.G.; Stewart, J.F.; Walker, G.D. Evaluation of the fungal endophyte *Clonostachys rosea* as an inoculant to enhance growth, fitness and productivity of crop plants. *Acta Hortic.* **2008**, *782*, 279–286. [CrossRef]
69. Manici, L.M.; Kelderer, M.; Caputo, F.; Saccà, M.L.; Nicoletti, F.; Topp, A.R.; Mazzola, M. Involvement of Dactylonectria and Ilyonectria spp. in tree decline affecting multi-generation apple orchards. *Plant Soil* **2018**, *425*, 217–230. [CrossRef]
70. Weber, R.W.S.; Entrop, A.-P. *Dactylonectria torresensis* as the Main Component of the Black Root Rot Complex of Strawberries and Raspberries in Northern Germany. *Erwerbs-Obstbau* **2017**, *59*, 157–169. [CrossRef]
71. Nigro, F.; Antelmi, I.; Sion, V.; Parente, P.; Pacifico, A. First Report of *Dactylonectria torresensis* Causing Foot and Root Rot of Olive Trees. *Plant Dis.* **2018**. [CrossRef]
72. Jankowiak, R.; Stępniewska, H.; Szwagrzyk, J.; Bilański, P.; Gratzer, G. Characterization of Cylindrocarpon-like species associated with litter in the old-growth beech forests of Central Europe. *For. Pathol.* **2016**, *46*, 582–594. [CrossRef]
73. Demers, J.E.; Gugino, B.K.; Jiménez-Gasco, M.d.M. Highly Diverse Endophytic and Soil *Fusarium oxysporum* Populations Associated with Field-Grown Tomato Plants. *Appl. Environ. Microbiol.* **2015**, *81*, 81–90. [CrossRef]
74. González-Teuber, M.; Vilo, C.; Bascuñán-Godoy, L. Molecular characterization of endophytic fungi associated with the roots of *Chenopodium quinoa* inhabiting the Atacama Desert, Chile. *Genom. Data* **2017**, *11*, 109–112. [CrossRef]
75. Rodriguez, R.J.; White, J.F., Jr.; Arnold, A.E.; Redman, R.S. Fungal endophytes: Diversity and functional roles: *Tansley review*. *New Phytol.* **2009**, *182*, 314–330. [CrossRef]
76. Castañares, E.; Stenglein, S.A.; Dinolfo, M.I.; Moreno, M.V. Fusarium tricinctum Associated with Head Blight on Wheat in Argentina. *Plant Dis.* **2010**, *95*, 496. [CrossRef]
77. Görke, C. Mykozönosen von Wurzel und Stamm von Jungbäumen unterschiedlicher Bestandsbegründungen. Available online: https://www.schweizerbart.de/publications/detail/isbn/9783443590758 (accessed on 28 February 2019).
78. Sieber, T.N. *Endophytische Pilze von Winterweizen (Triticum vulgare Vill.): Ein Vergleich Zwischen Weizen aus Gebeiztem und Solchem aus Ungebeiztem Saatgut*; ETH Zürich: Zürich, Switzerland, 1985.
79. Stark, C.; Babik, W.; Durka, W. Fungi from the roots of the common terrestrial orchid *Gymnadenia conopsea*. *Mycol. Res.* **2009**, *113*, 952–959. [CrossRef] [PubMed]

80. Petrini, O.; Fisher, P.; Petrini, L. Fungal endophytes of bracken (*Pteridium aquilinum*), with some reflections on their use in biological control. *Sydowia* **1992**, *44*, 282–293.
81. Cabral, A.; Rego, C.; Nascimento, T.; Oliveira, H.; Groenewald, J.Z.; Crous, P.W. Multi-gene analysis and morphology reveal novel Ilyonectria species associated with black foot disease of grapevines. *Fungal Biol.* **2012**, *116*, 62–80. [CrossRef] [PubMed]
82. Blumenstein, K.; Albrectsen, B.R.; Martín, J.A.; Hultberg, M.; Sieber, T.N.; Helander, M.; Witzell, J. Nutritional niche overlap potentiates the use of endophytes in biocontrol of a tree disease. *BioControl* **2015**, *60*, 655–667. [CrossRef]
83. Sánchez Márquez, S.; Bills, G.F.; Domínguez Acuña, L.; Zabalgogeazcoa, I. Endophytic mycobiota of leaves and roots of the grass *Holcus lanatus*. *Fungal Divers.* **2010**, *41*, 115–123. [CrossRef]
84. Nicoletti, R.; Fiorentino, A.; Scognamiglio, M. Endophytism of Penicillium _Species in Woody Plants. *TOMYCJ* **2014**, *8*, 1–26. [CrossRef]
85. Courtois, H. The mycoflora in the crown region and rhizosphere of Norway spruce (*Picea abies*), and its significance for interpreting forest decline symptoms. *Angew. Bot.* **1990**, *64*, 381–392.
86. Bezerra, J.D.P.; Santos, M.G.S.; Svedese, V.M.; Lima, D.M.M.; Fernandes, M.J.S.; Paiva, L.M.; Souza-Motta, C.M. Richness of endophytic fungi isolated from *Opuntia ficus-indica* Mill. (Cactaceae) and preliminary screening for enzyme production. *World J. Microbiol. Biotechnol.* **2012**, *28*, 1989–1995. [CrossRef] [PubMed]
87. Allen, T.R.; Millar, T.; Berch, S.M.; Berbee, M.L. Culturing and direct DNA extraction find different fungi from the same ericoid mycorrhizal roots. *New Phytol.* **2003**, *160*, 255–272. [CrossRef]
88. Menkis, A.; Vasaitis, R. Fungi in Roots of Nursery Grown Pinus sylvestris: Ectomycorrhizal Colonisation, Genetic Diversity and Spatial Distribution. *Microb. Ecol.* **2011**, *61*, 52–63. [CrossRef]
89. Summerbell, R.C. Microfungi associated with the mycorrhizal mantle and adjacent microhabitats within the rhizosphere of black spruce. *Can. J. Bot.* **1989**, *67*, 1085–1095. [CrossRef]
90. An, H.; Liu, Y.; Zhao, X.; Huang, Q.; Yuan, S.; Yang, X.; Dong, J. Characterization of cadmium-resistant endophytic fungi from *Salix variegata* Franch. in Three Gorges Reservoir Region, China. *Microbiol. Res.* **2015**, *176*, 29–37. [CrossRef]
91. Schlegel, M.; Dubach, V.; von Buol, L.; Sieber, T.N. Effects of endophytic fungi on the ash dieback pathogen. *FEMS Microbiol. Ecol.* **2016**, *92*. [CrossRef]
92. Kwaśna, H.; Mazur, A.; abędzki, A.; Kuźmiński, R.; akomy, P. Communities of fungi in decomposed wood of oak and pine. *For. Res. Pap.* **2016**, *77*, 261–275. [CrossRef]
93. Vaz, A.B.M.; Mota, R.C.; Bomfim, M.R.Q.; Vieira, M.L.A.; Zani, C.L.; Rosa, C.A.; Rosa, L.H. Antimicrobial activity of endophytic fungi associated with Orchidaceae in Brazil. *Can. J. Microbiol.* **2009**, *55*, 1381–1391. [CrossRef] [PubMed]
94. Xia, X.; Lie, T.K.; Qian, X.; Zheng, Z.; Huang, Y.; Shen, Y. Species Diversity, Distribution, and Genetic Structure of Endophytic and Epiphytic Trichoderma Associated with Banana Roots. *Microb. Ecol.* **2011**, *61*, 619–625. [CrossRef] [PubMed]
95. Werner, C.; Petrini, O.; Hesse, M. Degradation of the polyamine alkaloid aphelandrine by endophytic fungi isolated from *Aphelandra tetragona*. *FEMS Microbiol. Lett.* **1997**, *155*, 147–153. [CrossRef]
96. Kattner, D.; Schönhar, S. Occurrence of microscopic fungi in fine roots of apparently healthy Norway spruce on various sites. *Mitteilungen Verein Für Forstliche Standortskunde Forstpflanzenzüchtung* **1990**, *35*, 39–43.
97. Gan, H.; Churchill, A.C.L.; Wickings, K. Invisible but consequential: Root endophytic fungi have variable effects on belowground plant–insect interactions. *Ecosphere* **2017**, *8*, e01710. [CrossRef]
98. Schlegel, M.; Münsterkötter, M.; Güldener, U.; Bruggmann, R.; Duò, A.; Hainaut, M.; Henrissat, B.; Sieber, C.M.K.; Hoffmeister, D.; Grünig, C.R. Globally distributed root endophyte *Phialocephala subalpina* links pathogenic and saprophytic lifestyles. *BMC Genom.* **2016**, *17*, 1015. [CrossRef]
99. Bußkamp, J. Schadenserhebung, Kartierung und Charakterisierung des "Diplodia-Triebsterbens" der Kiefer, Insbesondere des Endophytischen Vorkommens in den Klimasensiblen Räumen und Identifikation von den in Kiefer (*Pinus sylvestris*) Vorkommenden Endophyten. Ph.D. Thesis, University of Kassel, Kassel,

3

Tree Diseases as a Cause and Consequence of Interacting Forest Disturbances

Richard C. Cobb [1,*] and Margaret R. Metz [2]

[1] Department of Plant Pathology, University of California, Davis, One Shields Ave, Davis, CA 95616, USA
[2] Department of Biology, Lewis & Clark College, 0615 S.W. Palatine Hill Road MSC 53, Portland, OR 97219, USA; mmetz@lclark.edu
* Correspondence: rccobb@ucdavis.edu

Academic Editors: Matteo Garbelotto and Paolo Gonthier

Abstract: The disease triangle is a basic and highly flexible tool used extensively in forest pathology. By linking host, pathogen, and environmental factors, the model provides etiological insights into disease emergence. Landscape ecology, as a field, focuses on spatially heterogeneous environments and is most often employed to understand the dynamics of relatively large areas such as those including multiple ecosystems (a landscape) or regions (multiple landscapes). Landscape ecology is increasingly focused on the role of co-occurring, overlapping, or interacting disturbances in shaping spatial heterogeneity as well as understanding how disturbance interactions mediate ecological impacts. Forest diseases can result in severe landscape-level mortality which could influence a range of other landscape-level disturbances including fire, wind impacts, and land use among others. However, apart from a few important exceptions, these disturbance-disease interactions are not well studied. We unite aspects of forest pathology with landscape ecology by applying the disease-triangle approach from the perspective of a spatially heterogeneous environment. At the landscape-scale, disturbances such as fire, insect outbreak, wind, and other events can be components of the environmental 'arm' of the disease triangle, meaning that a rich base of forest pathology can be leveraged to understand how disturbances are likely to impact diseases. Reciprocal interactions between disease and disturbance are poorly studied but landscape ecology has developed tools that can identify how they affect the dynamics of ecosystems and landscapes.

Keywords: forest pathogens; disease triangle; landscape ecology; disturbance interactions; fire; insect outbreak

---

1. Introduction

Forest diseases occur in an intricate environmental context that is a reflection of long host lifespan and fixed location. In contrast to other foci of disease ecology, environmental influences on forest health have been recognized as important factors influencing disease in individual trees since the emergence of forest pathology as a topic of scientific inquiry [1]. Environmental factors act on pathogens and tree hosts independently and, as a consequence, disease incidence may increase or decrease due to environmental variability. This environmental variation can include year-to-year climate variation (temperature, precipitation, humidity), environmental pollutants (ozone, acid deposition), and a range of biophysical or biotic disturbances such as fire, wind, herbivory or insect outbreak [2,3]. A substantial body of research has developed to elucidate environmental influences on both pathogens and their tree hosts, encompassing a diversity of environmental stresses and effects at scales ranging from cellular to landscape [4,5]. Integrating dynamics of both host and pathogen simultaneously with environmental effects on disease remains challenging, and yet is essential to understanding how forest diseases are likely to change in the future [6].

The disease triangle is a scale-free, flexible, and general model that has been used extensively to determine the etiology of many plant diseases and, more recently, zoonotic diseases. The model is a visualization of host, pathogen, and environment in a tripartite dynamic interaction framework that has been especially useful when applied at the individual-to-stand level. However, this application is in some contrast to the scale of the most problematic disease outbreaks which tend to occur across landscapes or regions [7–9].

The field of landscape ecology focuses on identifying mechanistic processes that lead to landscape-level patterns and the implications of cross-scale interactions in driving these patterns. Scale mismatch among mechanisms that underlie an emergent condition—including disease—is not a problem unique to forest pathology. Landscape ecology has developed understanding of the causes, consequences, and patterns of ecological changes that occur over large spatial scales such as plant community shifts resulting from fire, biotic agents including insects and disease, or land-use patterns [10–12]. Landscape ecology applies to a range of spatial extents, but is often employed to understand changes in collections of heterogeneous ecosystems (a landscape) or regions (a collection of landscapes). The field has developed a range of spatial analysis tools that pair with remote sensing technologies; these efforts have improved understanding of the causes and consequences of spatial patterns as well as increasingly accurate resource inventories.

Forest health researchers have applied landscape ecology approaches to gain insight into disease drivers acting at the spatial extent of a landscape or region [13,14]. These efforts have produced maps of spatiotemporal patterns of disease risk with great management application [10,15]. These advances are also important in that they apply epidemiological theory to large spatial extents although, at the landscape extent, forest pathology has tended to examine disease in isolation from other dynamics and processes. At the same time, understanding ecological dynamics at landscape-to-regional scales is increasingly focused on the role of interactions among disturbances which alter spatial structure and variation [16,17]. We suggest that this is a potential nexus of the two fields that can improve understanding of landscape dynamics as well as the mechanistic basis of disease emergence at broad spatial extents.

Disease forecasting and prediction of the resulting ecological impacts have obvious value to managers and policy makers. Shaping landscape-level structure, such as fuel levels, age-class, or species distribution, demands substantial economic investments. Disturbance events, including but not limited to disease, can challenge these goals and incur great ecological or management costs. Increasing interest within landscape ecology has focused on the overlap and potential interactions of landscape-level disturbances. Global environmental change, including altered fire regimes, changes in regional climate and weather, as well as increased introduction of exotic organisms, are all potential drivers of disturbance interactions that could increase the ecological impacts of these events [16–18]. Landscape-scale models and empirical studies of disease impacts are relatively infrequent compared to fire and land use [16–18]. At the same time, cellular-to-population level studies of disease illustrate considerable variation in intensity and underlying causes that emerge from the components and interactions encompassing the disease triangle. These insights can be leveraged into predictive epidemiological models that account for host susceptibility and transmission, characteristics that are distinct biologically from risk or contagion as applied in models of fire or land-use change. Put another way, disease is an emergent phenomenon with inherent mechanistic differences from other landscape-scale disturbances. Yet diseases are likely to affect, and be affected by, other landscape-level disturbances in ecologically important ways.

This paper examines linkages between forest diseases and landscape-scale disturbances to understand how these distinct events may interact to affect each other. Landscape ecology and forest pathology are notable for strong research foundations that suggest disease–disturbance interactions are likely to shape ecosystems. Yet, very few empirical examples of these interactions have been undertaken in spite of a strong emphasis within both fields on the role of environmental factors. This suggests potential for improving prediction of disease and disturbance impacts through an assessment of the strengths and research needs in each field. Although pathogens and insects have important differences in how they affect tree health, we examine inferences gained through the study of insect outbreak–disturbance interactions to frame hypotheses of disease–disturbance dynamics from ecosystem-to-landscape scales. Empirical research linking epidemiological process and ecosystem or landscape-level impacts of forest disease is generally lacking, so we also point to several studies aimed at addressing this knowledge gap. Landscape ecology has built metrics for quantifying landscape pattern, processing of remotely sensed data, and models of disturbance dynamics that can be placed within traditional forest pathology approaches. This structure provides a basis for rapid integration of ecosystem-to-landscape impacts and dynamics of disease into landscape models and associated theory. We link these bodies of knowledge by identifying mechanisms of interaction that are likely to determine if or when disease–disturbance interactions are ecologically important.

2. The Disease Triangle: A Primer for Landscape Ecologists

The disease triangle (Figure 1) is a valuable heuristic tool for envisioning and testing drivers of disease emergence. The model has a long history both within forest pathology and plant pathology more broadly [1]. The model frames disease as an emergent condition resulting from interactions of a pathogen with a susceptible host in suitable environmental conditions. Changes to any of the components can accelerate or dampen disease dynamics, while also altering the other components of the triangle. As a general, parsimonious, and dynamic conceptual model, the disease triangle also requires a researcher or manager to contextualize the disease system, including any unique characteristics of the focal host or pathogen.

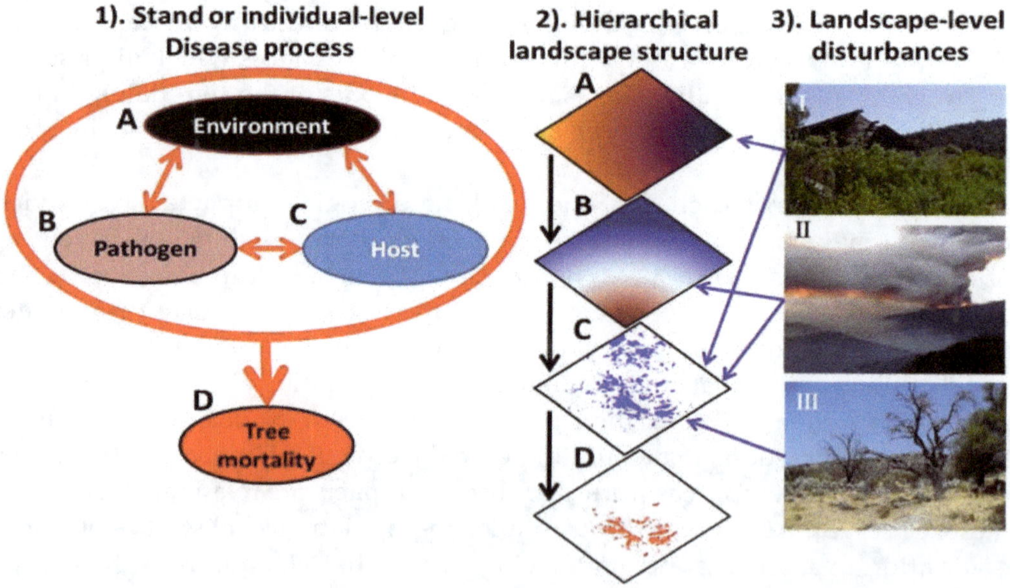

Figure 1. Linkages between forest pathology and landscape ecology with example landscape-level disturbances which can influence disease emergence and impacts. The traditional disease triangle (**1**) factors (A–C) are shown simultaneously in the spatial context of landscape ecology (**2**) along with landscape-level disturbances that are likely to interact with disease (**3**). Environment (1-A and 2-A) is shown on a gradient along with a waveform pathogen invasion process (1-B and 2-B) and realistic spatial heterogeneity of host distribution (1-C and 2-C) and mortality (1-D and 2-D). Examples of interactive disturbances are shown with arrows indicating the components of landscape structure and the disease triangle that are impacted: Invasive plants competitively inhibit forest host reestablishment following land abandonment with impacts to environmental conditions and host distribution (I—*Genista monspessulana* invasion of an old vineyard), fire can alter host and pathogen distribution (II—Soberanes Fire 2016, Big Sur—Photo credit K. Frangioso), and emergent insect outbreak-caused mortality of host populations (III—*Agrilus coxalis* mortality of coast live oak).

2.1. Pathogens

The major disease-causing pathogens of forest trees represent many branches on the tree of life, including fungi, oomycetes, or parasitic plants [3,19]. These pathogens can vary widely in their spatial distribution across a landscape. Some weak or facultative pathogens can be so widespread as to be practically ubiquitous, likely an evolutionary consequence of minimal impacts to host health or primarily saprotrophic energy acquisition [20,21]. Other more aggressive native pathogens such as *Heterobasidion* or *Dothistroma* species may be widespread within a region but only emerge as diseases following land use or climatic changes, respectively [22,23]. In contrast, diseases caused by invasive pathogens are often characterized by introduction foci that strongly control the spatial distribution of disease risk prior to equilibration of the pathogen in a new range. Introduced populations of *Phytophthora ramorum*, *P. cinnamomi*, or the beech bark disease insect-pathogen complex all share this strong signature of spatiotemporal variation in the distribution of disease risk [9,12,24].

2.2. Tree Hosts

Disease emergence or suppression may be driven by shifts in host presence or abundance. In forests, species abundance reflects a range of climate drivers, ecological processes, and cultural practices; several examples of documented or potential increases in disease with changes in host population are worth highlighting in light of these dynamics. Increased cultivation of *Hevea* species for rubber production is implicated in increased frequency and severity of South American leaf blight [19,25], a disease that rarely reaches inoculum levels needed for disease emergence under non-cultivated conditions. Increased tanoak importance in response to inadequate silvicultural investments has also been implicated in patterns of *P. ramorum* establishment risk and emergence of sudden oak death [26–28]. Afforestation in response to changes in land use or fire suppression has also been shown to increase *P. ramorum* hosts and subsequent patterns of disease emergence [29]. Vegetation changes in response to environmental and social or economic dynamics can also shift species abundances and spatial distributions with no significant changes in the represented taxa. For example, in northern forests of Quebec, human impacts reflecting changes in economic factors and silvicultural practices could influence future likelihood of insect outbreak [30]. Plant invasions can also alter host populations at broad spatial scales. In the southern hemisphere, several invasive pine species along with their mycorrhizal symbionts are aggressive invaders of native ecosystems. These invasions cause severe landscape-to-regional scale transformations of forest communities that also transform standing biomass, biogeochemical processes, and biodiversity [31,32].

Hosts also experience changes in susceptibility due to environmental responses at cellular-to-individual scales. Environmentally driven increases in host susceptibility to pathogens are commonly termed host "predisposition" to infection or disease. Tree hosts respond to physical or biotic stresses such as fire, drought, or insect and pathogen attack through a range of physical structures, chemical defenses, and biochemical pathways [4,5]. Thus, the physiological status of individual trees or entire populations can also determine the likelihood of disease emergence.

2.3. Environment

Within the context of the disease triangle, many environmental factors influence disease by limiting hosts or pathogen ranges and through direct influences on pathogens or their hosts. For example, temperature affects pathogen metabolic activity as well as host predisposition in response to environmental stress [33]. Global change-driven range expansions or contractions could make disease emergence more, or less, likely due to changes in the abundance or spatial distribution of either biological component [3,19,34]. Physical host wounds that act as a de facto surmounting of host physical defenses play an important role in infection [1,22,35,36]. Drought has been shown to cause physical damage to plant vascular tissues that may require multiple years to repair [37] and it is likely that other environmental stresses also cause multi-year changes in susceptibility.

Land management can be considered part of the environmental portion of the disease triangle, and is sometimes envisioned as a fourth component. This is warranted as land management policy can be more readily changed compared to an environmental event such as drought; policy and management can also affect environmental drivers of disease or landscape-level host distribution [3,38].

3. The Disease Triangle in the Context of Disturbance

In the absence of disease, disturbances play a critical role in structuring forest communities at local-to-landscape scales [17,30,39]. Ecological disturbances can alter any component of the disease triangle (Figure 1). This creates a dual context for disturbances: they are a component of the environmental portion of the model or also an external factor that can influence host composition or pathogen abundance. These effects in conjunction with environmental changes that determine host stresses or pathogens infection, can also create disturbance-feedbacks that will alter disease dynamics.

3.1. Environmental vs. Biotic Stress on Tree Hosts

An intricate system of plant hormone signaling has a strong influence on secondary chemical pathways and allocation of resources to repair physical damage from biotic and abiotic causes. Problematically for plant hosts, biochemical pathways that confer resistance to environmental stress can suppress those that confer resistance to insects and pathogens. The resulting hormonal interference of defensive pathways is known as plant hormonal cross-talk [4] and can be a significant factor leading to increased susceptibility (predisposition). Carbon- or nitrogen-based secondary chemistry can be important factors determining the degree of damage caused by insects or pathogens, but available carbon resources are also important for repairing tissue damage resulting from environmental stresses [5]. This suggests another resource dilemma for trees—whether to respond to environmental stress or biotic attack. Of course, physical damage, heat, and water stress may suppress defensive pathways for biotic attack via hormonal cross-talk [4] suggesting plant response to environmental factors may simply take precedence to biotic attack in some circumstances or species. It is likely that trees are sometimes caught in untenable "catch 22" traps where the host must respond to potentially overwhelming environmental or overwhelming biotic stresses and is likely to be overcome either way.

3.2. Example Disturbance Effects on Pathogens

Forest management and fire are especially important disturbances in forested landscapes that have been documented to influence disease dynamics in some circumstances. Each of these events can alter temperature and relative humidity which are particularly important controls on infection success for many important fungal pathogens (Figure 1). Tree harvesting can further alter stand characteristics that influence pathogen dispersal. For example, dwarf mistletoe dispersal is facilitated by increased canopy opening following thinning [40], and it is likely that increased air-flow and reduced tree density also increases the connectivity of susceptible hosts. Harvest and the resulting effects on substrate for pathogen survival, creation of infection pathways, and changes in host distribution or abundance have strong influences on disease emergence [14,20,22]. Increased input of below-ground organic matter, important for inoculum buildup of some root diseases, could occur following a variety of management actions [41,42]. Fire is somewhat unique in that it often causes substantial host and pathogen mortality [43] which should caution researchers against generalization without more examples. In contrast to fire impacts, fire suppression can lead to increased host density, which in turn may increase local inoculum build-up as well as alter microclimate in ways that influence pathogen establishment and the emergence of specific diseases [44].

Non-host plant invasions can also alter mutualistic host–microbial interactions, nutrient availability, and increase competition for other limiting resources such as light and water [45].

Most obviously, strong competitive interactions or competitive exclusion of host species by a novel invader could have powerful influences on disease emergence. However, this kind of shifting species dynamic is more likely when the invading plant amplifies pathogen populations and creates positive feedback between pathogen and invader [46,47]. Plant invasions can suppress or amplify pathogen populations and the relevance to disease will often depend on the interactions between a specific plant invader and pathogen.

3.3. *N Deposition, an Example of Changing Environmental Conditions at Broad Spatial Extents*

Environmental contamination has received less attention as environmental drivers of disease, but the spatial scales at which these events occur make them relevant to any discussion of disease–disturbance interactions. Atmospheric N deposition has potential to alter host populations by shifting species composition and dominance via changes in growth rates. N deposition has also been documented to increase pathogen attack [48,49] which may reflect mechanistic drivers from cellular-to-community scales. Studies of N-deposition on disease emergence are more common in understory or herbaceous species but whether this is due to an overall lack of study in overstory trees vs. greater susceptibility in these habitats is unclear. Increased N availability could alter a range of ecological interactions that influence tree host susceptibility such as phyllosphere host–microbial interactions and within-plant carbon allocation [33]. Species invasions, climate change, and other kinds of pollution could alter host or pathogen populations at broad spatial scales and are worth examining for disease-disturbance interactions.

3.4. *Disease as Disturbance*

Both native and exotic pathogens are substantial ecosystem disturbances that can cause a range of impacts with variable duration and intensities [50]. Root nematode pathogens, aerially dispersed *Phytophthora* pathogens, and native *Phellinus* root pathogens have each been shown to alter biogeochemical processes including nitrogen and carbon cycling [51–53]. Forest structural changes can be dramatic, and include shifts in size class distribution, selective removal of individuals, or changes in species composition [7,12,54,55]. The nature and extent of forest structural changes depend on specific biological characteristics of the pathogen such as the host range and virulence, as well as interactions with the biology of the tree host, including age- or size-related variation in susceptibility or host competency to transmit the pathogen. In the most extreme cases where host range is broad and many hosts suffer disease following infection, pathogens can drive a conversion from one ecosystem type to another. *Phytophthora cinnamomi* is an illustrative example, this pathogen causes the devastating disease Jarrah dieback which can convert woodland or forest to a low-statured heathland [9].

4. Disturbance Interactions: Perspectives from Landscape Ecology

Disturbance interactions are increasingly studied outside of the purview of forest pathology and the associated empirical and theoretical advances are relevant for bridging the fields (Table 1). Landscape ecology has made important advances in framing general theory and expectations regarding how a broad set of disturbances are likely to interact [16,17,56]. These efforts have incorporated differences in frequency, spatial pattern, and severity of physical and biological disturbances with the aim of assessing the relevance to landscape structure and management. Disturbance frequency, patterns, and impacts lie at the core of inquiry in landscape ecology, which stems from the importance of these events in driving spatial patterns [56]. In light of this, it is unsurprising that many of the intellectual frontiers of disturbance interactions (including but not limited to disease) are also firmly rooted within landscape ecology [13,14]. Systems at the landscape spatial scale are so often subject to variation in environmental stresses and stochastic events—such as lightning ignition sources or individual land-use actions—that understanding overlapping disturbances and their interactions is fundamental to mechanistic understanding and prediction over large areas.

Table 1. Example studies linking environmental changes and disease emergence or the interactive effects of biotic agents and landscape-level disturbance dynamics.

Biological Agent	Interacting Disturbance	Comments	Examples
	Landscape-level examples		
Native insects	Fire, wind, salvage harvest	Tested for interactive effects	[11,57,58]
Invasive pathogen	Fire	Tested for interactive effects	[18,43,59,60]
Root pathogens	Wind	Focused on environmental drivers of disease	[35,61]
Root pathogens	Management, fire suppression	Etiological investigation of landscape-level disease drivers	[22]
	Individual-level examples		
Insects or pathogens	Drought, salt stress, and heat	Synthesis, laboratory experiments	[4]
Pathogens	Drought	Synthesis	[5]

Discrete and intense disturbances—such as insect or disease interactions with fire—are notable for the relative ease of assessing population-level changes to vegetation. In these examples, mortality from either disturbance is rapid and easily recognized. The relative tractability of these study systems further leads to empirical and model based studies [11,56,59]. However, slowly accumulating and large-extent disturbance such as climate change and atmospheric N deposition are much more nuanced and difficult to assess. These events will have strong influences on landscape-level processes including disturbance frequency, extent, and impact but quantifying these dynamics with an empirically-backed and mechanistic framework poses many practical challenges [56].

How landscape scale dynamics affect ecological resiliency is also important for gauging the relevance of disturbance interactions to land management policy. Post-disturbance vegetation recovery rates and successional dynamics have a strong control over ecological resilience to interacting disturbances, for example, by affecting the likelihood of future disturbance events and their intensities [16]. Policy makers and managers face the dilemma that disturbance interactions are known to occur, or are highly likely, yet it is unclear how much these events will challenge local and regional management goals. This problem may be acute when multiple possible successional trajectories emerge from the local species pool and variation in disturbance impacts [17,50]. Despite important uncertainties, enough information is in hand to provide a point-of-departure for quantifying disease–disturbance interactions and incorporating these insights into land management policy.

4.1. Insect–Fire Interactions

Interactions between fire and insect outbreak have received attention among researchers due to the frequent overlap and relative importance of each disturbance [11,57,58]. Although biologically distinct from forest diseases, insect outbreaks provide insights into disturbance interactions and can help formulate expectations given changes to canopy structure, species composition, and changes in biomass distribution. Insect-driven mortality events have been dramatic and cause for alarm among land managers in fire-prone conifer forests of western North America. Many of these forests have been impacted by drought and fire suppression, which have influenced insect population dynamics and host physiological status [62,63]. In this respect, insect outbreaks are a useful model for disease in that tree mortality is driven by environmental changes including climate dynamics and land management. Severe insect-caused mortality often results in a short-term increase in highly flammable canopy fuels followed by an increase in ground fuels as dead material moves from the canopy to the forest floor. Somewhat counter to initial expectations, these mortality patterns have led to an overall decrease in fire impacts as measured by the likelihood of crown-fire, which are often the most damaging and dangerous forest fire conditions, for both bark beetle [11,57] and spruce budworm [58]. In these studies, the timing of fire and insect outbreak events was more important than the spatial overlap of each disturbance.

4.2. Disease–Disturbance Interactions

As previously noted, etiological studies have dominated most applications of the disease-triangle in forest pathology. For example, wind pressure on tree canopies can transfer a substantial amount of tension and compression force to root systems. The resulting wind-caused damage to fine roots is an important pathway of infection and subsequent creation of disease centers in balsam fir-spruce forests of the White Mountains, New Hampshire [61]. Historical management actions including fire suppression and cutting to control bark beetle outbreak have been demonstrated to influence patterns of *Heterobasidion* root disease centers in Yosemite Valley, California [22]. Historical patterns of conifer harvest have also likely influenced the landscape-level distribution of invasion risk for *P. ramorum* in coastal California and Oregon [24]. However, for each of these example systems, there has been little effort to inform prediction of future resource status, disturbance patterns, and rates (or capacity) of ecosystem recovery [17,56]. For example, wind-driven root infection decreases tree capacity to withstand wind events. Thus, wind can facilitate disease which then renders stands or ecosystems more vulnerable to wind-driven mortality [61]. Problematically, temporally delayed disease–disturbance interactions could be easily masked and result in misattribution of underlying mortality causes [50].

The few published studies focused on disease–disturbance interactions provide some insights that may help overcome this research challenge. The authors of this article along with colleagues conducted a series of studies examining interactive impacts of *P. ramorum* and fire in coastal California forests. Using a set of permanent study plots in disease-impacted forests that were surveyed before and after the 2008 Basin Fire, this work showed that fire impacts are contingent on the stages of disease progression, supporting comparisons to insect–fire interactions (compare with [57]). In our plots, fire intensity was positively related to the amount of fine canopy fuels associated with recently killed trees and decreased as these materials move from the canopy to the soil surface [18]. In contrast, fire-caused mortality of redwood, a species that is resilient to *P. ramorum* and typically also resilient to fire, increased with greater accumulation of ground fuels [59]. This pattern suggested that disease-generated fuels accumulated to the point that damage to tree root systems or cambium tissue was substantially increased. Ground fuel accumulation also impacted soil resources, losses of critical soil nutrients and carbon increased with increasing disease-related mortality [60], changes that may in turn alter post-fire vegetation succession. Although these studies are associated with an individual disease and fire (i.e., a single example), the mechanistic link between mortality, fuels, and ecosystem impacts suggests these disease–disturbance interactions could occur where other disease and fire events overlap.

5. Frontiers in the Study of Disease–Disturbance Interactions

Approaching a new study system or challenge with the most parsimonious model is often a recipe for rapid advancement. With some notable exceptions [11,17,56], most interactive disturbance studies have focused on two events. In the sudden oak death system, our sole explicit test of disease–disturbance interactions, we studied a single forest disease and a single fire. This study system spans multiple forest types which provides insight into variation in disease impacts, the importance of pre-disease species composition and structure [18,43,59,60] but ultimately provides only a partial view on the importance of interactive impacts. For example, our Big Sur study region is a fire-prone landscape that burned with a landscape-level wildfire only eight years following our initial study (the 2016 Soberanes Fire). How do repeated or subsequent disturbances modify or reflect previous disease–disturbance interactions? In the case of Big Sur and sudden oak death, *P. ramorum* was found to rapidly reinvade following the initial 2008 Basin wildfire [43], presumably impacts from the recent 2016 Soberanes wildfire reflect both the previously documented disease-fire interactions as well as an additional eight years of disease that occurred between wildfires. In Yosemite Valley, *Heterobasidion* disease centers are a serious public safety concern as several deaths have been associated with failures of infected trees [22]. This has motivated the aggressive removal of at risk trees, particularly in high-recreation use areas, as well as the greater application of prescribed burning in other parts of the valley. Yet, it is unclear how these management actions have influenced disease dynamics and furthered public safety beyond the immediate danger of tree failures. In both cases, disease–disturbance interactions could be reasonably argued to have increased or decreased ecological impacts under specific edaphic, vegetation, or management conditions.

Third and fourth-order disease–disturbance interactions remain minimally studied although these are certain to represent future management challenges. Fire-prone regions where multiple fires and disease events occur over time could be shaped by these factors. Paine et al. [56] and Johnstone et al. [17] suggested several general ecosystem trajectories resulting from disturbance interactions that tend to focus on transitions between stable ecosystem states or temporary states associated with recovery (Figure 2; examples 1 and 2). We recognize that these expectations may also fit disease–disturbance interactions. For example, disease-center formation in the White Mountains of New Hampshire resulting from wind impacts is a temporary ecosystem state [35,61]. Similarly, it has not yet been shown that sudden oak death–fire interactions will lead to long-term and permanent shifts in forest structure although it is plausible that increased tree mortality and soil nutrient loss will result in these changes. However, biotic disturbances are strongly influenced by the coincidence of environmental conditions favorable to establishment or outbreak suggesting that dynamic ecosystems should be expected in some circumstances (Figure 2; example 3). Oscillating dynamics where host populations recover and provide the basis for future outbreaks are a common pattern in many tree mortality events driven by native organisms [35,51,62]. Intensifying outbreaks resulting from climate-change-associated shifts in temperature, precipitation, and resulting host-stress appears likely given documentation of landscape-level declines in host physiological status and its impact to plant defense [3,4,37].

Hysteresis is a common challenge for much of ecology and the importance as well as the difficulty of capturing these effects is increased in studies at the landscape scale. The difficultly of scaling-up of results from the plot or stand scale where most ecological inquiry is made to the landscape scale is a common challenge in landscape ecology including the study of interactive disturbances [14,16]. An especially complex factor is that as the landscape is changed, the dynamics within the unit of study may become more dependent on processes and structure external to the measurement unit, often the plot or stand (see Meentemeyer et al. [14]). These changes may require many years to emerge as well as quantify, suggesting that surprises, often problematic to management goals, are likely to emerge without the tools to make judicious forecasts of disease–disturbance impacts. Several lines of inquiry can help prevent these problems: (1) increased efforts to disentangle three-way interactions that lead to disease at broad spatial scales. Disease-triangle applications are biased towards etiological descriptions of disease, yet disease emergence could be a secondary or contributing factor influencing tree mortality and disturbance interactions [19,37].

Understanding if disease is a cause (primary agent) or a symptom (secondary agent) of a forest health problem is essential to any response. (2) Increase focus on multi-pathogen systems. Tree mortality from one event, such as the invasion of an exotic pathogen is likely to alter the conditions for the emergence of a second pathogen. For example, native Armillaria pathogens are widespread in forests at risk to *P. ramorum* invasion and the substantial addition of buried coarse roots is likely to increase local Armillaria biomass and associated disease [8,20]. (3) Increase the scope of disease–disturbance studies to pathogens with different epidemiological dynamics, plant parts which are attacked, and impacts to host health. At present, studies of fire–insect outbreak has examined insects which attack both the bole as well as the canopy of hosts [57,58]. A broad range of pathogens attack roots and leaves which may alter the timing and duration of dead fine fuels in the canopy; previous work has shown this to be a critical factor influencing disease–fire relationships [18]. Contrasting these classes of pathogens with bole-canker pathogens, vascular wilts, and a range of other biological differences among pathogens [5] could help link pathogenicity with effects on other disturbances.

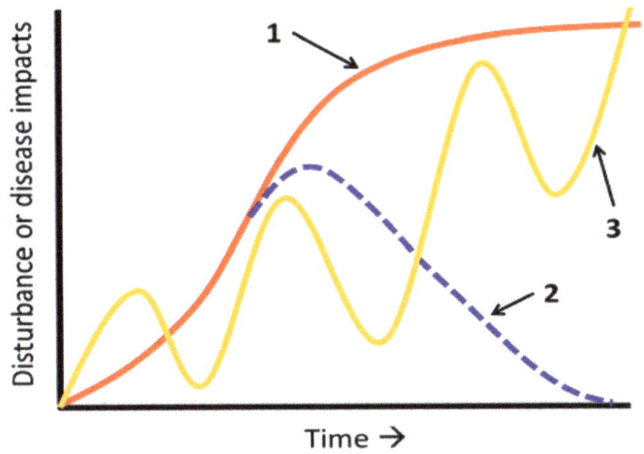

Figure 2. Several examples of disturbance and disease interactive outcomes over time. Impacts may increase and reach a new stable state (1—red); increases may gradually recover to pre-outbreak levels (2—blue dashed); or feedbacks between disease and disturbance could create fluctuating dynamics over time, here depicted as an oscillating, but increasing impact (3—yellow).

6. Conclusions

Landscape-scale disturbances such as wind, fire, or land use can (i) modify the impacts of a disease [22,61]; or (ii) can be influenced by disease in a manner which increases or decreases the ecological impacts of these disturbances [59]. In both cases, the potential to magnify the ecological impacts of either disturbance can have important management implications [16,17], but the mechanistic processes driving these impacts are quite different. The disease triangle is a valuable model for improving understanding of disease–disturbance interactions but to date, the approach has been primarily employed to study environmental influences on disease emergence. This body of work provides the mechanistic foundation for landscape ecologists to model disease emergence at broad spatial scales along with providing the basis to address associated management goals and challenges. However, the influence of disease on other disturbance processes is less well studied, and although the disease triangle is fully reciprocal, little effort has focused on disturbance changes stemming from ecosystem-to-landscape level impacts of disease. A nexus of interest and research efforts is developing between the fields [13,14]; forest pathology can provide epidemiological foundations unique to understanding disease emergence while landscape ecology can provide broad spatial extent models and empirical assessment of ecosystem and landscape dynamics. Of course, more data on ecological impacts of disease and collaboration between forest pathologists and landscape ecologists is needed to develop and test predictions of disease–disturbance interactions that will inform management decisions. However, each field is primed to take advantage of knowledge gained from the other and determine when, if, or how disease–disturbance interactions can be incorporated into land management policy and actions.

Acknowledgments: This work was funded by NSF Grant DEB EF-0622770 as part of the joint NSF-NIH Ecology of Infectious Disease program, the Gordon and Betty Moore Foundation, Cal Fire Green House Gas Reduction Fund, and the USDA Forest Service Forest Health Protection and Pacific Southwest Research Station. The authors thank David Rizzo and two anonymous reviewers for helpful comments on earlier versions of this manuscript.

Author Contributions: R.C.C. and M.R.M. conceived the research, conducted the analysis, and wrote the paper.

References

1. Hartig, R. *Text-Book of the Diseases of Trees*; Macmillan and Company: London, UK, 1894.
2. Desprez-Loustau, M.-L.; Marçais, B.; Nageleisen, L.-M.; Piou, D.; Vannini, A. Interactive effects of drought and pathogens in forest trees. *Ann. For. Sci.* **2006**, *63*, 597–612. [CrossRef]

3. Pautasso, M.; Schlegel, M.; Holdenrieder, O. Forest health in a changing world. *Microb. Ecol.* **2014**, *69*, 826–842. [CrossRef] [PubMed]
4. Bostock, R.M.; Pye, M.F.; Roubtsova, T.V. Predisposition in plant disease: Exploiting the nexus in abiotic and biotic stress perception and response. *Annu. Rev. Phytopathol.* **2014**, *52*, 517–549. [CrossRef] [PubMed]
5. Oliva, J.; Stenlid, J.; Martínez-Vilalta, J. The effect of fungal pathogens on the water and carbon economy of trees: Implications for drought-induced mortality. *New Phytol.* **2014**, *203*, 1028–1035. [CrossRef] [PubMed]
6. Stenlid, J.; Oliva, J. Phenotypic interactions between tree hosts and invasive forest pathogens in the light of globalization and climate change. *Philos. Trans. R. Soc. B* **2016**, *371*, 20150455. [CrossRef] [PubMed]
7. Paillet, F.L. Chestnut: History and ecology of a transformed species. *J. Biogeogr.* **2003**, *29*, 1517–1530. [CrossRef]
8. Rizzo, D.M.; Garbelotto, M.; Hansen, E.M. Phytophthora ramorum: Integrative research and management of an emerging pathogen in California and Oregon forests. *Annu. Rev. Phytopathol.* **2005**, *43*, 309–335. [CrossRef] [PubMed]
9. Shearer, B.L.; Crane, C.E.; Barrett, S.; Cochrane, A. Phytophthora cinnamomi invasion, a major threatening process to conservation of flora diversity in the South-west Botanical Province of Western Australia. *Aust. J. Bot.* **2007**, *55*, 225–238. [CrossRef]
10. Orwig, D.A.; Foster, D.R.; Mausel, D.L. Landscape patterns of hemlock decline in New England due to the introduced hemlock woolly adelgid. *J. Biogeogr.* **2002**, *29*, 1475–1487. [CrossRef]
11. Kulakowski, D.; Veblen, T.T. Effect of prior disturbances on the extent and severity of wildfire in Colorado subalpine forests. *Ecology* **2007**, *88*, 759–769. [CrossRef] [PubMed]
12. Garnas, J.R.; Ayres, M.P.; Liebhold, A.M.; Evans, C. Subcontinental impacts of an invasive tree disease on forest structure and dynamics. *J. Ecol.* **2011**, *99*, 532–541. [CrossRef]
13. Holdenrieder, O.; Pautasso, M.; Weisberg, P.J.; Lonsdale, D. Tree diseases and landscape processes: The challenge of landscape pathology. *Trends Ecol. Evol.* **2004**, *19*, 446–452. [CrossRef] [PubMed]
14. Meentemeyer, R.K.; Haas, S.E.; Václavík, T. Landscape epidemiology of emerging infectious diseases in natural and human-altered ecosystems. *Annu. Rev. Phytopathol.* **2012**, *50*, 379–402. [CrossRef] [PubMed]
15. Cunniffe, N.J.; Gilligan, C.A. A theoretical framework for biological control of soil-borne plant pathogens: Identifying effective strategies. *J. Theor. Biol.* **2011**, *278*, 32–43. [CrossRef] [PubMed]
16. Buma, B. Disturbance interactions: Characterization, prediction, and the potential for cascading effects. *Ecosphere* **2015**, *6*, 70. [CrossRef]
17. Johnstone, J.F.; Allen, C.D.; Franklin, J.F.; Frelich, L.E.; Harvey, B.J.; Higuera, P.E.; Mack, M.C.; Meentemeyer, R.K.; Metz, M.R.; Perry, G.L.; et al. Changing disturbance regimes, ecological memory, and forest resilience. *Front. Ecol. Environ.* **2016**, *14*, 369–378. [CrossRef]
18. Metz, M.R.; Frangioso, K.M.; Meentemeyer, R.K.; Rizzo, D.M. Interacting disturbances: Wildfire severity affected by stage of forest disease invasion. *Ecol. Appl.* **2011**, *21*, 313–320. [CrossRef] [PubMed]
19. Desprez-Loustau, M.-L.; Aguayo, J.; Dutech, C.; Hayden, K.J.; Husson, C.; Jakushkin, B.; Marçais, B.; Piou, D.; Robin, C.; Vacher, C. An evolutionary ecology perspective to address forest pathology challenges of today and tomorrow. *Ann. For. Sci.* **2016**, *73*, 45–67. [CrossRef]
20. Baumgartner, K.; Rizzo, D.M. Ecology of *Armillaria* spp. in Mixed-Hardwood Forests of California. *Plant Dis.* **2001**, *85*, 947–951. [CrossRef]
21. Stergiopoulos, I.; Gordon, T.R. Cryptic fungal infections: The hidden agenda of plant pathogens. *Plant-Microbe Interact.* **2014**, *5*, 506. [CrossRef] [PubMed]
22. Slaughter, G.; Rizzo, D. Past forest management promoted root disease in Yosemite Valley. *Calif. Agric.* **1999**, *53*, 17–24. [CrossRef]
23. Woods, A.J.; Martín-García, J.; Bulman, L.; Vasconcelos, M.W.; Boberg, J.; La Porta, N.; Peredo, H.; Vergara, G.; Ahumada, R.; Brown, A.; et al. Dothistroma needle blight, weather and possible climatic triggers for the disease's recent emergence. *For. Pathol.* **2016**, *46*, 443–452. [CrossRef]
24. Cunniffe, N.J.; Cobb, R.C.; Meentemeyer, R.K.; Rizzo, D.M.; Gilligan, C.A. Modeling when, where, and how to manage a forest epidemic, motivated by sudden oak death in California. *Proc. Natl. Acad. Sci. USA* **2016**, *113*, 5640–5645. [CrossRef] [PubMed]
25. Lieberei, R. South American Leaf Blight of the Rubber Tree (*Hevea* spp.): New Steps in Plant Domestication using Physiological Features and Molecular Markers. *Ann. Bot.* **2007**, *100*, 1125–1142. [CrossRef] [PubMed]
26. Bowcutt, F. Tanoak target: The rise and fall of herbicide use on a common native tree. *Environ. Hist.* **2011**, *16*, 197–225. [CrossRef]

27. Cobb, R.C.; Rizzo, D.M.; Hayden, K.J.; Garbelotto, M.; Filipe, J.A.N.; Gilligan, C.A.; Dillon, W.W.; Meentemeyer, R.K.; Valachovic, Y.S.; Goheen, E.; et al. Biodiversity Conservation in the Face of Dramatic Forest Disease: An Integrated Conservation Strategy for Tanoak (*Notholithocarpus densiflorus*) Threatened by Sudden Oak Death. *Madroño* **2013**, *60*, 151–164. [CrossRef]
28. Meentemeyer, R.K.; Cunniffe, N.J.; Cook, A.R.; Filipe, J.A.N.; Hunter, R.D.; Rizzo, D.M.; Gilligan, C.A. Epidemiological modeling of invasion in heterogeneous landscapes: Spread of sudden oak death in California (1990–2030). *Ecosphere* **2011**, *2*, 1–24. [CrossRef]
29. Meentemeyer, R.K.; Rank, N.E.; Anacker, B.L.; Rizzo, D.M.; Cushman, J.H. Influence of land-cover change on the spread of an invasive forest pathogen. *Ecol. Appl.* **2008**, *18*, 159–171. [CrossRef] [PubMed]
30. Danneyrolles, V.; Arseneault, D.; Bergeron, Y. Pre-industrial landscape composition patterns and post-industrial changes at the temperate–boreal forest interface in western Quebec, Canada. *J. Veg. Sci.* **2016**, *27*, 470–481. [CrossRef]
31. Nuñez, M.A.; Horton, T.R.; Simberloff, D. Lack of belowground mutualisms hinders Pinaceae invasions. *Ecology* **2009**, *90*, 2352–2359. [CrossRef] [PubMed]
32. Dickie, I.A.; John, S.G.M.; Yeates, G.W.; Morse, C.W.; Bonner, K.I.; Orwin, K.; Peltzer, D.A. Belowground legacies of Pinus contorta invasion and removal result in multiple mechanisms of invasional meltdown. *AoB Plants* **2014**, *6*, plu056. [CrossRef] [PubMed]
33. Burdon, J.J.; Thrall, P.H.; Ericson, L. The current and future dynamics of disease in plant communities. *Annu. Rev. Phytopathol.* **2006**, *44*, 19–39. [CrossRef] [PubMed]
34. Ghelardini, L.; Pepori, A.L.; Luchi, N.; Capretti, P.; Santini, A. Drivers of emerging fungal diseases of forest trees. *For. Ecol. Manag.* **2016**, *381*, 235–246. [CrossRef]
35. Rizzo, D.M.; Harrington, T.C. Root movement and root damage of red spruce and balsam fir on subalpine sites in the White Mountains, New Hampshire. *Can. J. For. Res.* **1988**, *18*, 991–1001. [CrossRef]
36. Etheridge, D.E.; Craig, H.M. Factors influencing infection and initiation of decay by the Indian paint fungus (*Echinodontiumtinctorium*) in western hemlock. *Can. J. For. Res.* **1976**, *6*, 299–318. [CrossRef]
37. Anderegg, W.R.L.; Plavcová, L.; Anderegg, L.D.L.; Hacke, U.G.; Berry, J.A.; Field, C.B. Drought's legacy: Multiyear hydraulic deterioration underlies widespread aspen forest die-off and portends increased future risk. *Glob. Chang. Biol.* **2013**, *19*, 1188–1196. [CrossRef] [PubMed]
38. Dillon, W.W.; Meentemeyer, R.K.; Vogler, J.B.; Cobb, R.C.; Metz, M.R.; Rizzo, D.M. Range-wide threats to a foundation tree species from disturbance interactions. *Madroño* **2013**, *60*, 139–150. [CrossRef]
39. Thompson, J.R.; Carpenter, D.N.; Cogbill, C.V.; Foster, D.R. Four centuries of change in Northeastern United States forests. *PLoS ONE* **2013**, *8*, e72540. [CrossRef] [PubMed]
40. Mehl, H.K.; Mori, S.R.; Frankel, S.J.; Rizzo, D.M. Mortality and growth of dwarf mistletoe-infected red and white fir and the efficacy of thinning for reducing associated losses. *For. Pathol.* **2013**, *43*, 193–203. [CrossRef]
41. Garbelotto, M.; Gonthier, P. Biology, epidemiology, and control of heterobasidion species worldwide. *Annu. Rev. Phytopathol.* **2013**, *51*, 39–59. [CrossRef] [PubMed]
42. Rizzo, D.M.; Whiting, E.C.; Elkins, R.B. Spatial distribution of armillaria mellea in pear orchards. *Plant Dis.* **1998**, *82*, 1226–1231. [CrossRef]
43. Beh, M.M.; Metz, M.R.; Frangioso, K.M.; Rizzo, D.M. The key host for an invasive forest pathogen also facilitates the pathogen's survival of wildfire in California forests. *New Phytol.* **2012**, *196*, 1145–1154. [CrossRef] [PubMed]
44. Hawkins, A.E.; Henkel, T.W. Native forest pathogens facilitate persistence of Douglas-fir in old-growth forests of Northwestern California. *Can. J. For. Res.* **2011**, *41*, 1256–1266. [CrossRef]
45. Ehrenfeld, J.G. Ecosystem Consequences of Biological Invasions. *Annu. Rev. Ecol. Evol. Syst.* **2010**, *41*, 59–80. [CrossRef]
46. Cobb, R.C.; Meentemeyer, R.K.; Rizzo, D.M. Apparent competition in canopy trees determined by pathogen transmission rather than susceptibility. *Ecology* **2010**, *91*, 327–333. [CrossRef] [PubMed]
47. Holt, R.D.; Dobson, A.P.; Begon, M.; Bowers, R.G.; Schauber, E.M. Parasite establishment in host communities. *Ecol. Lett.* **2003**, *6*, 837–842. [CrossRef]
48. Mitchell, C.E.; Reich, P.B.; Tilman, D.; Groth, J.V. Effects of elevated CO_2, nitrogen deposition, and decreased species diversity on foliar fungal plant disease. *Glob. Chang. Biol.* **2003**, *9*, 438–451. [CrossRef]
49. Strengbom, J.; Englund, G.; Ericson, L. Experimental scale and precipitation modify effects of nitrogen addition on a plant pathogen. *J. Ecol.* **2006**, *94*, 227–233. [CrossRef]

50. Eviner, V.T.; Likens, G.E. Effects of pathogens on terrestrial ecosystem function. In *Infectious Disease Ecology. Effects of Ecosystems on Disease and Disease on Ecosystems*; Ostfeld, R.S., Keesing, F., Eviner, V.T., Eds.; Princeton University Press: Princeton, NJ, USA, 2008; pp. 260–283.
51. Matson, P.A.; Boone, R.D. Natural Disturbance and Nitrogen Mineralization: Wave-Form Dieback of Mountain Hemlock in the Oregon Cascades. *Ecology* **1984**, *65*, 1511–1516. [CrossRef]
52. Hobara, S.; Tokuchi, N.; Ohte, N.; Koba, K.; Katsuyama, M.; Kim, S.J.; Nakanishi, A. Mechanism of nitrate loss from a forested catchment following a small-scale, natural disturbance. *Can. J. For. Res.* **2001**, *31*, 1326–1335. [CrossRef]
53. Cobb, R.C.; Eviner, V.T.; Rizzo, D.M. Mortality and community changes drive sudden oak death impacts on litterfall and soil nitrogen cycling. *New Phytol.* **2013**, *200*, 422–431. [CrossRef] [PubMed]
54. Hansen, E.M.; Goheen, E.M. Phellinus weirii and other native root pathogens as determinants of forest structure and process in western North America. *Annu. Rev. Phytopathol.* **2000**, *38*, 515–539. [CrossRef] [PubMed]
55. Metz, M.R.; Frangioso, K.M.; Wickland, A.C.; Meentemeyer, R.K.; Rizzo, D.M. An emergent disease causes directional changes in forest species composition in coastal California. *Ecosphere* **2012**, *3*, 86. [CrossRef]
56. Paine, R.T.; Tegner, M.J.; Johnson, E.A. Compounded perturbations yield ecological surprises. *Ecosystems* **1998**, *1*, 535–545. [CrossRef]
57. Simard, M.; Romme, W.H.; Griffin, J.M.; Turner, M.G. Do mountain pine beetle outbreaks change the probability of active crown fire in lodgepole pine forests? *Ecol. Monogr.* **2010**, *81*, 3–24. [CrossRef]
58. James, P.M.A.; Robert, L.-E.; Wotton, B.M.; Martell, D.L.; Fleming, R.A. Lagged cumulative spruce budworm defoliation affects the risk of fire ignition in Ontario, Canada. *Ecol. Appl.* **2016**, *27*, 532–544. [CrossRef] [PubMed]
59. Metz, M.R.; Varner, J.M.; Frangioso, K.M.; Meentemeyer, R.K.; Rizzo, D.M. Unexpected redwood mortality from synergies between wildfire and an emerging infectious disease. *Ecology* **2013**, *94*, 2152–2159. [CrossRef] [PubMed]
60. Cobb, R.C.; Meentemeyer, R.K.; Rizzo, D.M. Wildfire and forest disease interaction lead to greater loss of soil nutrients and carbon. *Oecologia* **2016**, *182*, 265–276. [CrossRef] [PubMed]
61. Worrall, J.J.; Harrington, T.C. Etiology of canopy gaps in spruce–fir forests at Crawford Notch, New Hampshire. *Can. J. For. Res.* **1988**, *18*, 1463–1469. [CrossRef]
62. Negrón, J.F.; McMillin, J.D.; Anhold, J.A.; Coulson, D. Bark beetle-caused mortality in a drought-affected ponderosa pine landscape in Arizona, USA. *For. Ecol. Manag.* **2009**, *257*, 1353–1362. [CrossRef]
63. Allen, C.D.; Breshears, D.D.; McDowell, N.G. On underestimation of global vulnerability to tree mortality and forest die-off from hotter drought in the Anthropocene. *Ecosphere* **2015**, *6*, 129. [CrossRef]

4

Diversity of *Phytophthora* Species Associated with *Quercus ilex* L. in Three Spanish Regions Evaluated by NGS

Beatriz Mora-Sala [1,*], David Gramaje [2], Paloma Abad-Campos [1] and Mónica Berbegal [1]

1 Instituto Agroforestal Mediterráneo, Universitat Politècnica de València, Camino de Vera s/n, 46022 Valencia, Spain; pabadcam@eaf.upv.es (P.A.-C.); mobermar@etsia.upv.es (M.B.)
2 Instituto de Ciencias de la Vid y del Vino, Consejo Superior de Investigaciones Científicas–Universidad de la Rioja–Gobierno de La Rioja, Ctra. de Burgos km. 6, 26007 Logroño, Spain; david.gramaje@icvv.es
* Correspondence: beamosa@upvnet.upv.es

Abstract: The diversity of *Phytophthora* species in declining *Fagaceae* forests in Europe is increasing in the last years. The genus *Quercus* is one of the most extended *Fagaceae* genera in Europe, and *Q. ilex* is the dominant tree in Spain. The introduction of soil-borne pathogens, such as *Phytophthora* in *Fagaceae* forests modifies the microbial community present in the rhizosphere, and has relevant environmental and economic consequences. A better understanding of the diversity of *Phytophthora* spp. associated with *Q. ilex* is proposed in this study by using Next Generation Sequencing (NGS) in six *Q. ilex* stands located in three regions in Spain. Thirty-seven *Phytophthora* phylotypes belonging to clades 1 to 12, except for clades 4, 5 and 11, are detected in this study, which represents a high diversity of *Phytophthora* species in holm oak Spanish forests. *Phytophthora chlamydospora, P. citrophthora, P. gonapodyides, P. lacustris, P. meadii, P. plurivora, P. pseudocryptogea, P. psychrophila* and *P. quercina* were present in the three regions. Seven phylotypes could not be associated with known *Phytophthora* species, so they were putatively named as *Phytophthora* sp. Most of the detected phylotypes corresponded to terrestrial *Phytophthora* species but aquatic species from clades 6 and 9 were also present in all regions.

Keywords: forest disease monitoring; oomycetes; natural ecosystems; holm oak decline

---

1. Introduction

The genus *Phytophthora* comprises nowadays more than 150 species with a broad host range, and includes well-known plant pathogens that are devastating natural ecosystems [1–3]. *Fagaceae* forests, with *Phytophthora* hosts such as *Quercus, Castanea, Lithocarpus* or *Fagus* species, are an example of declining forests affected by *Phytophthora* worldwide [4–26]. In Europe, *Castanea, Fagus* and *Quercus* are the most extended *Fagaceae* genera [13]. The diversity of *Phytophthora* species found in the last thirty years associated with the rhizosphere of these genera, reveals a complex syndrome in the decline of these forests [10,13,27–31]. On one hand, there is the difficulty of diagnosing which *Phytophthora* species present is the primary pathogen. On the other hand, it is also known that depending on which *Phytophthora* species is established first, the damage on the host varies [32]. Furthermore, all *Phytophthora* species present in the rhizosphere compromise forests regeneration, as they cause seedling damping off [13]. Within the genus *Phytophthora* we can find typical species of riparian ecosystems and/or water bodies that belong to clades 6 and 9. These species behave as saprotrophs or opportunistic organisms, causing in some cases tree decline, nevertheless its role is not well understood [32–35]. The remaining *Phytophthora* species included in the other clades (1 to 12, except 6 and 9) have a terrestrial life cycle, although they often end up in watercourses by runoff [13,35–37]. Either aquatic or terrestrial *Phytophthora* species are related to *Fagaceae* forests' decline [13].

Global trade increases the risk of unnoticed introductions of alien species into natural ecosystems [38,39]. The introduction of soil-borne pathogens, such as *Phytophthora* in *Fagaceae* forests modifies the microbial community present in the rhizosphere, and has relevant environmental and economic consequences [3]. Soil properties, land use, environmental conditions, the host plants and/or the microbial background determine the microbiota composition from a site [40]. Introduced pathogens have to compete with other microorganisms for available resources, which can lead to a decrease of the native microbiota. Hosts that co-evolved with soil microorganisms can adapt more easily to biotic and abiotic stresses, but a shift in the microbiota structure can trigger the host decline [3,40].

Next generation sequencing (NGS) technologies have stood out as an essential tool for environmental and ecological studies [41,42]. Pyrosequencing the Internal Transcribed Spacer (ITS1) is an efficient and accurate NGS technique for the detection and identification of *Phytophthora* spp. in environmental samples [29,35,43–45]. A better understanding of the diversity of *Phytophthora* spp. associated with a host can be provided by metagenomics, even if these *Phytophthora* spp. are not the most prevalent pathogens [3,45–47]. Nevertheless, a holistic study, combining biological and molecular identification tools, can substantially improve the diagnosis, because in many cases establishing species boundaries via molecular methods it is not an easy task [2,39,43]. In this context, it is interesting to note that samples from the current study were previously subjected to traditional isolation, baiting and real-time polymerase chain reaction (PCR) methods to detect *Phytophthora* spp. [48]. The objective of the present study was to unravel the *Phytophthora* community present in these previously studied areas using NGS technology. Thus, the diversity and abundance of *Phytophthora* spp. in six holm oak forests located in southwestern and eastern Spain were investigated using an amplicon pyrosequencing approach and the implications are further discussed.

2. Materials and Methods

2.1. Study Site and Sampling

The six study areas are holm oak forests located in three different Spanish regions: Holm oak rangelands ("dehesas") in Extremadura in southwestern Spain (province of Cáceres), four holm oak stands in the Comunidad Valenciana region in eastern Spain (two in Valencia province, one in Castellón province and one in Alicante province) and one holm oak stand in Cataluña in northeastern Spain (Barcelona province). Samplings were conducted in the autumn (fall) and winter season in different years (2012–2015), and in some areas, it was repeated for two consecutive years.

Soil samples (0.5–1 kg approx.) were collected in all the surveys, consisting in a mixture of four subsamples taken from four different points 1 m around the selected holm oak. In the surveys conducted from 2013 to 2015, along with the soil samples, roots from the rhizosphere were also taken to be analyzed. Soil and roots samples were conserved at 5 °C until DNA extraction was performed. Baiting with leaflets of *Rhododendron* sp., *Viburnum tinus* L, *Quercus ilex* L, *Quercus suber* L, *Ceratonia siliqua* L. and/or *Dianthus caryophyllus* L. petals was performed as described in a previous study [48], and the vegetal material from the baitings was conserved at -80 °C until DNA extraction.

In the Extremadura region in 2012, 60 soil samples from the declining and non-declining *Q. ilex* rhizosphere were collected from 10 "dehesas" during the autumn. In 2013, 54 soil and root samples and 216 baiting samples from 15 "dehesas" were analyzed. In the Comunidad Valenciana region, during 2014 and 2015, holm oak stands were surveyed, generating 26 soil and roots samples and 104 baiting samples from declining trees. In the Cataluña region in 2013, 15 soil and root samples and 45 baiting samples were processed from declining holm oaks.

2.2. DNA Extraction, PCR, and Preparation of the Amplicon Libraries

Each soil sample was passed through a 2 mm sieve to remove the organic matter and gravel. Once it was homogenized, 50–80 g per subsample was lyophilized overnight and pulverized using a FRITSCH Variable Speed Rotor Mill-PULVERISETTE 14 (ROSH, Oberstein, Germany). DNA from each sample was extracted by duplicate with the Power Soil DNA Isolation Kit (MO BIO Laboratories, Carlsbad, CA, USA) following the manufacturer´s protocol. Roots and baiting vegetal material samples were firstly ground using a mortar and pestle under liquid nitrogen.

Healthy leaflets of the different vegetal species used as baitings were included as negative controls. Once homogenized, DNA extraction was performed from 60–80 mg of material per sample using the Power Plant Pro DNA Isolation Kit (MO BIO Laboratories, Carlsbad, CA, USA).

Amplicon libraries were generated with a nested polymerase chain reaction (PCR) according to the methodology described by Català et al. [35]. The expected size of the amplicons ranged from 280 up to 450 bp. The duplicate amplicons obtained from each sample were pooled if they were positive. Negative samples were retried, firstly diluting 10 times the first round PCR product for the second round PCR. If the retried samples kept on being negative, DNA was diluted ten times, and nested PCR was performed without diluting the first round product. Positive amplicons were double purified using the Agencourt AMPureXP Bead PCR Purification protocol (Beckman Coulter Genomics, MA, USA). Samples were sequenced on the GS Junior 454 system (Roche 454 Life Science, Brandford, CT, USA) at the genotyping service facility from the Universitat de València (Burjassot, Spain).

2.3. Bioinformatics Processing and Statistical Analysis

The sequence reads were processed as described in Català et al. [35]. MOTU (Molecular Operational Taxonomic Unit) clustering of the reads was done with the 90% length coverage threshold and a 99% score coverage threshold using BLASTCLUST software [49]. The consensus sequences of the MOTUs were identified using BLAST tool in the *Phytophthora* Database [50], the MegaBLAST tool from the GenBank database [49] and a customized database. Afterwards, our query sequences from each survey and type of material (soil/vegetal) were aligned separately using MUSCLE [51]. Phylogenetic analyses consisting of maximum likelihood were performed in MEGA 6.06 [52] with the suggested suitable model, where gaps and missing data were treated as complete deletions. The robustness of the topology was evaluated by 1000 bootstrap replications [53].

A *Phytophthora* operational taxonomic unit (OTU) table was created excluding other genus reads, unaligned sequences and singletons. OTUs represented globally by less than five reads were removed [54]. The resulting quality-filtered dataset was used for the assessment of diversity and richness. The relationship in OTU composition among samples were investigated by calculating Bray Curtis metrics, and visualized by means of PCoA plots. Samples without *Phytophthora* reads were excluded for the diversity analyses. Biodiversity indices and principle statistics analyses on taxonomic profiles were analyzed in the tool MicrobiomeAnalyst [55].

3. Results

3.1. Sequencing Results

In the Extremadura region, 66,732 total ITS1 sequences were generated from the pyrosequenced soil samples collected in 2012. These had an average length of 308 bp, and only 61,576 high quality sequences were considered for the analysis after trimming, and excluding singletons; 48.7% of the sequences were identified from declining trees, and 51.3% from non-declining trees.

From samples collected in 2013 in the Extremadura region, 377,799 ITS1 sequences reads were obtained with an average length of 302 bp. After trimming and excluding singletons, 317,961 high quality sequences were used for the analysis; 103,333 from the root samples (45.05% from declining trees and 54.95% from non-declining trees), 56,112 sequences from the soil samples (45.3% from declining trees and 54.7% from non-declining trees) and 158,516 sequences from the baiting samples (51.68% from declining trees and 48.32% from non-declining trees). No *Phytophhtora* phylotypes are detected from the negative baiting controls.

In the Cataluña region, 63,105 total ITS1 sequences were generated from the pyrosequenced samples. These had an average length of 283 bp, and only 55,230 high quality sequences were considered for the analysis after trimming and excluding singletons (28,381 from roots, 9887 from soils and 16,962 from baitings).

In the Comunidad Valenciana region, 177,398 ITS1 raw sequences with an average length of 309 bp were obtained. After trimming and excluding singletons, 78,962 high quality sequences were considered for the analysis (3,977 from sampled roots, 18,861 from soils and 56,124 from baitings).

3.2. *Identification of Phytophthora Phylotypes*

3.2.1. Extremadura Region

A total of 33 different *Phytophthora* phylotypes are identified in the Extremadura region during the surveys conducted in 2012 and 2013 (Table 1). Detected phylotypes belong to all clades except for clades 4, 5 and 11, the clade 6 having the highest number of *Phytophthora* phylotypes detected (Table 1). In 2012, 20 *Phytophthora* phylotypes were obtained, and 15 of them identified to the species level: *Phytophthora bilorbang* Aghighi, G.E. Hardy, J.K. Scott & T.I. Burgess, *Phytophthora botryosa* Chee, *Phytophthora cambivora* (Petri) Buisman, *Phytophthora chlamydospora* Brasier and Hansen, *Phytophthora cinnamomi* Rands., *Phytophthora cryptogea* Pethybr. and Laff., *Phytophthora gemini* Man in 't Veld, Rosendahl, Brouwer and de Cock, *Phytophthora gonapodyides* (H.E. Petersen) Buisman, *Phytophthora hydropathica* Hong and Gallegly, *Phytophthora lacustris* Brasier, Cacciola, Nechwatal, Jung and Bakonyi, *Phytophthora lagoariana*, *Phytophthora multivora* Scott and Jung, *Phytophthora plurivora* Jung and Burgess, *Phytophthora psychrophila* Jung and Hansen, *Phytophthora quercina* Jung and *Phytophthora riparia* Reeser, Sutton and Hansen.

One phylotype belong to complex *Phytophthora uliginosa-europaea* as the ITS1 region used in the assay could not resolve its identity. Three phylotypes do not correspond to any *Phytophthora* sequence included in databases; therefore, these putative new species are named *Phytophthora* sp.1, *Phytophthora* sp.2 and *Phytophthora* sp.3.

In 2013, 25 *Phytophthora* phylotypes were recovered in Extremadura and 22 of them were identified to the species level: *Phytophthora* taxon *ballota*, *P. bilorbang*, *P. chlamydospora*, *P. cinnamomi*, *Phytophthora citrophthora* (R.E. Sm. and E.H. Sm.) Leonian, *Phytophthora clandestina* Taylor, Pascoe and Greenhalgh, *P. cryptogea*, *Phytophthora gallica* Jung and Nechwatal, *P. gonapodyides*, *P. hydropathica*, *Phytophthora insolita* Ann and Ko, *Phytophthora* sp. *kelmania*, *P. lacustris*, *Phytophthora lactucae* Bertier, Brouwer and de Cock, *P. lagoariana*, *Phytophthora meadii* McRae, *Phytophthora megasperma* Drechsler, *Phytophthora* sp. *palustris*, *P. plurivora*, *Phytophthora pseudocryptogea* Safaiefarahani, Mostowfizadeh, G.E. Hardy, and T.I. Burgess, *P. psychrophila*, *P. quercina* and *Phytophthora rosacearum* Hansen and Wilcox. The ITS1 region used in the assay could not resolve the identity of one *Phytophthora* phylotype complex: *Phytophthora uliginosa-europaea*.

One *Phytophthora* phylotype does not match with any *Phytophthora* sequence included in the databases, therefore this putative new species is named as *Phytophthora* sp.4.

Table 1. *Phytophthora* phylotypes detected by next generation sequencing (NGS), based on the Internal Transcribed Spacer (ITS1) region.

Phylotypes	Clade	Spanish Regions					
		Extremadura	Cataluña	Comunidad Valenciana			
				Font Roja	Hunde	Pina	Alcublas
BAL	1	√		√	√		
BIL	6	√					
BOT	2	√					
CAM	7	√					
CHL	6	√	√			√	√
CIN	7	√	√				
CIP	2	√	√			√	
CLA	1	√					
CRY	8	√					
GAL	10	√					
GEM	6	√					
GON	6	√	√				√
HYD	9	√	√				
INS	9	√	√				
KEL	8	√				√	
LAC	6	√	√			√	√
LCT		√					
LAG	9	√	√				
MEA	2	√	√			√	
MEG	6	√					√
MUL	2	√					
PAS	9	√					
PLU	2	√	√			√	√
PSC	8	√	√		√	√	√
PSY	3	√	√	√	√		√
QUE	12	√	√	√	√	√	
RIP	6	√					
ROS	6	√					
TEN	1						√
ULIG-EUR	7	√		√			
SP.1	6	√					
SP.2	7	√					
SP.3	8	√					
SP.4	1	√					
SP.5	7		√				
SP.6	3			√			
SP.7	1						√

BAL, *P.* taxon *ballota*; BIL, *P. bilorbang*; BOT, *P. botryosa*; CAM, *P. cambivora*; CHL, *P. chlamydospora*; CIN, *P. cinnamomi*; CIP, *P. citrophthora*; CLA, *P. clandestina*; CRY, *P. cryptogea*; GAL, *P. gallica*; GEM, *P. gemini*; GON, *P. gonapodyides*; HYD, *P. hydropathica*; INS, *P. insolita*; KEL, *P. sp. kelmania*; LAC, *P. lacustris*; LCT, *P. lactucae*; LAG, *P. lagoariana*; MEA, *P. meadii*; MEG, *P. megasperma*; MUL, *P. multivora*; PAS, *P.* sp. *palustris*; PLU, *P. plurivora*; PSC, *P. pseudocryptogea*; PSY, *P. psychrophila*; QUE, *P. quercina*; RIP, *P. riparia*; ROS, *P. rosacearum*; TEN, *P. tentaculata*; ULIG-EUR, *P. uliginosa-P. europaea*; SP.1-SP.7, new phylotypes found not identified to the species level.

3.2.2. Cataluña Region

Fourteen *Phytophthora* phylotypes are identified in the Cataluña region during the survey conducted in 2013, belonging to clades 2, 3, 6, 7, 8, 9 and 12 (Table 1). Thirteen phylotypes are identified to the species level: *P. chlamydospora*, *P. cinnamomi*, *P. citrophthora*, *P. gonapodyides*, *P. hydropathica*, *P. insolita*, *P. lacustris*, *P. lagoariana*, *P. meadii*, *P. plurivora P. pseudocryptogea*, *P. psychrophila* and *P. quercina*. One *Phytophthora* phylotype does not match with any *Phytophthora* sequence included in the databases, therefore this putative new species is named *Phytophthora* sp.5.

3.2.3. Comunidad Valenciana Region

The *Phytophthora* phylotypes identified in Comunidad Valenciana during the surveys conducted in the period 2014-2015 is shown in Table 1. Sixteen *Phytophthora* phylotypes are identified in Comunidad Valenciana belonging to clades 1, 2, 3, 6, 7, 8 and 12 (Table 1). Thirteen phylotypes are identified to the species level: *P.* taxon *ballota*, *P. chlamydospora*, *P. citrophthora*, *P. gonapodyides*, *P.* sp. *kelmania*, *P. lacustris*, *P. meadii*, *P. megasperma*, *P. plurivora P. pseudocryptogea*, *P. psychrophila*, *P. quercina* and *Phytophthora tentaculata* Kröber and Marwitz. The ITS1 region used in the assay could not resolve the identity of one *Phytophthora* phylotype complex: *P. uliginosa-europaea*. Two *Phytophthora* phylotypes do not match with any *Phytophthora* sequences included in the databases and thus, these putative new species are named as *Phytophthora* sp.6. and *Phytophthora* sp.7.

3.2.4. Phytophthora Diversity

The relative abundance of *Phytophthora* species detected from 2013 to 2015 in the three studied regions is shown in Figure 1A. The Extremadura region presents the greatest diversity of *Phytophthora* species. According to the type of material used for *Phytophthora* isolation, baiting material has the highest relative abundance of *Phytophthora* species, followed by roots and soils (Figure 1B).

Regarding symptomatology, non-declining holm oak samples in the Extremadura present a higher diversity of *Phytophthora* species either in the 2013 survey or analyzing 2012 and 2013 together, than samples from declining trees (Figure 2A,C), but this diversity was slightly higher in declining samples in the 2012 survey (Figure 2B).

The ANOVA shows that the alpha diversity was significantly higher in the Extremadura region, followed by Comunidad Valenciana and Cataluña (Figure 3). Both the Chao 1 estimator (p value < 0.05) and the Shannon index (p value < 0.01) were significant according to the factor region (Table 2). Regarding the factor material, the Shannon index was significant (p value $= 0.033$), showing greater diversity of *Phytophthora* species when using baiting material, while the Chao 1 estimator was non-significant (p value $= 0.091$) (Table 2).

Table 2. Analysis of variance (ANOVA) table for the alpha diversity of *Phytophthora* species detected in the study.

	α Diversity		
	Region	**Material**	**Region × Material**
Chao 1 estimator	$F_{2,245} = 10.8$ $p \leq 0.01$	$F_{2,196} = 1.9$ $p = 0.091$	$F_{4,188} = 5.5$ $p \leq 0.01$
Shannon index	$F_{2,245} = 7.1$ $p \leq 0.01$	$F_{2,196} = 2.3$ $p = 0.033$	$F_{4,188} = 4.7$ $p \leq 0.01$

In the Extremadura region where the factor symptomatology was studied, in 2012 alpha diversity measured by the Chao1 estimator results was significant for the factor symptomatology (p value $= 0.0218$), but the Shannon Index was non-significant (p value $= 0.1303$). Nevertheless, in 2013 the ANOVA shows that both estimators were significant for the factor material and non-significant for the factor symptomatology (Table 3).

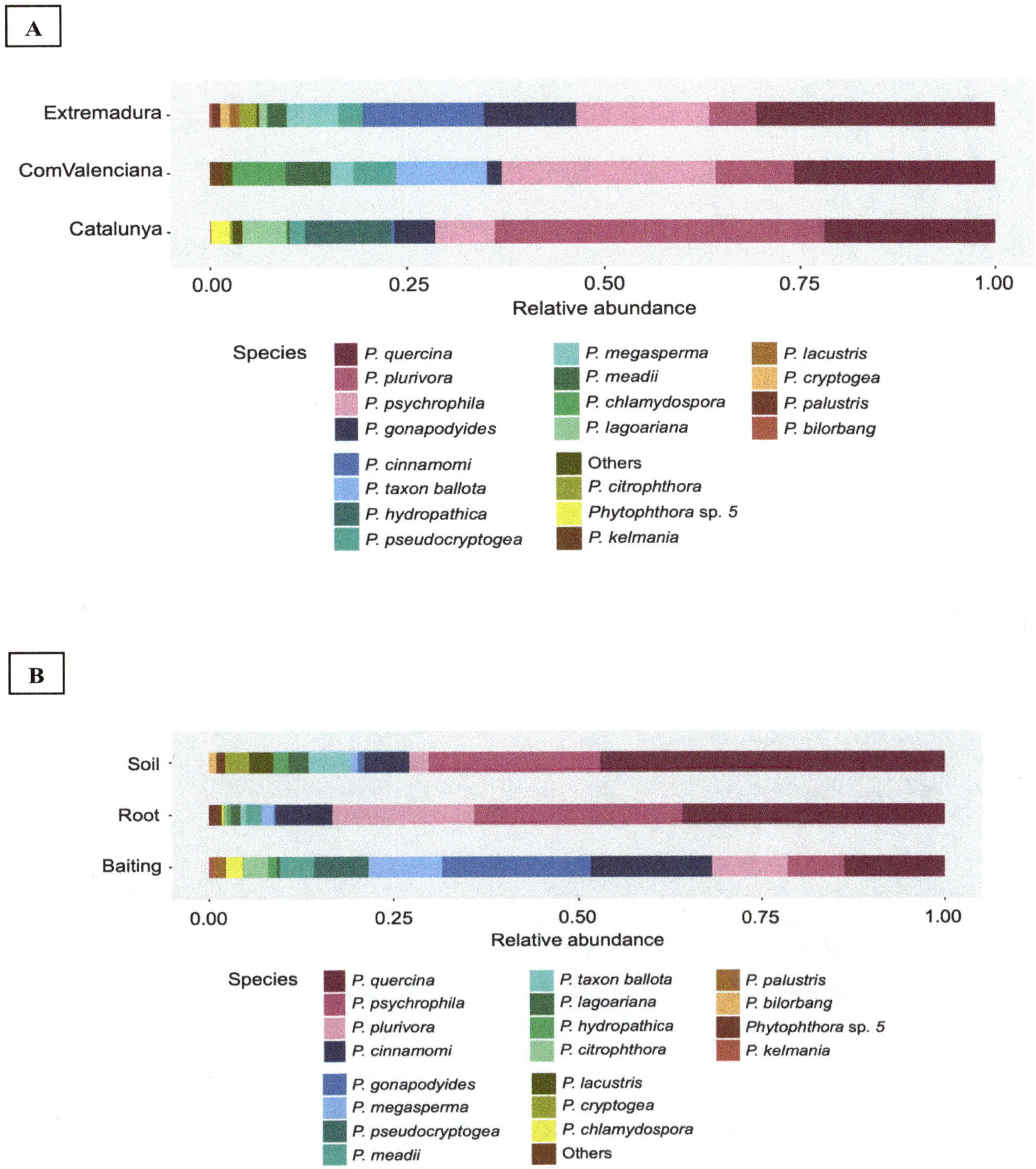

Figure 1. Relative abundance of *Phytophthora* species detected in the three studied regions from 2013 to 2015 showing more than 1% relative abundance of all reads. Species representing less than 1% of the total reads are grouped in "others". (**A**): According to the factor region; Extremadura, Comunidad Valenciana (ComValenciana) and Cataluña (Catalunya). (**B**): According to the factor type of material; soils, roots and baitings. Results of the survey performed in Extremadura in 2012 are excluded in this analysis, since only soil samples were included.

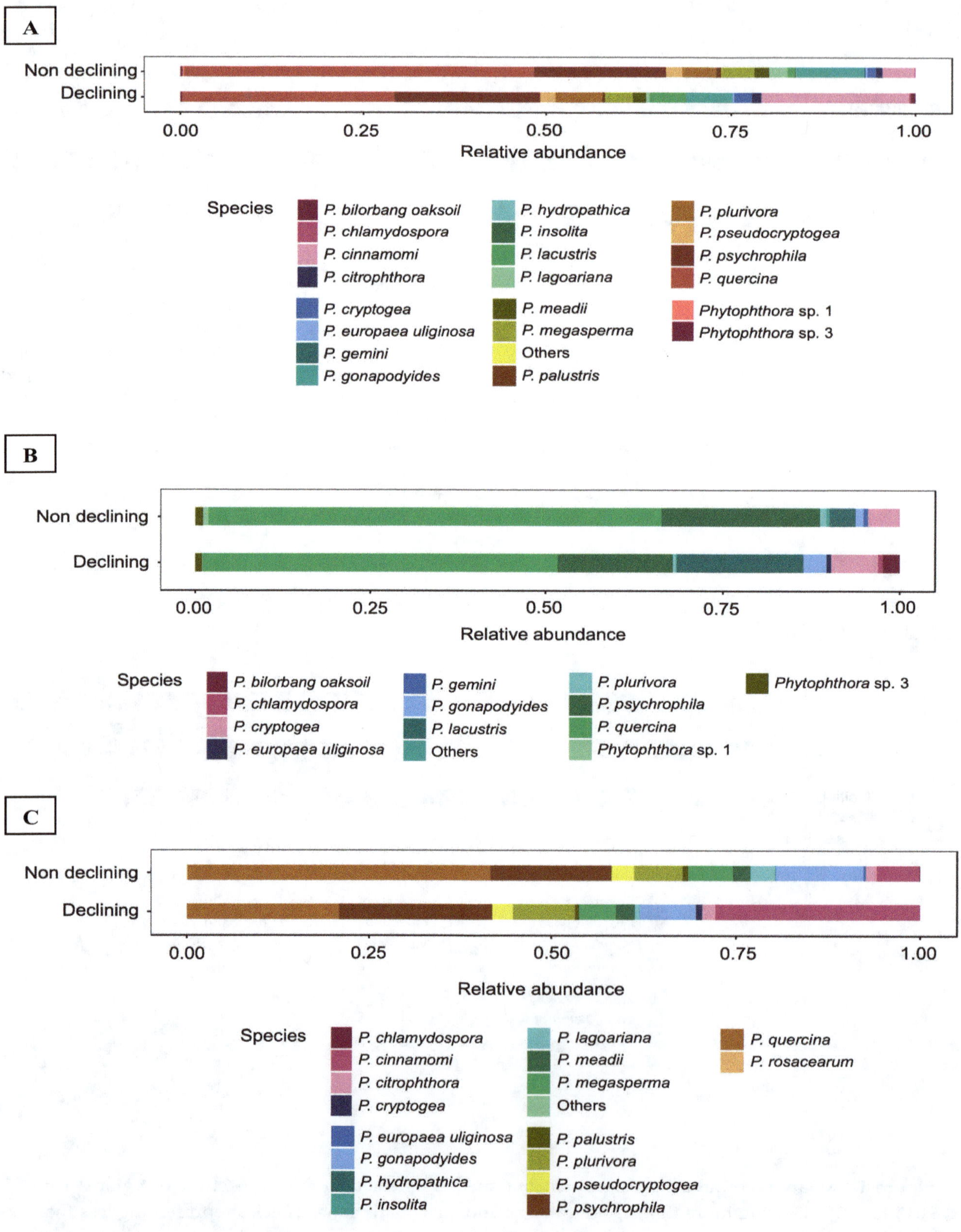

Figure 2. Relative abundance of *Phytophthora* species showing more than 1% relative abundance of all reads in Extremadura region according to the factor symptomatology. Species representing less than 1% of the total reads are grouped in others. (**A**): During the surveys conducted in 2012 and 2013. (**B**): Survey conducted in 2012. (**C**): Survey conducted in 2013.

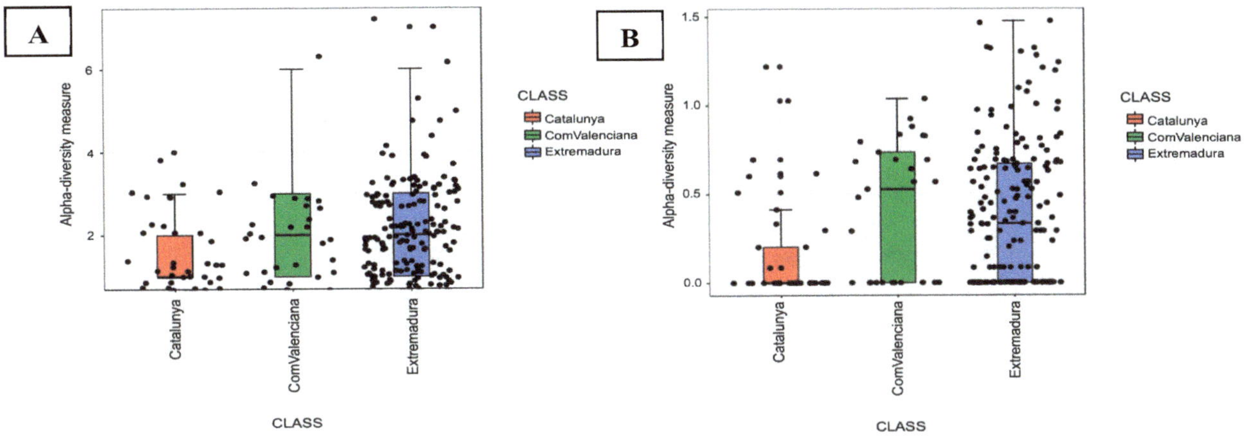

Figure 3. Boxplot showing alpha diversity measures of the *Phytophthora* diversity according to the factor region: Extremadura, Comunidad Valenciana (ComValenciana) and Cataluña (Catalonia). (**A**): Chao 1 estimator (p value < 0.05). (**B**): Shannon index (p value < 0.01).

Table 3. Analysis of variance (ANOVA) tables for the alpha diversity of *Phytophthora* species detected in Extremadura dehesas during the surveys conducted in 2012 and 2013.

	α Diversity					
Survey	**Chao 1 Estimator**			**Shannon Index**		
	Material	**Symptoms**	**Material × Symptoms**	**Material**	**Symptoms**	**Material × Symptoms**
2012	ndt	$F_{1,47} = 2.39$ $p = 0.0218$	ndt	ndt	$F_{1,47} = 1.54$ $p = 0.1303$	ndt
2013	$F_{2,126} = 6.03$ $p = 0.0029$	$F_{1,127} = 1.23$ $p = 0.2199$	$F_{2,123} = 2.7$ $p = 0.023$	$F_{2,126} = 4.95$ $p = 0.0081$	$F_{1,127} = 0.83$ $p = 0.4085$	$F_{2,123} = 2.4$ $p = 0.041$

ndt = not determined.

The principal coordinates analysis (PCoAs) of Bray Curtis show that the *Phytophthora* population in Extremadura had more variation in the number of species present, although there were similarities with the populations of the other two regions (Figure 4A). There are no significant differences in the diversity of *Phytophthora* species detected in the type of material used or the symptomatology of the sampled trees (Figure 4B,C).

Nine *Phytophthora* phylotypes were detected in the three regions (24.3%): *P. citrophthora*, *P. meadii*, *P. plurivora*, *P. psychrophila*, *P. quercina*, *P. chlamydospora*, *P. gonapodyides*, *P. lacustris* and *P. pseudocryptogea* (Figure 5A). Regarding the type of material, 11 *Phytophthora* phylotypes were common to root, soil and baiting material (28.9%): *P. plurivora*, *P. psychrophila*, *P. quercina*, *P. chlamydospora*, *P. gonapodyides*, *P. lacustris*, *P. megasperma*, *P. cinnamomi*, *P.* uliginosa - europaea, *P. insolita* and *P. lagoariana* (Figure 5B). Finally, 20 *Phytophthora* phylotypes were present in declining and non-declining holm oak trees, which represent 54.05% of the detected *Phytophthora* phylotypes (Figure 5C).

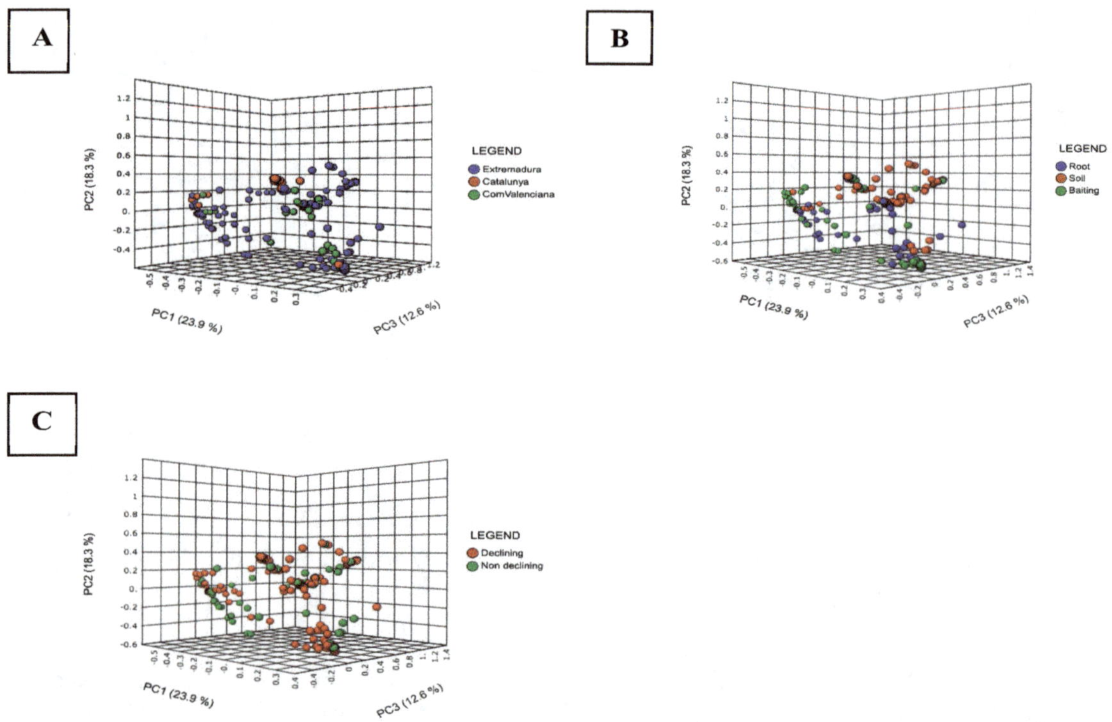

Figure 4. Principal Coordinate Analysis (PCoA) based on Bray Curtis dissimilarity metrics, showing the distance in the *Phytophthora* spp. composition according to the different factors studied. (**A**): Regions surveyed; Extremadura, Comunidad Valenciana (ComValenciana) and Cataluña (Catalunya). (**B**): Type of material used for the detection; roots, soils and baitings. (**C**): Symptomatology of the holm oaks sampled: declining and non-declining.

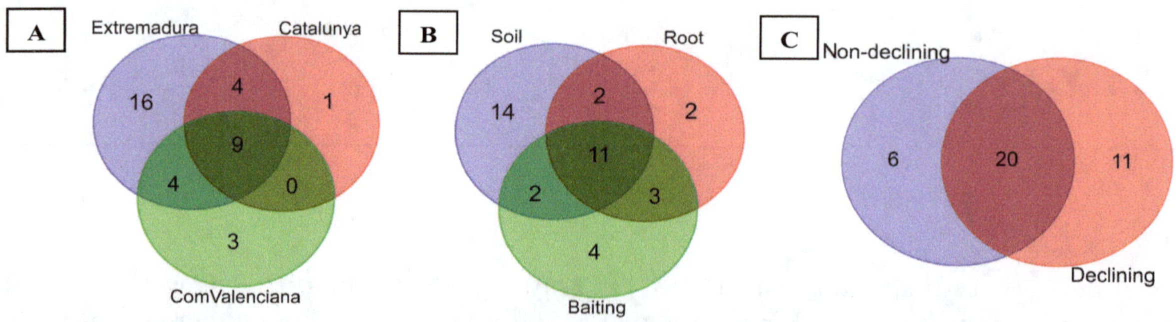

Figure 5. Venn diagram showing the overlap of operational taxonomic units (OTUs) identified in the *Phytophthora* population present in the holm oaks surveyed in the study according to the different factors studied. (**A**): Regions surveyed; Extremadura, Comunidad Valenciana (ComValenciana) and Cataluña (Catalunya), (**B**): Type of material used for the detection; roots, soils and baitings. (**C**): Symptomatology of the holm oaks sampled: declining and non-declining.

4. Discussion

Thirty-seven *Phytophthora* phylotypes belonging to clades 1 to 12, except for clades 4, 5 and 11, were detected in this study, which represents a high diversity of *Phytophthora* species in holm oak Spanish forests compared to previous Spanish NGS studies [44,56]. *Phytophthora chlamydospora*, *P. citrophthora*, *P. gonapodyides*, *P. lacustris*, *P. meadii*, *P. plurivora*, *P. pseudocryptogea*, *P. psychrophila* and *P. quercina* were present in the three regions. Seven of these phylotypes cannot be associated with known *Phytophthora* species, so they are putatively named as *Phytophthora* sp. Most of the detected phylotypes correspond to terrestrial *Phytophthora* species, but aquatic species from clades 6 and 9 were also present in all regions. In general, the most abundant phylotypes in the study are *P. quercina*, followed by *P. psychrophila*, *P. cinnamomi* and *P. plurivora*; all of them terrestrial species. Our results concur with Català et al. [44], which reported *P. quercina* and *P. psychrophila* as the most relevant species associated with *Q. ilex* in eastern Spain. Nevertheless, they differ from Ruiz Gómez et al. [56] describing *P. plurivora*, *P. quercina*, *P. cinnamomi* and *P. cactorum* as the most frequent detected species in Andalucía holm oak rangelands (dehesas) located in southern Spain.

In the Extremadura region, all *Phytophthora* clades are present with the exception of clades 4, 5 and 11. Terrestrial species dominate the *Phytophthora* community as *P. quercina* 30%, *P. psychrophila* 17% and *P. cinnamomi* 15%, are the most abundant species. Nevertheless, also aquatic species are identified, such as *P. gonapodyides* 12% and *P. megasperma* 7%, which follow in abundance. Hence, a mixture of *Q. ilex* pathogenic terrestrial and aquatic species is identified in the Extremadura region. In the Cataluña region, the *Phytophthora* community associated with the studied *Q. ilex* stand was dominated again by terrestrial species (*P. plurivora* 42%, *P. quercina* 22% and *P. psychrophila* 8%), followed by aquatic species (*P. hydropathica* 11%, *P. lagoariana* 6% and *P. gonapodyides* 5%). Clades 1, 4, 5, 8, 10 and 11 were not detected in the studied holm oaks. In the Comunidad Valenciana region, the *Phytophthora* community is made up primarily by terrestrial *Phytophthora* species from clades 1, 2, 3, 7, 8 and 12 (*P. psychrophila* 27%, *P. quercina* 26%, *P.* taxon *ballota* 11%, *P. plurivora* 10%); although there are also present in lower abundance aquatic species from clade 6 such as *P. chlamydospora* 7%, *P. gonapodyides* 2% or *P. megasperma* 3%.

The weather conditions of the Extremadura region were more favorable for *Phytophthora* development in 2013 than in 2012, since in 2012 Extremadura received less precipitation than in 2013, as reported in 2018 by Mora-Sala et al. [48]. Redondo et al. [57] state that there is a decrease in the diversity of terrestrial *Phytophthora* communities when temperature and precipitation decreases, precipitation being the main driving factor, except for clades 2 and 8, in which temperature was more conditioning. According to Redondo et al. [57], aquatic *Phytophthora* species are inversely conditioned by temperature and precipitation. The *Phytophthora* diversity found in Extremadura in 2013 was higher than in 2012, which fits with the parameters stated in the study of Redondo et al. [57]. Moreover, more types of sampling materials (roots and baitings) were used for DNA extraction in 2013 than in 2012, which might potentially increase the phylotype diversity. Redondo et al. [57] conclude that the land use is only significant for aquatic species, having more diversity in urban or agricultural sites than in forests. In our study, we found more species belonging to the clade 6 in Extremadura oak rangelands, such as *P. gonapodyides* or *P. megasperma*, which are man-made forests, although they are also identified in lower abundance in Cataluña and Comunidad Valenciana. Clade 9 species (*P. hydropathica* and *P. lagoariana*) are also detected in Extremadura and in Cataluña.

Phytophthora cinnamomi is an important and devastating *Quercus* pathogen in Spain [27,58–61], although it is not the most abundant species detected in the *Quercus* forests studied. The pathogen is significantly more abundant in declining trees than in non-declining trees, as previously reported by Mora-Sala et al. using qPCR and traditional isolation methods [48], and that also corroborates the metabarcoding study of Ruiz Gómez et al. [56]. *Phytophthora cinnamomi* is not detected in Comunidad Valenciana.

This is in agreement with previously published reports, and may be due to the unsuitable conditions for the pathogen development in the area consisting on primarily calcareous soils with a high pH [26,44,48,62].

Phytophthora quercina is a specific oak pathogen, present in the three studied regions, that probably has co-evolved with *Q. ilex* in Spain, rotting the trees slowly and progressively. As *P. quercina* was not described until 1999 [8], it could not be associated with the studies of the decline of Spanish holm oak conducted in the past. In addition, *P. quercina* is a difficult pathogen to isolate, due to its slow growth, thus its detection was not always possible in previous studies [8,15,26,30]. Moreover, as generally *P. cinnamomi* was present, and *P. quercina* is not a fast growing pathogen, the decline was associated with *P. cinnamomi*. Although recent studies report the presence of *P. quercina*, it is thought of as not being as frequent as *P. cinnamomi*. Mora-Sala et al. demonstrate that *P. quercina* is more frequent than

P. cinnamomi, not only in Comunidad Valenciana, but also in the Extremadura and Cataluña regions by qPCR [48]. Ruiz Gómez et al. [56] report *P. quercina* as the fourth-most abundant oomycete in Andalucía holm oak rangelands, while *P. cinnamomi* stands in the ninth position. The present study based on amplicon pyrosequencing supports our previous results using qPCR and verifies that this pathogen is highly abundant in the studied *Quercus* Spanish regions.

Phytophthora psychrophila is detected in high abundance in the three regions sampled, especially in Comunidad Valenciana, either in declining or in non-declining trees. This species was previously related to *Quercus* spp. dieback in Spain where it is apparently well distributed and it is only able to be isolated during the winter and spring/autumn periods [10,26,56]. As Pérez-Sierra et al. report in 2013, it is perfectly adapted to xeric Mediterranean conditions due to the thick wall of its resting structures that enables it to overcome unsuitable environmental conditions. Pathogenicity tests with *Q. ilex* seedlings demonstrate that *P. psychrophila* is considered an aggressive pathogen, causing dieback of the root system, mainly the fine roots, necrotic lesions and open cankers [26,48].

Phytophthora plurivora is significantly detected in roots and it is even present in all regions, resulting more abundantly in the Cataluña region, where it is the most frequent detected species. *Phytophthora plurivora* is a widespread species in Europe, which had been already detected in *Q. ilex* in Spain [44,56], and its pathogenicity to *Q. ilex* is demonstrated in a previous study [63] causing the absence of fine roots, necrotic lesions, open cankers, dieback of the whole root system and collar rot. Ruiz Gómez et al. [56] report *P. plurivora* as the most abundant oomycete detected in the Spanish oak rangelands surveyed in their study. *Phytophthora plurivora* in our study seems to be more associated with declining *Q. ilex* trees, as Ruiz Gómez et al. [56] report. The homothallic behavior of *Phytophthora* species, as *P. plurivora* or *P. psychrophila*, facilitates its reproduction and establishment in new areas, boosting the risk of forest decline [10,26,29,36,57,64].

From the aquatic species detected, *P. gonapodyides* was previously detected affecting *Q. ilex* in Spain [26,33,48]. *Phytophthora megasperma* is reported to reduce the root system [26,63] although this species is considered an opportunistic oak pathogen present in oak forests [9,15,16]. The role of the remaining aquatic *Phytophthora* species detected in the *Q. ilex* decline is still unknown.

The sequencing results support the presence of *P.* taxon *ballota* in two forests of the Comunidad Valenciana and in Extremadura regions. This uncultured phylotype was previously detected in oak forests in Comunidad Valenciana [44], but it is the first time that it has been detected outside the Comunidad Valenciana. The designation of new phylotypes from environmental DNA remains a committed issue, but it is reported in other studies [65–67]. Both in previous studies and the present study, no *Phytophthora* culture was isolated that coincided with this proposed phylotype using traditional methods. Further surveys targeting this organism should be performed to try the isolation.

This study detected the presence of *P. pseudocryptogea* in all three regions, and it has also been isolated from other *Fagaceae* such as *Castanea sativa* in the North of Spain (Mora-Sala and Català unpublished), suggesting that it is well established in Spain and probably in the past many isolates identified as *P. cryptogea*, were actually *P. pseudocryptogea*. In 2018, *P. pseudocryptogea* was firstly reported in Extremadura region from *Q. ilex* rhizosphere [48] and it has just been reported in Sicily affecting *Q. ilex* [36], so it seems to be well established in the Mediterranean basin.

The pathogenicity on *Q. ilex* has not been tested, but its pathogenicity on other *Fagaceae* (Mora-Sala and Català, unpublished), suggests that this species could contribute to the oak decline.

Phytophthora community diversity recovered from *Fagaceae* forests in Europe has increased in the last years [8–10,13,16,20,21,23,24,26,29,30,36,44,64,68]. The implementation of NGS technologies to forests surveys helps to improve the knowledge about the *Phytophthora* spp. diversity associated with *Fagaceae* forests and to identify possible new introduced species [3,29,44,45,47]. *Phytophthora* spp. can adapt to a wide variety of environmental conditions [38]. In Spain, the oak decline due to *Phytophthora* spp. is related to the effect of the water stress; seasonal droughts followed by floods enhance the root damage induced by *Phytophthora* species. The versatility of these species to cope with the changing scenarios and the increase of extreme weather conditions that are occurring nowadays, focuses the attention on this destructive genus. The composition of *Phytophthora* species in these ecosystems is changing because of their adaptation to new environmental conditions and new species introduction. New technologies help us improve knowledge about species diversity in these new scenarios. In the present study, NGS reveals a higher *Phytophthora* species diversity than previously detected by traditional isolation, baiting and qPCR. However, the best approach should combine all available methodologies for a correct *Phytophthora* diagnosis, facilitating a quick answer facing the potential introduction of new and/or quarantine organisms.

5. Conclusions

The use of amplicon pyrosequencing reveals a high diversity of *Phytophthora* species associated with *Q. ilex*. Baiting material has the highest relative abundance of *Phytophthora* species. The highest alpha diversity is obtained in the region of Extremadura, the diversity of Comunidad Valenciana and Cataluña being lower following this order. In general terms, the *Phytophthora* diversity is highest in non-declining *Q. ilex* than in declining trees. The implementation of molecular tools in *Phytophthora* forests monitoring, complement and help to overcome the limitations of traditional methods, being useful to improve the knowledge about the real composition of the species present in these ecosystems.

Author Contributions: B.M.-S. conducted field sampling, performed experimental work and data analysis, discussed the results and wrote the paper. D.G. performed the data analysis. M.B. participated in the data analyses, discussed the results and revised the manuscript. P.A.C. designed the study, participated in the experimental work and revised the manuscript.

Acknowledgments: We would like to thank M. León from the Instituto Agroforestal Mediterráneo-UPV (Spain) for its technical assistance. This research was supported by funding from the project AGL2011-30438-C02-01 (Ministerio de Economía y Competitividad, Spain) and Euphresco [Instituto Nacional de Investigación y Tecnología Agraria y Agroalimentaria (EUPHESCO-CEP: "Current and Emerging Phytophthoras: Research Supporting Risk Assesssment and Risk Management")].

References

1. Erwin, D.C.; Ribeiro, O.K. *Phytophthora Diseases Worldwide*; APS Press: St. Paul, MN, USA, 1996; p. 562. ISBN 0-89054-212-0.
2. Mideros, M.F.; Turissini, D.A.; Guayazán, N.; Ibarra-Avila, H.; Danies, G.; Cárdenas, M.; Lagos, L.E.; Myers, K.; Tabima, J.; Goss, E.M.; et al. *Phytophthora betacei*, a new species within *Phytophthora* clade 1c causing late blight on *Solanum betaceum* in Colombia. *Persoonia* **2018**, *41*, 39–55. [CrossRef] [PubMed]
3. Tremblay, É.D.; Duceppe, M.O.; Bérubé, J.A.; Kimoto, T.; Lemieux, C.; Bilodeau, G.J. Screening for exotic forest pathogens to increase survey capacity using metagenomics. *Phytopathology* **2018**, *108*, 1509–1521. [CrossRef] [PubMed]
4. Shearer, B.L.; Tippett, J.T. *Jarrah Dieback: The Dynamics and Management of Phytophthora Cinnamomi in the Jarrah (Eucalyptus Marginata) Forests of South-Western Australia*; Lewis, M., Ed.; Department of Conservation and Land Management: Como, WA, Australia, 1989; ISSN 1032-8106.
5. Brasier, C.M. Oak tree mortality in Iberia. *Nature* **1992**, *360*, 539. [CrossRef]
6. Brasier, C.M. *Phytophthora cinnamomi* as a contributory factor on European oak declines. In *Recent Advances in Studies on Oak Decline, Proceedings of the International Congress, Brindisi, Italy, 13–18 September 1992*; Luisi, N., Lerario, P., Vannini, A., Eds.; Università degli Studi: Bari, Italy, 1992; pp. 49–58.
7. Jung, T.; Blaschke, H.; Neumann, P. Isolation, identification and pathogenicity of *Phytophthora* species from declining oak stands. *Eur. J. Forest Pathol.* **1996**, *26*, 253–272. [CrossRef]
8. Jung, T.; Cooke, D.E.L.; Blaschke, H.; Duncan, J.M.; Oßwald, W. *Phytophthora quercina* sp. nov., causing root rot of European oaks. *Mycol. Res.* **1999**, *103*, 785–798. [CrossRef]
9. Jung, T.; Blaschke, H.; Oßwald, W. Involvement of soilborne *Phytophthora* species in Central European oak decline and the effect of site factors on the disease. *Plant Pathol.* **2000**, *49*, 706–718. [CrossRef]

10. Jung, T.; Hansen, E.M.; Winton, L.; Oßwald, W.; Delatour, C. Three new species of *Phytophthora* from European oak forests. *Mycol. Res.* **2002**, *106*, 397–411. [CrossRef]
11. Jung, T.; Nechwatal, J.; Cooke, D.E.L.; Hartmann, G.; Blaschke, M.; Oßwald, W.; Duncan, J.M.; Delatour, C. *Phytophthora pseudosyringae* sp. nov., a new species causing root and collar rot of deciduous tree species in Europe. *Mycol. Res.* **2003**, *107*, 772–789. [CrossRef]
12. Jung, T.; Hudler, G.W.; Jensen-Tracy, S.L.; Griffiths, H.M.; Fleischmann, F.; Oßwald, W. Involvement of *Phytophthora* species in the decline of European beech in Europe and the USA. *Mycologist* **2005**, *19*, 159–166. [CrossRef]
13. Jung, T.; Jung Horta, M.; Cacciola, S.O.; Cech, T.; Bakonyi, J.; Seress, D. Multiple new cryptic pathogenic Phytophthora species from Fagaceae forests in Austria, Italy and Portugal. *IMA Fungus* **2017**, *8*, 219–244. [CrossRef]
14. Robin, C.; Desprez-Loustau, M.L.; Capron, G.; Delatour, C. First record of *Phytophthora cinnamomi* on cork and holm oaks in France and evidence of pathogenicity. *Ann. Scie. Forest.* **1998**, *55*, 869–883. [CrossRef]
15. Hansen, E.; Delatour, C. *Phytophthora* species in oak forests of north-east France. *Ann. For. Sci.* **1999**, *56*, 539–547. [CrossRef]
16. Vettraino, A.M.; Barzanti, G.P.; Bianco, M.C.; Ragazzi, A.; Capretti, P.; Paoletti, E.; Vannini, A. Occurrence of *Phytophthora* species in oak stands in Italy and their association with declining oak trees. *For. Pathol.* **2002**, *32*, 19–28. [CrossRef]
17. Vettraino, A.M.; Morel, O.; Perlerou, C.; Robin, C.; Diamandis, S.; Vannini, A. Occurrence and distribution of *Phytophthora* species in European chestnut stands, and their association with Ink Disease and crown decline. *Eur. J. Plant Pathol.* **2005**, *111*, 169–180. [CrossRef]
18. Rizzo, D.M.; Garbelotto, M.; Davidson, J.M.; Slaughter, G.W.; Koike, S.T. *Phytophthora ramorum* as the cause of extensive mortality of *Quercus* spp. and *Lithocarpus densiflorus* in California. *Plant Dis.* **2002**, *86*, 205–214. [CrossRef]
19. Rizzo, D.M.; Garbelotto, M. Sudden Oak Death: Endangering California and Oregon Forest Ecosystems. *Front. Ecol. Environ.* **2003**, *1*, 197–204. [CrossRef]
20. Balci, Y.; Halmschlager, E. Incidence of *Phytophthora* species in oak forests in Austria and their possible involvement in oak decline. *For. Pathol.* **2003**, *33*, 157–174. [CrossRef]
21. Balci, Y.; Halmschlager, E. *Phytophthora* species in oak ecosystems in Turkey and their association with declining oak trees. *Plant Pathol.* **2003**, *52*, 694–702. [CrossRef]
22. Balci, Y.; Balci, S.; Eggers, J.; MacDonald, W.L.; Juzwik, J.; Long, R.P.; Gottschalk, K.W. *Phytophthora* species associated with forest soils in Eastern and North-central U.S. oak ecosystems. *Plant Dis.* **2007**, *91*, 705–710. [CrossRef]
23. Balci, Y.; Balci, S.; MacDonald, W.L.; Gottschalk, K.W. Relative susceptibility of oaks to seven species of *Phytophthora* isolated from oak forest soils. *For. Pathol.* **2008**, *38*, 394–409. [CrossRef]
24. Balci, Y.; Balci, S.; Blair, J.E.; Park, S.Y.; Kang, S.; MacDonald, W.L. *Phytophthora quercetorum* sp. nov., a novel species isolated from eastern and north-central U.S. oak forest soils. *Mycol. Res.* **2008**, *112*, 906–916. [CrossRef] [PubMed]
25. Vannini, A.; Vettraino, A.M. *Phytophthora cambivora* . *For. Phytophthoras* **2011**, *1*. [CrossRef]
26. Pérez-Sierra, A.; López-García, C.; León, M.; García-Jiménez, J.; Abad-Campos, P.; Jung, T. Previously unrecorded low-temperature *Phytophthora* species associated with *Quercus* decline in a Mediterranean forest in eastern Spain. *For. Pathol.* **2013**, *43*, 331–339. [CrossRef]
27. Brasier, C.M. *Phytophthora cinnamomi* and oak decline in southern Europe-Environmental constraints including climate change. *Ann. Sci. For.* **1996**, *53*, 347–358. [CrossRef]
28. Jung, T.; Orlikowski, L.; Henricot, B.; Abad-Campos, P.; Aday, A.G.; Aguín Casal, O.; Bakonyi, J.; Cacciola, S.O.; Cech, T.; et al. Widespread *Phytophthora* infestations in European nurseries put forest, semi-natural and horticultural ecosystems at high risk of *Phytophthora* diseases. *For. Pathol.* **2016**, *46*, 134–163. [CrossRef]
29. Vannini, A.; Bruni, N.; Tomassini, A.; Franceschini, S.; Vettraino, A.M. Pyrosequencing of environmental soil samples reveals biodiversity of the *Phytophthora* resident community in chestnut forests. *FEMS Microbiol. Ecol.* **2013**, *85*, 433–442. [CrossRef] [PubMed]
30. Jankowiak, R.; Stępniewska, H.; Bilański, P.; Kolařík, M. Occurrence of *Phytophthora plurivora* and other *Phytophthora* species in oak forests of southern Poland and their association with site conditions and the health status of trees. *Folia Microbiol.* **2014**, *59*, 531. [CrossRef]

31. Scanu, B.; Linaldeddu, B.T.; Deidda, A.; Jung, T. Diversity of *Phytophthora* species from declining Mediterranean maquis vegetation, including two new species, *Phytophthora crassamura* and *P. ornamentata* sp. nov. *PLoS ONE* **2015**, *10*, e0143234. [CrossRef]
32. Corcobado, T.; Miranda-Torres, J.J.; Martín-García, J.; Jung, T.; Solla, A. Early survival of *Quercus ilex* subspecies from different populations after infections and co-infections by multiple *Phytophthora* species. *Plant Pathol.* **2017**, *66*, 792–804. [CrossRef]
33. Corcobado, T.; Cubera, E.; Pérez-Sierra, A.; Jung, T.; Solla, A. First report of *Phytophthora gonapodyides* involved in the decline of *Quercus ilex* in xeric conditions in Spain. *New Dis. Rep.* **2010**, *22*, 33. [CrossRef]
34. Hansen, E.M.; Reeser, P.W.; Sutton, W. *Phytophthora* beyond agriculture. *Annu. Rev. Phytopathol.* **2012**, *50*, 359–378. [CrossRef] [PubMed]
35. Català, S.; Pérez-Sierra, A.; Abad-Campos, P. The use of genus-specific amplicon pyrosequencing to assess *Phytophthora* species diversity using eDNA from soil and water in northern Spain. *PLoS ONE* **2015**, *10*, e0119311. [CrossRef] [PubMed]
36. Jung, T.; La Spada, F.; Pane, A.; Aloi, F.; Evoli, M.; Horta Jung, M.; Scanu, B.; Faedda, R.; Rizza, C.; Puglisi, I.; et al. Diversity and Distribution of *Phytophthora* Species in Protected Natural Areas in Sicily. *Forests* **2019**, *10*, 259. [CrossRef]
37. Jung, T.; Pérez-Sierra, A.; Durán, A.; Jung, M.H.; Balci, Y.; Scanu, B. Canker and decline diseases caused by soil- and airborne *Phytophthora* species in forests and woodlands. *Persoonia* **2018**, *40*, 182–220. [CrossRef]
38. Brasier, C.M. The biosecurity threat to the UK and global environment from international trade in plants. *Plant Pathol.* **2008**, *57*, 792–808. [CrossRef]
39. O'Brien, P.; Williams, N.; Hardy, G. Detecting *Phytophthora*. *Crit. Rev. Microbiol.* **2009**, *35*, 169–181. [CrossRef]
40. Berlanas, C.; Berbegal, M.; Elena, G.; Laidani, M.; Cibriain, J.F.; Sagües, A.; Gramaje, D. The Fungal and Bacterial Rhizosphere Microbiome Associated with Grapevine Rootstock Genotypes in Mature and Young Vineyards. *Front. Microbiol.* **2019**, *10*, 1142. [CrossRef]
41. Taberlet, P.; Coissac, E.; Hajibabaei, M.; Rieseberg, L.H. Environmental DNA. *Mol. Ecol.* **2012**, *21*, 1789–1793. [CrossRef]
42. Oulas, A.; Pavloudi, C.; Polymenakou, P.; Pavlopoulos, G.A.; Papanikolaou, N.; Kotoulas, G.; Arvanitidis, C.; Iliopoulos, L. Metagenomics: Tools and Insights for Analyzing Next-Generation Sequencing Data Derived from Biodiversity Studies. *Bioinform. Biol. Insights* **2015**, *9*, BBI-S12462. [CrossRef]
43. Vettraino, A.M.; Bonats, P.; Tomassini, A.; Bruni, N.; Vannini, A. Pyrosequencing as a tool for detection of K species: Error rate and risk of false MOTUs. *Lett. Appl. Microbiol.* **2012**, *55*, 390–396. [CrossRef]
44. Català, S.; Berbegal, M.; Pérez-Sierra, A.; Abad-Campos, P. Metabarcoding and development of new real-time specific assays reveal *Phytophthora* species diversity in Holm Oak forests in eastern Spain. *Plant Pathol.* **2017**, *66*, 115–123. [CrossRef]
45. Prigigallo, M.I.; Abdelfattah, A.; Cacciola, S.O.; Faedda, R.; Sanzani, S.M.; Cooke, D.E.L.; Schena, L. Metabarcoding analysis of *Phytophthora* diversity using genus-specific primers and 454 pyrosequencing. *Phytopathology* **2016**, *106*, 305–313. [CrossRef] [PubMed]
46. Scibetta, S.; Schena, L.; Chimento, A.; Cacciola, S.O.; Cooke, D.E.L. A molecular method to assess *Phytophthora* diversity in environmental samples. *J. Microbiol. Methods* **2012**, *88*, 356–368. [CrossRef] [PubMed]
47. Burgess, T.I.; McDougall, K.L.; Scott, P.M.; Hardy, G.E.; Garnas, J. Predictors of *Phytophthora* diversity and community composition in natural areas across diverse Australian ecoregions. *Ecography* **2018**, *42*, 565–577. [CrossRef]
48. Mora-Sala, B.; Berbegal, M.; Abad-Campos, P. The use of qPCR reveals a high frequency of *Phytophthora quercina* in two Spanish holm oak areas. *Forests* **2018**, *9*, 697. [CrossRef]
49. Altschul, S.F.; Madden, T.L.; Scheaffer, A.A.; Zhang, J.; Zhang, Z.; Miller, W.; Lipman, D.J. Gapped BLAST and PSIBLAST: A new generation of protein database search programs. *Nucleic Acids Res.* **1997**, *25*, 3389–3402. [CrossRef]
50. Park, J.; Park, B.; Veeraraghavan, N.; Jung, K.; Lee, Y.H.; Rossman, A.; Farr, D.; Cline, E.; Grünwald, N.J.; Luster, D.; et al. *Phytophthora* database: A forensic database supporting the identification and monitoring of *Phytophthora*. *Plant Dis.* **2008**, *92*, 966–972. [CrossRef] [PubMed]
51. Edgar, R.C. MUSCLE: Multiple sequence alignment with high accuracy and high throughput. *Nucleic Acids Res.* **2004**, *32*, 1792–1797. [CrossRef]

52. Tamura, K.; Stecher, G.; Peterson, D.; Filipski, A.; Kumar, S. MEGA6: Molecular Evolutionary Genetics Analysis version 6.0. *Mol. Biol. Evol.* **2013**, *30*, 2725–2729. [CrossRef]
53. Felsenstein, J. Confidence limits on phylogenies: An approach using the bootstrap. *Evolution* **1985**, *39*, 783–791. [CrossRef]
54. Glynou, K.; Nam, B.; Thines, M.; Maciá-Vicente, J.G. Facultative root-colonizing fungi dominate endophytic assemblages in roots of nonmycorrhizal *Microthlaspi* species. *New Phytol.* **2018**, *217*, 1190–1202. [CrossRef] [PubMed]
55. Dhariwal, A.; Chong, J.; Habib, S.; King, I.L.; Agellon, L.B.; Xia, J. MicrobiomeAnalyst: A web-based tool for comprehensive statistical, visual and meta-analysis of microbiome data. *Nucleic Acids Res.* **2017**, *45*, W180–W188. [CrossRef]
56. Ruiz Gómez, F.J.; Navarro-Cerrillo, R.M.; Pérez deLuque, A.; Oβwald, W.; Vannini, A.; Morales-Rodríguez, C. Assessment of functional and structural changes of soil fungal and oomycete communities in holm oak declined dehesas through metabarcoding analysis. *Sci. Rep.* **2019**, *9*, 5315. [CrossRef] [PubMed]
57. Redondo, M.A.; Boberg, J.; Stenlid, J.; Oliva, J. Contrasting distribution patterns between aquatic and terrestrial *Phytophthora* species along a climatic gradient are linked to functional traits. *ISME J.* **2018**, *12*, 2967–2980. [CrossRef] [PubMed]
58. Brasier, C.M.; Robredo, F.; Ferraz, J.F.P. Evidence for *Phytophthora cinnamomi* involvement in Iberian oak decline. *Plant Pathol.* **1993**, *42*, 140–145. [CrossRef]
59. Gallego, F.J.; Perez de Algaba, A.; Fernandez-Escobar, R. Etiology of oak decline in Spain. *Eur. J. For. Pathol.* **1999**, *29*, 17–27. [CrossRef]
60. Sánchez, M.E.; Caetano, P.; Ferraz, J.; Trapero, A. *Phytophtora* disease of *Quercus ilex* in south-western Spain. *For. Pathol.* **2002**, *32*, 5–18. [CrossRef]
61. Camilo-Alves, C.S.P.; da Clara, M.I.E.; de Almeida Ribeiro, N.M.C. Decline of Mediterranean oak trees and its association with *Phytophthora cinnamomi*: A review. *Eur. J. Forest Res.* **2013**, *132*, 411–432. [CrossRef]
62. Serrano, M.S.; de Vita, P.; Fernández-Rebollo, P.; Sánchez, M.E. Calcium fertilizers induce soil suppressiveness to *Phytophthora cinnamomi* root rot of *Quercus ilex*. *Eur. J. Plant Pathol.* **2012**, *132*, 271–279. [CrossRef]
63. Mora-Sala, B.; Abad-Campos, P.; Berbegal, M. Response of *Quercus ilex* seedlings to *Phytophthora* spp. root infection in a soil infestation test. *Eur. J. Plant Pathol.* **2018**, *154*, 2115–2225. [CrossRef]
64. Jung, T.; Burgess, T.I. Re-evaluation of *Phytophthora citricola* isolates from multiple woody hosts in Europe and North America reveals a new species, *Phytophthora plurivora* sp. nov. *Persoonia* **2009**, *22*, 95–110. [CrossRef] [PubMed]
65. Ioos, R.; Laugustin, L.; Rose, S.; Tourvieille, J.; Tourvieille de Labrouhe, D. Development of a PCR test to detect the downy mildew causal agent *Plasmopara halstedii* in sunflower seeds. *Plant Pathol.* **2007**, *56*, 209–218. [CrossRef]
66. Zhao, J.; Wang, X.J.; Chen, C.Q.; Huang, L.L.; Kang, Z.S. A PCR-based assay for detection of *Puccinia striiformis* f. sp. *tritici* in wheat. *Plant Dis.* **2007**, *91*, 1669–1674. [CrossRef] [PubMed]
67. Alaei, H.; Baeyen, S.; Maes, M.; Höfte, M.; Heungens, K. Molecular detection of *Puccinia horiana* in *Chrysanthemum* × *morifolium* through conventional and real-time PCR. *J. Microbiol. Methods* **2009**, *76*, 136–145. [CrossRef]
68. Mrázková, M.; Černý, K.; Tomosovsky, M.; Strnadová, V.; Gregorová, B.; Holub, V.; Panek, M.; Havrdová, L.; Hejná, M. Occurrence of *Phytophthora multivora* and *Phytophthora plurivora* in the Czech Republic. *Plant Prot. Sci.* **2013**, *49*, 155–164. [CrossRef]

5

Diversity and Distribution of *Phytophthora* Species in Protected Natural Areas in Sicily

Thomas Jung [1,2], Federico La Spada [3], Antonella Pane [3], Francesco Aloi [3,4], Maria Evoli [3], Marilia Horta Jung [1,2], Bruno Scanu [5], Roberto Faedda [3], Cinzia Rizza [3], Ivana Puglisi [3], Gaetano Magnano di San Lio [6], Leonardo Schena [6] and Santa Olga Cacciola [3,*]

1. Phytophthora Research Centre, Mendel University in Brno, Zemědělská 1, 61300 Brno, Czech Republic; dr.t.jung@t-online.de (T.J.); marilia.horta@mendelu.cz (M.H.J.)
2. Phytophthora Research and Consultancy, Am Rain 9, 83131 Nussdorf, Germany
3. Department of Agriculture, Food and Environment (Di3A), University of Catania, Via Santa Sofia 100, 95123 Catania, Italy; federicolaspada@yahoo.it (F.L.S.); apane@unict.it (A.P.); francescoaloi88@gmail.com (F.A.); marevoli@gmail.com (M.E.); rfaedda@unict.it (R.F.); cinziarizza@libero.it (C.R.); ipuglisi@unict.it (I.P.)
4. Department of Agricultural, Food and Forest Sciences, University of Palermo, Viale delle Scienze Ed., 4, 90128 Palermo, Italy
5. Dipartimento di Agraria, Sezione di Patologia vegetale ed Entomologia (SPaVE), Università degli Studi di Sassari, Viale Italia 39, 07100 Sassari, Italy; bscanu@uniss.it
6. Dipartimento di Agraria, Mediterranean University of Reggio Calabria, località Feo di Vito, 89122 Reggio Calabria, Italy; gmagnano@unirc.it (G.M.d.S.L.); lschena@unirc.it (L.S.)

* Correspondence: olgacacciola@unict.it.

Abstract: The aim of this study was to investigate the occurrence, diversity, and distribution of *Phytophthora* species in Protected Natural Areas (PNAs), including forest stands, rivers, and riparian ecosystems, in Sicily (Italy), and assessing correlations with natural vegetation and host plants. Fifteen forest stands and 14 rivers in 10 Sicilian PNAs were studied. *Phytophthora* isolations from soil and stream water were performed using leaf baitings. Isolates were identified using both morphological characters and sequence analysis of the internal transcribed spacer (ITS) region. A rich community of 20 *Phytophthora* species from eight phylogenetic clades, including three new *Phytophthora* taxa, was recovered (17 species in rhizosphere soil from forest stands and 12 species in rivers). New knowledge about the distribution, host associations, and ecology of several *Phytophthora* species was provided.

Keywords: soilborne pathogens; invasive species; natural ecosystems; streams; vegetation type; baiting; ITS region

---

1. Introduction

Due to its location in the central Mediterranean Sea and its vast area of 25.708 km^2, Sicily is one of the most important biodiversity areas in Europe and in the Mediterranean basin [1,2] harboring more than 3000 plant species [3], 321 of which are endemic to Sicily [4]. Sicily's outstanding floristic and ecological diversity was acknowledged by the establishment of numerous Protected Natural Areas (PNAs; Italian National Law 394/91), including three Regional Parks, 72 Regional Natural Reserves and 223 Sites of Community Importance (Habitats Directive 92/43/EEC).

During a recent monitoring of the health conditions of oak and beech trees in forests of the Etna, Madonie, and Nebrodi Regional Parks in Sicily (southern Italy), severe symptoms of crown decline were observed, indicating fine root losses caused by soilborne pathogens from the genus *Phytophthora* [5]. With more than 150 described species grouped in twelve multigenic phylogenetic

Clades [6], this oomycete genus comprises some of the most aggressive plant pathogens of forests and other natural ecosystems [7–16]. Several studies highlighted the diversity of *Phytophthora* species in native vegetation and their potential impact on natural ecosystems [17–26]. The presence of exotic, potentially invasive *Phytophthora* species often represents a threat for the survival of native plant species and may alter the stability of the entire ecosystem. In Sardinia, a survey in the National Park of the La Maddalena archipelago demonstrated the involvement of exotic *Phytophthora* species in the widespread mortality of *Quercus ilex* trees and Mediterranean maquis vegetation [25,27,28]. Outplanting of infected nursery stock is considered a primary pathway for the introduction of non-native *Phytophthora* species into forest ecosystems [10,29–36]. In recent years, great attention has been paid to surface water as a source of *Phytophthora* inoculum in natural ecosystems. Surveys of rivers, streams, and riparian ecosystems in several continents have revealed a huge diversity of *Phytophthora* species, including primarily aquatic species which are considered as opportunistic pathogens, but also soilborne and airborne primary pathogens [8,20,21,37–41]. However, all *Phytophthora* species have the potential to be disturbance factors in natural ecosystems, in particular, those of exotic origin, provided that the environmental conditions are conducive to disease development [9,10,32]. The number of species known in the genus *Phytophthora* has increased dramatically during the past decade, mainly due to extensive surveys in previously unexplored ecosystems such as natural forests, riparian ecosystems, streams, and irrigation systems [6,20,21,25,42,43]. The ecological role of most of these new species and their distribution in natural ecosystems are still largely unknown, although the knowledge of the *Phytophthora* community and its potential impact on native vegetation is a prerequisite for proper management of PNAs. Despite the large number of *Phytophthora* species reported from nurseries and agricultural crops in Sicily, their occurrence and ecology in natural environments have received little attention. The aims of this study were to examine (i) the diversity and distribution of *Phytophthora* species in forest stands and river systems of Sicilian PNAs and (ii) their association with natural vegetation and potential host plants.

2. Materials and Methods

2.1. Sampling and Phytophthora Isolation

Ten PNAs in northern and eastern Sicily, including the three Regional Parks (RP), five Regional Nature Reserves (RNR), and two Sites of Community Importance (SCI), characterized by different ecological conditions were included in this study (Table S1, Figure S1). Twenty sites in 15 characteristic Sicilian forest stands (FS) in seven PNAs and 14 rivers running through nine PNAs were included in the survey of distribution and diversity of *Phytophthora* species (Table 1 and 2, Figures 1 and 2). Sampling activities were carried out during the spring of 2013 and 2015.

In total, 83 rhizosphere soil samples from mature specimens of 17 tree species were collected in the 15 forest stands (Table 1). Soil sampling and isolation methodologies were performed according to Jung [9]. Subsamples of ca. 200 mL soil were used for baiting tests at 18–20 °C in a walk-in growth chamber with 12 h natural daylight. Young leaves of native species (mainly *Ceratonia siliqua* and *Quercus* spp.) were used as baits floated over flooded soil. Necrotic segments (2 × 2 mm) from Infected leaves were plated onto selective PARPNH-agar [11]. Petri dishes were incubated at 20 °C in the dark. Outgrowing *Phytophthora* hyphae were transferred onto V8-juice agar (V8A) under the stereomicroscope. *Phytophthora* isolations from rivers were performed using an in situ baiting technique [21]. At each site, 10 non-wounded young leaves of *C. siliqua*, *Quercus* spp. and *Citrus* spp. were placed in a mesh-bag styrofoam raft (25 × 30 cm) [21] rigged to float on the water surface. In total 35 rafts were placed in 14 rivers (Table 2, Figure 2) and collected after 3–5 days. Isolations from necrotic leaf lesions were carried out as described before. All obtained isolates were maintained on V8A and stored at 6 °C in the dark.

Table 1. Vegetation, geological substrate, municipality, geographic coordinates, and altitude of the 15 forest stands sampled in seven Protected Natural Areas in Sicily, tree species sampled and *Phytophthora* taxa isolated.

Forest Stand (FS) No.	Protected Natural Area [a]	Vegetation (Natura 2000 Code, Forest Stand Type, Phytocoenosis) [b,c,d]	Geological Substrate	Municipality	Sampling Site No.	Geographic Coordinates (DATUM WGS84)	Altitude (m a.s.l.)	Sampled Tree Species (No. of *Phytophthora*-Positive Soil Samples/Sampled Trees)	*Phytophthora* spp. (No. of Positive Soil Samples) [f]
FS-1	Etna RP	Natura 2000 CODE: 9340. Forest stand type: Meso-Mediterranean evergreen oak forest. Phytocoenosis: *Teucrio siculi-Quercetum ilicis* subass. *Teucrietosum siculi.*	Volcanic (Alkali Basalt-Na)	Zafferana Etnea (CT)	I	37°41′44.53″ N–15°05′00.04″ E	1030	*Quercus ilex* L. (3/5)	MUL (1); QUE (2)
					II	37°41′05.94″ N–15°05′13.04″ E	890	*Quercus pubescens* Willd. s. l. (4/5)	PSY (4) [g]
					V	37°41′53.92″ N–15°06′01.05″ E	660	*Q. pubescens* s. l. (4/5)	CAM (1); QUE (3)
FS-2	Etna RP	Natura 2000 CODE: 91M0. Forest stand type: Supra-Mediterranean turkey oak forest. Phytocoenosis: *Vicio cassubicae-Quercetum cerridis.*	Volcanic (Alkali Basalt-Na)	Sant'Alfio (CT)	III	37°46′26.02″ N–15°05′37.23″ E	1345	*Q. pubescens* s. l. (3/6)	PSY (3) [h]
FS-3	Etna RP	EUNIS CODE: G1.916 [e]. Forest stand type: Supra-Mediterranean birch forest. Phytocoenosis: Aggregation with *Betula aetnensis.*	Volcanic (Alkali Basalt-Na)	Sant'Alfio (CT)	IV	37°46′14.90″ N–15°03′34.56″ E	1667	*Betula aetnensis* Raf. (0/1)	-
FS-4	Etna RP	Natura 2000 CODE: 9220. Forest stand type: Supra-Mediterranean beech forest. Phytocoenosis: *Epipactido meridionalis-Fagetum sylvaticae.*	Volcanic (Alkali Basalt-Na)	Castiglione di Sicilia (CT)	X	37°48′50.94″ N–15°01′24.42″ E	1874	*Fagus sylvatica* L. (1/1)	VUL (1)
FS-5	Nebrodi RP	Natura 2000 CODE: 9210. Forest stand type: Supra-Mediterranean beech forest. Phytocoenosis: *Anemono apenninae-Fagetum sylvaticae.*	Sedimentary–M. Soro Flysh (Marly claystones and limestones, grading upward to quarzarenites)	Militello Rosmarino (ME)	VI	37°56′22.20″ N–14°40′15.49″ E	1450	*F. sylvatica* (5/7)	CAM (4); MEG (1) [i,j]
								Q. pubescens s. l. (1/1)	CAM (1)
				Cersarò (ME)	IX	37°55′40.90″ N–14°41′35.48″ E	1783	*F. sylvatica* (1/3)	CAM (1)
FS-6	Nebrodi RP	Natura 2000 CODE: 9340. Forest stand type: Meso-Mediterranean evergreen oak forest. Phytocoenosis: *Teucrio siculi-Quercetum ilicis.*	Sedimentary–M. Soro Flysh (Marly claystones and limestones, grading upward to quarzarenites)	San Fratello (ME)	VII	37°57′16.38″ N–14°37′18.34″ E	1050	*Q. ilex* (3/5)	CAM (1); GON (2); PSY (1) [i]
FS-7	Nebrodi RP	Natura 2000 CODE: 91M0. Forest stand type: Meso-Mediterranean turkey oak forest. Phytocoenosis: *Arrhenathero nebrodensis-Quercetum cerridis.*	Sedimentary–M. Soro Flysh (Marly claystones and limestones, grading upward to quarzarenites)	Randazzo (CT)	VIII	37°56′40.81″ N–14°54′17.89″ E	1420	*F. sylvatica* (1/1)	CAM (1) [i]
								Quercus cerris L. (1/1)	CAM (1) [i]

Table 1. *Cont.*

Forest Stand (FS) No.	Protected Natural Area [a]	Vegetation (Natura 2000 Code, Forest Stand Type, Phytocoenosis) [b,c,d]	Geological Substrate	Municipality	Sampling Site No.	Geographic Coordinates (DATUM WGS84)	Altitude (m a.s.l.)	Sampled Tree Species (No. of *Phytophthora*-Positive Soil Samples/Sampled Trees)	*Phytophthora* spp. (No. of Positive Soil Samples) [f]
FS-8	Nebrodi RP	Natura 2000 CODE: 9330. Forest stand type: Meso-Mediterranean cork oak forest. Phytocoenosis: *Genisto aristatae-Quercetum suberis*.	Sedimentary–Numidian Flysch (quarzarenites and clays)	Geraci Siculo (PA)	XVII	37°53′22.33″ N–14° 8′10.77″ E	710	*Quercus suber* L. (2/2)	GON (2); MEG (1)
FS-9	Madonie RP	Natura 2000 CODE: 9380. Forest stand type: Supra-Mediterranean holly forest. Phytocoenosis: *Ilici aquifoliae-Quercetum austrotyrrhenicae*.	Sedimentary–Numidian Flysh (quarzarenites and clays)	Petralia Sottana (PA)	XVIII	37°53′46.39″ N–14°3′55.22″ E	1390	*Ilex aquifolium* L. (1/1)	CAM (1)
FS-10	Madonie RP	Natura 2000 CODE: 91AA. Forest stand type: Meso-Mediterranean *Quercus pubescens* forest. Phytocoenosis: *Qurcetum leptobalani*.	Sedimentary–Numidian Flysh (quarzarenites and claystones)	Castelbuono (PA)	XIX	37°53′51.02″ N–14°3′58.77″ E	1412	*Q. pubescens* s. l. (1/3)	CAM (1)
FS-11	Madonie RP	Natura 2000 CODE: 9380. Forest stand type: Meso Mediterranean evergreen oak and holly forest. Phytocoenosis: *Geranio versicoloris-Quercetum ilicis*.	Sedimentary–Numidian Flysh (quarzarenites and claystones)	Castelbuono (PA)	XX	37°54′20.46″ N–14°4′29.39″ E	1110	*I. aquifolium* (0/3)	- [g]
								Q. ilex (2/4)	QUE (1); TYR (1); [g]
					XXI	37°54′50.19″ N–14°4′40.07″ E	850	*Castanea sativa* Mill. (1/2)	PLU (1)
FS-12	Pantalica RNR	Natura 2000 CODE: 92C0. Forest stand type: Thermo-Mediterranean riparian plane tree forest. Phytocoenosis: *Platano-Salicetum pedicellatae*.	Sedimentary (algal calcarenites and calcirudites)	Sortino (SR)	XI	37°07′48.0″ N–15°01′26.5″ E	236	*Populus nigra* L. (1/1)	PSC (1)
								Salix pedicellata Desf. (1/1)	PSC (1)
								Q. ilex + *Fraxinus oxycarpa* Bieb., mixed sample (1/1)	PSC (1); PLU (1); LAC (1)
								Platanus orientalis L. (1/1)	CAC (1); PLU (1)
								Ostrya carpinifolia Scop. (1/1)	LAC (1); PLU (1)
								P. orientalis + *Q. ilex*, mixed sample (1/1)	LAC (1); PLU (1)
					XII, XIV	37°08′19.3″ N–15°02′13.3″ E	221	*P. nigra* (1/1)	CAC (1); PLU (1) [k]
								Populus alba L. (1/1)	PSC (1); LAC (1); KEL (1)
								S. pedicellata (1/1)	PSC (1); LAC (1)
								Nerium oleander L. (1/1)	PLU (1)
								Celtis australis L. (1/1)	POL (1)
								Q. ilex (1/1)	PSC (1); PLU (1)
								P. orientalis (2/2)	CIP (1); LAC (2)
FS-13	Ciane RNR	Natura 2000 CODE: 92A0. Forest stand type: Thermo-Mediterranean riparian willow, poplar, and ash forest. Phytocoenosis: *Salicetum albo–pedicellatae*.	Alluvial sediments (loam and sandy limestone)	Siracusa (SR)	XIII	37°02′40.3″ N–15°14′40.7″ E	4	*F. oxycarpa* (4/4)	CRA (1); PSC (3); LAC (2); MEG (1); PLU (2);

Table 1. *Cont.*

Forest Stand (FS) No.	Protected Natural Area [a]	Vegetation (Natura 2000 Code, Forest Stand Type, Phytocoenosis) [b,c,d]	Geological Substrate	Municipality	Sampling Site No.	Geographic Coordinates (DATUM WGS84)	Altitude (m a.s.l.)	Sampled Tree Species (No. of *Phytophthora*-Positive Soil Samples/Sampled Trees)	*Phytophthora* spp. (No. of Positive Soil Samples) [f]
FS-14	Cavagrande RNR	Natura 2000 CODE: 92C0. Forest stand type: Thermo-Mediterranean riparian plane tree forest. Phytocoenosis: *Platano-Salicetum pedicellatae.*	Alluvial sediments (loam and sandy limestone)	Siracusa (SR)	XV	36°57′2.62″ N–15°11′8.15″ E	8	*Salix caprea* L. (2/2)	LAC (2); POL (1)
								P. orientalis (3/3)	CAC (1); PSC (3); MUL (2); PLU (1);
FS-15	Irminio SCI	Natura 2000 CODE: 92C0. Thermo-Mediterranean riparian plane tree forest. Phytocoenosis: *Platano-Salicetum pedicellatae.*	Sedimentary (calcarenites and marns)	Ragusa (RG)	XVI	37°00′1.9″ N–14°46′31.5″ E	430	*F. oxycarpa* (1/1)	PLU (1); PSC (1)
								Q. pubescens s. l. (2/2)	CIT (1); PLU (2)
								P. orientalis (1/1)	PLU (1)

[a] Etna RP = Etna Regional Park; Nebrodi-RP = Nebrodi Regional Park; Madonie-RP = Madonie Regional Park; Pantalica RNR = Pantalica, Valle dell'Anapo e Torrente Cavagrande Regional Natural Reserve (RNR); Ciane RNR = Fiume Ciane e Saline di Siracusa RNR; Cavagrande RNR = Cavagrande del Cassibile RNR; Irminio SCI = ITA080002—Alto corso del Fiume Irminio Site of Community Importance (SCI). [b] Vegetation features were in accordance with Natura 2000 sites data and respective management plans: ftp://ftp.minambiente.it/PNM/Natura2000/TrasmissioneCE_2016/schede_mappe/Sicilia/SIC_schede/. [c] Natura 2000 habitats: http://www.minambiente.it/sites/default/files/archivio/allegati/rete_natura_2000/int_manual_eu28.pdf. [d] *Teucrio siculi-Quercetum ilicis* subass. *Teucrietosum siculi*: Meso-Mediterranean acidophilous oak stand characterized by *Quercus ilex* L. mixed with calcifuge downy oaks (*Quercus dalechampii* and *Quercus congesta*). *Vicio cassubicae-Quercetum cerridis*: Supra-Mediterranean deciduous turkey oak stand characterized by *Quercus cerris* mixed with downy oaks (*Q. congesta* and *Q. delachampii*) *Fraxinus ornus* and *Acer obtusatum*. Aggregation with *Betula aetnensis*: Supra-Mediterranean pioneer vegetation dominated by the endemic *B. aetnensis* mixed with beech, turkey oak, and *Pinus nigra* subsp. *calabrica*. *Epipactido meridionalis-Fagetum sylvaticae*: Supra-Mediterranean beech forest dominated by *Fagus sylvatica*. *Anemono apenninae-Fagetum sylvaticae*: Acidophilous Supra-Mediterranean beech forest characterized by *F. sylvatica in* association in *Ilex aquifolium*. *Teucrio siculi-Quercetum ilicis*: Meso-Mediterranean acidophilous oak forest stand characterized by *Q. ilex* mixed with deciduous oaks (*Quercus virgiliana* and *Q. congesta*). *Arrhenathero nebrodensis-Quercetum cerridis*: Meso-Mediterranean acidophilous turkey oak forest stand typified by *Q. cerris;* at the higher altitude (ca. 1400 m) it is mixed with beech forest stands (*Anemono apenninae-Fagetum sylvaticae*). *Genisto aristatae-Quercetum suberis*: Meso-Mediterranean acidophilous cork oak forest stand. *Ilici aquifoliae-Quercetum austrotyrrhenicae*: Acidophilous supra-Mediterranean forest community dominated by arborescent *Ilex aquifolium* mainly associated with *Quercus petraea* subsp. *austrotyrrhenica* and other plant species (*Acer obtusatum, Acer campestre, Ulmus glabra*). *Quercetum leptobalani*: Acidophilous meso-Mediterranean deciduous community typified by *Quercus leptobalanos* growing together with other oak species (*Q. dalechampii, Q. congesta, Quercus amplifolia, Q. ilex*). *Geranio versicoloris-Quercetum ilicis*: Meso-Mediterranean forest of *Q. ilex* growing on flysch at an altitude of 900-1200 m. This acidophilous plant community, is characterized by the dominance of *Q. ilex*, growing together with *I. aquifolium*. *Platano-Salicetum pedicellatae*: Thermo-Mediterranean Hyblean plateau riparian forest dominated by *Platanus orientalis* growing in association with *Salix* spp., *Populus* spp., *Fraxinus oxycarpa*, and *Nerium oleander*. *Salicetum albo—pedicellatae*: Thermo-Mediterranean riparian forest communities growing on soils with a high water table. It is characterized by *Salix* spp. and *Populus* spp. in association with *F. oxycarpa*. [e] EUNIS habitat: http://eunis.eea.europa.eu/habitats/1176. [f] CAC = *Phytophthora cactorum;* CAM = *P.* ×*cambivora* (previously *P. cambivora*); CIP = *P. citrophthora;* CIT = *Phytophthora citricola* 12; CRA = *P. crassamura;* GON = *P. gonapodyides;* KEL = *Phytophthora* sp. kelmania; LAC = *P. lacustris;* MEG = *P. megasperma;* MUL = *P. multivora;* PLU = *P. plurivora;* POL = *P. polonica;* QUE = *P. quercina;* PSC = *P. pseudocryptogea;* PSY = *P. psychrophila;* TYR = *P. tyrrhenica;* VUL = *P. vulcanica*. [g] *Pythium* sp. alternatum-like isolated. [h] *Gibberella moniliformis* also isolated. [i] *Pythium* and *Phytopythium* spp. also isolated. [j] *Elongisporangium anandrum* also isolated. [k] *Pythium* sp. JN5-like also isolated.

Table 2. Vegetation and geological features of drainage basins of 14 rivers surveyed in nine Protected Natural Areas in Sicily; location of sites with baiting rafts, and *Phytophthora* taxa isolated.

River	Protected Natural Area [a]	Location of Drainage Basin	Forest Vegetation in Drainage Basin (Natura 2000 Code, Forest Stand Type, Phytocoenosis) [b,c,d]	Geological Features of Drainage Basin	Raft No.	Municipality	Geographic Coordinates (DATUM WGS84	Altitude (m a.s.l.)	*Phytophthora* spp. [e]
Anapo	Pantalica RNR	Northern area of eastern sector of the Hyblean plateau	Natura 2000 CODE:92C0 Forest stand types: Riparian forests. Phytocoenosis: *Platano-Salicetum pedicellatae*	Limestone (algal calcarenites and calcirudites)	1	Sortino (SR)	37°07′48.0″ N–15°01′26.5″ E	294	LAC
					2	Sortino (SR)	37°07′48.0″ N–15°01′26.5″ E	294	LAC
					3	Sortino (SR)	37°08′19.3″ N–15°02′13.3″ E	219	CIP, LAC, PSC
					4	Sortino (SR)	37°08′19.3″ N–15°02′13.3″ E	219	-
Ciane	Ciane RNR	Eastern area of eastern sector of the Hyblean plateau	Natura 2000 CODE: 92A0 Forest stand types: Riparian forests. Phytocoenosis: *Salicetum albo-pedicellatae*	Alluvial sediments (derived from loam and sandy limestone)	5	Siracusa (SR)	37°02′34.4″ N–15°13′37.5″ E	4	KEL, LAC, PSC
					6	Siracusa (SR)	37°02′34.4″ N–15°13′37.5″ E	4	FRI, LAC
					7	Siracusa (SR)	37°02′34.4″ N–15°13′37.5″ E	4	LAC, MUL
					8	Siracusa (SR)	37°02′34.4″ N–15°13′37.5″ E	4	LAC, PSC
					9	Siracusa (SR)	37°02′34.4″ N–15°13′37.5″ E	4	LAC
					10	Siracusa (SR)	37°02′34.4″ N–15°13′37.5″ E	4	LAC
					11	Siracusa (SR)	37°02′34.4″ N–15°13′37.5″ E	4	LAC
					12	Siracusa (SR)	37°02′34.4″ N–15°13′37.5″ E	4	LAC
Cassibile	CavagrandeRNR	Eastern area of western sector of the Hyblean plateau	Natura 2000 CODE: 92C0 Forest stand types: Riparian forests. Phytocoenosis: *Platano-Salicetum pedicellatae*	Limestone (algal calcarenites and calcirudites)	13	Siracusa (SR)	36°57′2.05″N–15°11′11.22″E	8	HYD, LAC, PSC
					14	Siracusa (SR)	36°57′2.05″N–15°11′11.22″E	8	HYD, LAC
Irminio	Irminio SCI	Northwestern area of western sector of the Hyblean plateau	Natura 2000 CODE: 92C0 Forest stand types: Riparian forests. Phytocoenosis: *Platano-Salicetum pedicellatae*	Limestone and claystone (calcarenites and marns)	15	Ragusa (RG)	37°00′23.3″ N–14°46′45.1″ E	400	-
					16	Ragusa (RG)	37°00′23.3″ N–14°46′45.1″ E	400	-
					17	Ragusa (RG)	36°57′20.7″ N–14°46′06.2″ E	300	LAC [f]
Alcantara	Nebrodi RP	Southeastern area of Nebrodi mountains	Natura 2000 CODE:92A0 Forest stand types: Riparian forests. Phytocoenosis: *Salicetum albo-purpurae*	Numidian Flysch (quarzarenites and claystones)	18	Randazzo (CT)	37°52′50.4″ N–14°56′49.6″ E	718	GON, LAC [g]
Fiume di Troina	Nebrodi RP	Southeastern area of Nebrodi mountains	Natura 2000 CODE:91AA, 91M0, 92A0 Forest stand types: Woodlands and riparian forests. Phytocoenosis: *Erico-Quercetum virgilianae; Arrhenathero nebrodensis-Quercetum cerridis; Salicetum albo-purpureae.*	Numidian Flysch (quarzarenites and claystones)	19	San Teodoro (ME)	37°48′32.2″ N–14°41′53.1″ E	605	LAC [h]
					26	San Teodoro (ME)	37°48′32.2″ N–14°41′53.1″ E	605	LAC

Table 2. *Cont.*

River	Protected Natural Area [a]	Location of Drainage Basin	Forest Vegetation in Drainage Basin (Natura 2000 Code, Forest Stand Type, Phytocoenosis) [b,c,d]	Geological Features of Drainage Basin	Raft No.	Municipality	Geographic Coordinates (DATUM WGS84	Altitude (m a.s.l.)	*Phytophthora* spp. [e]
Flascio	Nebrodi RP	Southeastern area of Nebrodi mountains	Natura 2000 CODE:92A0 Forest stand types: Riparian forests Phytocoenosis: *Salicetum albo-purpureae*	Numidian Flysch (quarzarenites and claystones)	20	Randazzo (CT)	37°52′51.4″ N–14°52′50.6″ E	856	LAC [h,i]
					21	Randazzo (CT)	37°52′51.4″ N–14°52′50.6″ E	856	LAC [i]
Della Saracena	Nebrodi RP	Southeastern area of Nebrodi mountains	Natura 2000 CODE:91AA, 92A0 Forest stand types: Woodlands and riparian forests. Phytocoenosis: *Erico-Quercetum virgilianae; Salicetum albo-purpureae.*	Numidian Flysch (quarzarenites and claystones)	22	Bronte (CT)	37°52′07.3″ N–14°50′56.2″ E	811	CAM, GON, LAC, POL [i,j]
					25	Maniace (CT)	37°51′02.6″ N–14°48′04.3″ E	624	GON, LAC
Martello	Nebrodi RP	Southeastern area of Nebrodi mountains	Natura 2000 CODE:91M0, 92A0 Forest stand types: Woodlands and riparian forests. Phytocoenosis: *Arrhenathero nebrodensis-Quercetum cerridis; Salicetum albo-purpurae.*	Numidian Flysch (quarzarenites and claystones)	23	Maniace (CT)	37°51′27.7″ N–14°47′29.8″ E	676	LAC
Cutò	Nebrodi RP	Southern area of Nebrodi mountains	Natura 2000 CODE:92A0 Forest stand types: Riparian forests Phytocoenosis: *Salicetum albo-purpurae*	Numidian Flysch (quarzarenites and claystones)	24	Maniace (CT)	37°51′57.9″ N–14°46′00.4″ E	708	LAC
Sciambro	Etna RP	Northeastern area of Volcano Etna	Natura 2000 CODE:9530 Forest stand types: Woodland Phytocoenosis: *Junipero hemisphaericae-Pinetum calabricae.*	Volcanic (Alcali-Basalt-Na)	27	Linguaglossa (CT)	37°46′58.9″ N–15°3′04.7″ E	1656	GON
					28	Linguaglossa (CT)	37°46′58.4″ N–15°3′02.5″ E	1669	-
					29	Linguaglossa (CT)	37°46′57.0″ N–15°2′01.8″ E	1682	GON
Fiumefreddo	FiumefreddoRNR	Northeastern boundary of Volcano Etna	Natura 2000 CODE:92A0 Forest stand types: Riparian Forests Phytocoenosis: *Populetalia albae* The following were also present: (i) citrus groves, (ii) nurseries; (iii) artificial forest of *Eucalyptus* and *Carya cordiformis*.	Alluvial sediments (derived from loam, sandy limestone, and volcanic rocks).	30	Fiumefreddo di Sicilia (CT)	37°47′22.15″ N–15°13′55.63″ E	6	LAC, MUL, PLU, PSC, THE
					31	Fiumefreddo di Sicilia (CT)	37°47′25.98″ N–15°14′3.89″ E	6	LAC, PSC, THE

Table 2. *Cont.*

River	Protected Natural Area [a]	Location of Drainage Basin	Forest Vegetation in Drainage Basin (Natura 2000 Code, Forest Stand Type, Phytocoenosis) [b,c,d]	Geological Features of Drainage Basin	Raft No.	Municipality	Geographic Coordinates (DATUM WGS84	Altitude (m a.s.l.)	*Phytophthora* spp. [e]
Fiumara d'Agrò	Agrò SCI	Southeastern area of Peloritani mountains	Natura 2000 CODE:91AA, 91E0 Forest stand types: Woodlands and Riparian Forests. Phytocoenosis: *Erico-Quercetum virgilianae; Spartio-Nerietum oleandri.*	Metamorphic (Phyllites)	32	Limina (ME)	37°57′22.4″ N–15°16′20.8″ E	202	LAC, PLU, PSC
					33	Limina (ME)	37°57′22.4″ N–15°16′20.8″ E	202	-
Fiumedinisi	Fiumedinisi RNR	Southeastern area of Peloritani mountains	Natura 2000 CODE:91AA, 92A0; 92C0. Forest stand types: Woodlands and Riparian Forests. Phytocoenosis: *Erico-Quercetum virgilianae; Platano-Salicetum gussonei; Salicetum albo-purpuree.*	Metamorphic (mainly green shists and amphibolites)	34	Fiumedinisi (ME)	38°01′47.8″ N–15°22′21.3″ E	214	-
					35	Fiumedinisi (ME)	38°01′47.8″ N–15°22′21.3″ E	214	CIP, LAC [g]

[a] Etna RP = Etna Regional Park; Nebrodi RP = Nebrodi Regional Park; Madonie RP = Madonie Regional Park; Pantalica RNR = Pantalica, Valle dell'Anapo e Torrente Cavagrande Regional Natural Reserve (RNR); Ciane RNR = Fiume Ciane e Saline di Siracusa RNR; Cavagrande RNR = Cavagrande del Cassibile RNR; Fiumedinisi RNR = Fiume Fiumedinisi e Monte Scuderi RNR; Agrò SCI = ITA030019—Tratto Montano del Bacino della Fiumara di Agrò—Site of Community Importance (SCI); Irminio SCI = ITA080002—Alto corso del Fiume Irminio SCI. [b] Forest vegetation features were in accordance with Natura 2000 sites data and respective management plans: ftp://ftp.minambiente.it/PNM/Natura2000/TrasmissioneCE_2016/schede_mappe/Sicilia/SIC_schede/. [c] Natura 2000 Habitats: http://www.minambiente.it/sites/default/files/archivio/allegati/rete_natura_2000/int_manual_eu28.pdf [d] *Platano-Salicetum pedicellatae*: Thermo-Mediterranean Hyblean plateau riparian forest dominated by *Platanus orientalis* growing in association with *Salix* spp., *Populus* spp., *Fraxinus oxycarpa*, and *Nerium oleander*. *Salicetum albo-pedicellatae*: Thermo-Mediterranean riparian forest communities that grow on soils with a high water table. It is characterized by *Salix* spp. and *Populus* spp. in association with *F. oxycarpa*. *Salicetum albo-purpurea*: thermo-meso-Mediterranean riparian forest dominated by *Salix purpurea*, *Salixalba*, and *Salix pedicellata* in association with *Populus* spp. *Arrhenathero nebrodensis-Quercetum cerridis*: meso-Mediterranean acidophilous turkey oak forest stand typified by *Quercus cerris*; at higher altitudes (ca. 1400 m) it is mixed with beech forests (*Anemono apenninae-Fagetum sylvaticae*). *Populetalia albae*: riparian forests characterized by communities of *S. alba* and *Populus alba*. *Erico-Quercetum virgilianae*: meso-Mediterranean acidophilous woodland dominated by *Quercus dalechampii* in association with *Fraxinus ornus*. *Spartio-Nerietum oleandri*: thermo-Mediterranean community characteristic of Sicilian "Fiumara" streams, dominated by *N. oleander* in association with *Salix* spp. and *Populus* spp. *Platano-Salicetum gussonei*: thermo-Mediterranean community characteristic of Sicilian "Fiumara" streams, typified by *P. orientalis* and *Salix gussonei*. *Junipero hemisphaericae-Pinetum calabricae*: supra-Mediterranean Calabrian laricio pine forest with a dense structure. [e] CAM = *Phytophthora* ×*cambivora*; CIP = *P. citrophthora*; FRI = *P. frigida*, GON = *P. gonapodyides*; HYD = *P. hydropathica*; KEL = *P.* sp. kelmania; LAC = *P. lacustris*; MUL = *P. multivora*; PLU = *P. plurivora*; POL = *P. polonica*, PSC = *P. pseudocryptogea*; THE = *P. thermophila*. [f] *Pythium* sp. JN6-like also isolated. [g] *Pythium* sp. strain 1-9-like also isolated. [h] *Pythium* sp. F-1509-like also isolated. [i] *Pythium* sp. dissotocum-like also isolated. [j] *Pythium* sp. FL-2016d-like also isolated.

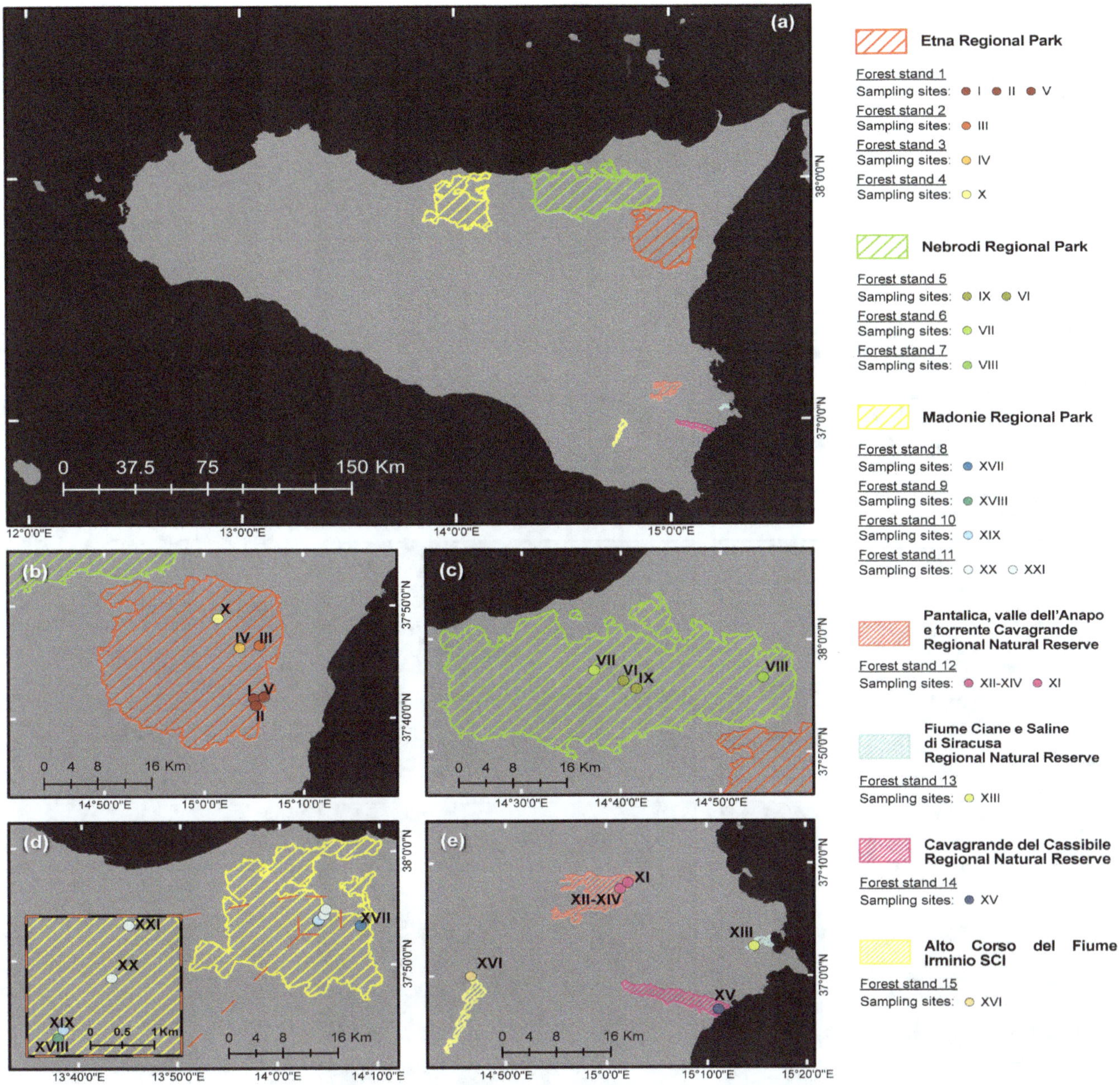

Figure 1. Geographical location of the 15 forest stands and the seven Protected Natural Areas included in the *Phytophthora* survey of natural forests in Sicily, projected using the Universal Transverse Mercator (UTM) (**a**). Location of the sampled forest sites within the Etna (**b**), Nebrodi (**c**), and Madonie (**d**) Regional Parks; and in the "Pantalica, valle dell'Anapo e torrente Cavagrande", "Fiume Ciane e Saline di Siracusa", "Cavagrande del Cassibile" Regional Natural Reserves and the "ITA080002—Alto Corso del Fiume Irminio" Site of Community Importance (SCI) (**e**).

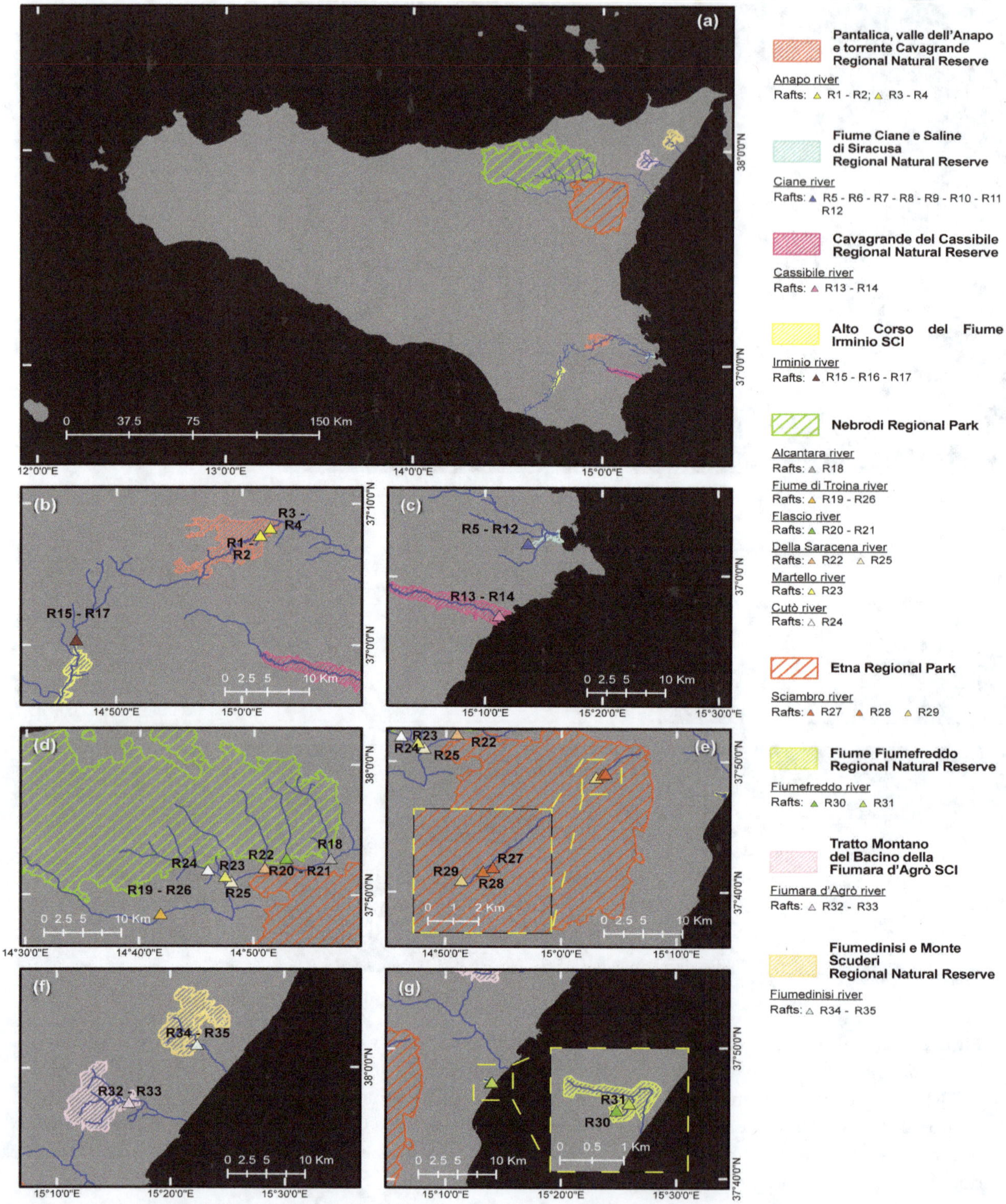

Figure 2. Geographical location of the nine Protected Natural Areas and the 14 river systems included in the *Phytophthora* survey of rivers in Sicily, projected using the Universal Transverse Mercator (UTM) (**a**). Riparian sampling sites (R) along the river systems running through: "Pantalica, valle dell'Anapo e Torrente Cavagrande" Regional Natural Reserve (RNR) and "ITA080002—Alto Corso del Fiume Irminio" Site of Community Importance (SCI) (**b**); "Fiume Ciane e Saline di Siracusa" and "Cavagrande del Cassibile" RNRs (**c**); "Nebrodi" (**d**) and "Etna" (**e**) Regional Parks; "ITA030019—Tratto Montano del Bacino della Fiumara d'Agrò" SCI and "Fiumedinisi e Monte Scuderi" RNR (**f**); and "Fiume Fiumefreddo" RNR (**g**).

2.2. *Morphological Characterization of Isolates*

Seven-days-old cultures grown at 20 °C in the dark on V8A were used to group all obtained isolates into morphotypes on the basis of their colony growth patterns. In addition, morphological features of sporangia, oogonia, antheridia, chlamydospores, hyphal swellings, and aggregations were examined [20,42] and compared with species descriptions in the literature.

2.3. *Molecular Identification of Isolates*

Molecular analyses were performed with 387 (184 from soil and 203 from rivers) of the 841 obtained isolates, representative of all morphotypes, soil samples, and baiting rafts. DNA was extracted from pure cultures grown on V8A using the PowerPlant® Pro DNA Isolation Kit (MO BIO Laboratories, Inc., Carlsbad, CA, USA), following the manufacturer's protocol. DNA was stored at −20 °C until further use.

The identification of *Phytophthora* species was performed by sequence analysis of the internal transcribed spacer (ITS) region of ribosomal DNA (rDNA). For amplification, forward primers ITS6 or ITS1 [44] and reverse primer ITS4 were used [45]. The PCR amplification mix and thermocycler conditions were as in [44]. PCR products were purified and sequenced by Macrogen Europe (Amsterdam, The Netherlands) in both directions with the primers used for amplification. Sequences were analyzed using FinchTV v.1.4.0 (https://digitalworldbiology.com/FinchTV). For species identification, blast searches in GenBank (http://www.ncbi.nlm.nih.gov/BLAST/) and in a local database containing sequences of ex-type or key isolates from published studies were performed. Isolates were assigned to a species when their sequences were at least 99% identical to a reference isolate. ITS sequences from representative isolates of this study were deposited at GenBank (www.ncbi.nlm.nih.gov/genbank; accession numbers are given in Table S2).

3. Results

Morphological and ITS sequence analyses revealed the occurrence of multiple *Phytophthora* species in each of the sampled PNAs. ITS sequence analyses showed that 351 of the 387 (90.7%) analyzed isolates (162 from forest soils and 189 from rivers) matched with 99–100% identity reference sequences of 16 known *Phytophthora* species and the designated *Phytophthora* sp. kelmania [46]. Nine isolates belonged to two species recently described as *Phytophthora vulcanica* and *Phytophthora tyrrhenica* [6] from Clade 7a, and to a new, yet undescribed species from Clade 2, while 27 isolates (7.0%) were assigned to other oomycete genera (Tables 1 and 2).

3.1. *Phytophthora Diversity and Distribution in Forest Stands*

In all oak and beech forests sampled, the majority of trees showed disease symptoms including thinning and dieback of crowns, fine root losses and, in some cases, bleeding stem cankers, whereas in riparian forests diseased trees had a scattered distribution. Noteworthy, in the riparian forest FS-13 along the Ciane river, is the fact that almost all *Fraxinus oxycarpa* Bieb. trees showed severe dieback and mortality. Overall, in all seven selected PNAs, *Phytophthora* species were found in 14 of 15 sampled forest stands (93.3%). In total, 17 *Phytophthora* species from eight of the 12 known phylogenetic clades [6] were isolated from 61 of the 83 (73.5%) soil samples collected from 16 of the 17 tree species sampled (94%) (Table 1, Figure 3a,c, Figure 4a,b, and Figure S2a,c,e). Only in one forest stand (FS-3) could no *Phytophthora* isolates be obtained from the only tested tree species *Betula aetnensis* Raf.

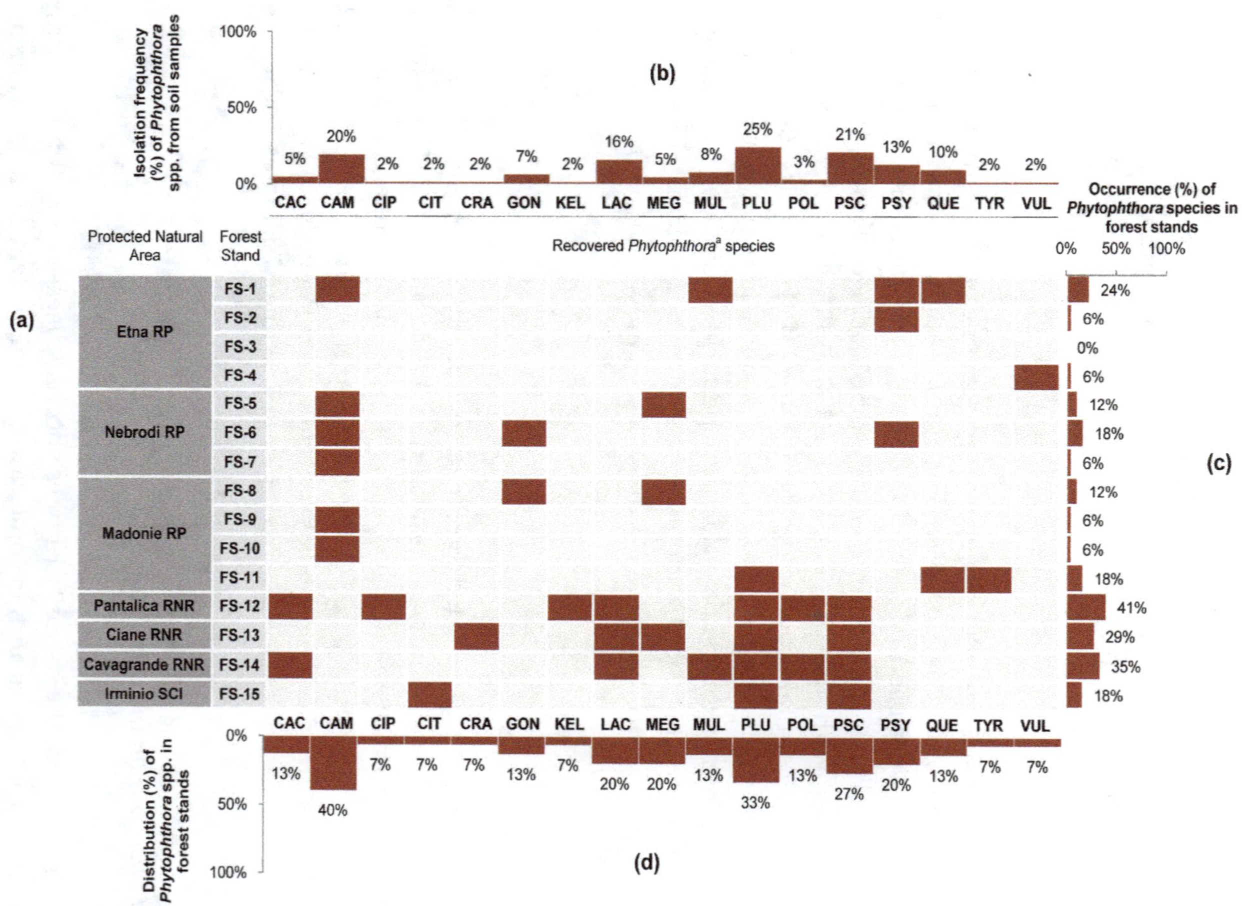

CAC = *Phytophthora cactorum*; CAM = *Phytophthora ×cambivora*; CIP = *Phytophthora citrophthora*; CIT = *Phytophthora citricola* 12; CRA = *Phytophthora crassamura*; GON = *Phytophthora gonapodyides*; KEL = *Phytophthora* sp. kelmania; LAC = *Phytophthora lacustris*; MEG = *Phytophthora megasperma*; MUL = *Phytophthora multivora*; PLU = *Phytophthora plurivora*; POL = *Phytophthora polonica*; PSC = *Phytophthora pseudocryptogea*; PSY = *Phytophthora psychrophila*; QUE = *Phytophthora quercina*; TYR = *Phytophthora tyrrhenica*; VUL = *Phytophthora vulcanica*.

Figure 3. Distribution and diversity of *Phytophthora* species in sampled forest stands from Protected Natural Areas in Sicily. (**a**) Etna RP = Etna Regional Park; Nebrodi RP = Nebrodi Regional Park; Madonie RP = Madonie Regional Park; Pantalica RNR = Pantalica, Valle dell'Anapo e Torrente Cavagrande Regional Natural Reserve (RNR); Ciane RNR = Fiume Ciane e Saline di Siracusa RNR; Cavagrande RNR = Cavagrande del Cassibile RNR; Irminio SCI = ITA080002—Alto corso del Fiume Irminio Site of Community Importance (SCI), (**b**) isolation frequency (%) of *Phytophthora* species from *Phytophthora*-positive soil samples, (**c**) occurrence (%) of *Phytophthora* species in sampled forest stands, (**d**) distribution (%) of *Phytophthora* species in the sampled forest stands.

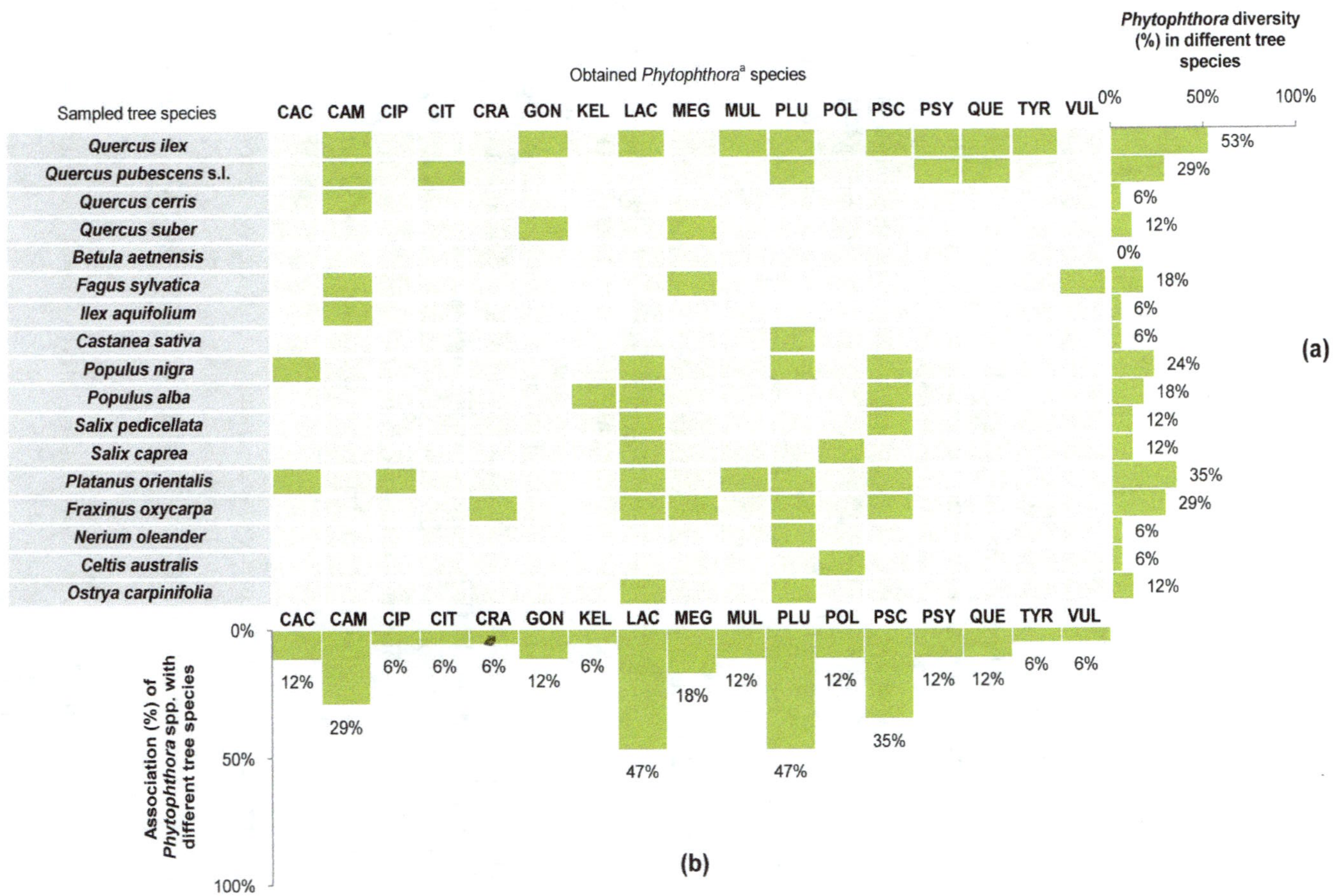

CAC = *Phytophthora cactorum*; CAM = *Phytophthora ×cambivora*; CIP = *Phytophthora citrophthora*; CIT = *Phytophthora citricola* 12; CRA = *Phytophthora crassamura*; GON = *Phytophthora gonapodyides*; KEL = *Phytophthora* sp. kelmania; LAC = *Phytophthora lacustris*; MEG = *Phytophthora megasperma*; MUL = *Phytophthora multivora*; PLU = *Phytophthora plurivora*; POL = *Phytophthora polonica*; PSC = *Phytophthora pseudocryptogea*; PSY = *Phytophthora psychrophila*; QUE = *Phytophthora quercina*; THY = *Phytophthora tyrrhenica*; VUL = *Phytophthora vulcanica*.

Figure Association of *Phytophthora* species with different tree species in Protected Natural Areas in Sicily. Dark-green color represents a *Phytophthora*—host tree association, (**a**) diversity of *Phytophthora* species in different tree species (in % of all *Phytophthora* species found), (**b**) association of *Phytophthora* species with the sampled tree species (in % of all tree species sampled).

Species from Clade 7, i.e., *Phytophthora* ×*cambivora* (previously *P. cambivora*), *Phytophthora vulcanica*, and *Phytophthora tyrrhenica*, were isolated from 53% of the sampled forest stands (Figure 3 and Figure S2b) in three of the seven protected natural areas (Etna, Nebrodi, and Madonie RPs) (Table 1, Figure 3a, and Figure S2a,d). *Phytophthora* ×*cambivora* was isolated from all sampled meso-, and supra-Mediterranean forest stands: In the Etna RP (FS-1) from *Quercus pubescens* Willd. *sensu latu* (s.l.); in the Nebrodi RP from *Fagus sylvatica* L., *Quercus cerris* L., *Quercus ilex* L., and *Q. pubescens* s.l. and all sampled forest stands (FS-5 to FS-7); and in the Madonie RP (FS-9, FS-10) from *Ilex aquifolium* L. and *Q. pubescens* s.l. (Table 1, Figure 3a,b,d, and Figure 4). *Phytophthora* ×*cambivora* occurred in an altitude range between 660 and 1780 m above sea level (a.s.l.). *Phytophthora vulcanica* and *P. tyrrhenica* were recovered from *F. sylvatica* in FS-4 and from *Q. ilex* in FS-11, respectively (Table 1).

Four Clade 6 species, *Phytophthora gonapodyides*, *Phytophthora megasperma*, *Phytophthora lacustris*, and *Phytophthora crassamura*, were found in 40% of the sampled forest stands in five PNAs (Figure 3 and Figure S2b). *Phytophthora gonapodyides* occurred between 700 and 1000 m a.s.l. in the rhizosphere of *Q. ilex* and *Quercus suber* L. in meso-Mediterranean evergreen (FS-6) and cork oak (FS-8) woodlands, respectively (Table 1, Figure 4). *Phytophthora megasperma* was isolated from supra-, meso-, and thermo-Mediterranean forest stands in three PNAs: In the Nebrodi RP from *F. sylvatica* (FS-5); in the Madonie RP from *Q. suber* (FS-8); and in the Fiume Ciane e Saline di Siracusa RNR (Ciane RNR) from *Fraxinus oxycarpa* (FS-13). This *Phytophthora* species inhabited a wide altitudinal range between 4 and 1450 m a.s.l. (Table 1, Figure 3a,b,d, and Figure 4). *Phytophthora lacustris* was isolated from the rhizosphere of eight different tree species between 4 and 236 m a.s.l. in three thermo-Mediterranean riparian forest stands (FS-12 to FS-14) located in three PNAs (Table 1, Figure 3a,b,d, and Figure 4). *Phytophthora crassamura* only occurred in the rhizosphere of *F. oxycarpa* in the Ciane RNR (FS-13) (Table 1, Figure 4).

Species from *Phytophthora* Clade 2 were present in six of the seven monitored PNAs (Figure S2a,b,d). *Phytophthora plurivora* was most widespread, occurring in 25% of the *Phytophthora*-positive soil samples taken from eight different tree species in 33% of the sampled forest stands (FS-11 to FS-15) and in five PNAs (Table 1, Figure 3a,b,d, and Figure 4). Interestingly, this pathogen was recovered from seven trees species in the riparian thermo-Mediterranean plane tree stand (FS-12) of the Pantalica RNR. The altitudinal distribution of *P. plurivora* ranged from an altitude of 4 to 850 m a.s.l. *Phytophthora multivora* was associated with *Q. ilex* in the Etna RP (FS-1) and with *Platanus orientalis* L. in the Cassibile RNR (FS-14) (Table 1, Figure 3a,b,d). *Phytophthora citrophthora* was only found in the rhizosphere of *P. orientalis* in FS-12 (Table 1). A previously unknown species from the '*Phytophthora citricola* complex', informally designated here as *P. citricola* 12, was recovered from the rhizosphere of *Q. pubescens* s.l. in riparian stand FS-15 in Irminio SCI (Table 1).

The Clade 1 species *Phytophtora cactorum* occurred in two riparian thermo-Mediterranean plane tree forests in two PNAs (Table 1, Figure S2a,b,d, and Figure 3a,b,d). In the Pantalica RNR (FS-12) and the Cassibile RNR (FS-14), *P. cactorum* was associated with *P. orientalis* and *Populus nigra* L., respectively (Figure 4).

The Clade 3 species *Phytophthora psychrophila* was found associated with *Q. pubescens* s.l. in two forest stands (FS-1, FS-2) of the Etna RP and with *Q. ilex* in stand FS-6 of the Nebrodi RP. The altitudinal distribution ranged from 890 to 1345 m a.s.l. (Table 1).

Phytophthora pseudocryptogea from Clade 8 was frequently isolated at an altitude between 4 and 240 m from rhizosphere soil of six different tree species in four riparian thermo-Mediterranean forest stands (FS-12 to FS-15) located in four distinct PNAs (Table 1, Figure 3a,b,d). Another Clade 8 taxon, *Phytophthora* sp. kelmania, was detected in only one soil sample from *Populus alba* L. in stand FS-12 (Table 1, Figure 4).

Phytophthora polonica from Clade 9 was associated with *Celtis australis* L. and *Salix caprea* L. in two riparian thermo-Mediterranean forest stands, FS-12 in the Pantalica RNR and FS-14 in the Cavagrande RNR, respectively (Table 1, Figure 3a,b,d, and Figure 4).

The oak-specific pathogen *Phytophthora quercina* from the recently described Clade 12 [6] was recovered between 660 and 1110 m a.s.l. from *Q. ilex* and *Q. pubescens* s.l. at two sites of FS-1 in the Etna RP and from *Q. ilex* in the Madonie RP (FS-11) (Table 1, Figure 3a,b,d and Figure 4).

3.2. *Phytophthora* Diversity and Distribution in Rivers within PNAs

In total 12 *Phytophthora* species from five phylogenetic clades were detected in all monitored rivers running through all nine selected PNAs (Table 2, Figure 5a,c, and Figure S3a–f); 29 of the 35 baiting rafts (83%) were *Phytophthora*-positive.

Most common were mainly aquatic *Phytophthora* species from ITS Clade 6 that were recovered from all monitored river systems and PNAs (Figure S3a,d,e). *Phytophthora lacustris* occurred between 4 and 850 m a.s.l. in 77% of the baiting rafts and in all watercourses except for the Sciambro river (Table 2, Figure 5a,b,d), a torrential high altitude stream, which only flows seasonally during snowmelt. In five rivers, *P. lacustris* was the only *Phytophthora* species detected. *Phytophthora gonapodyides* was found in an altitudinal range between ca. 700 and 1700 m a.s.l. in the Alcantara and Della Saracena rivers (Nebrodi RP) and in the Sciambro river (Etna RP); in the latter it was the only *Phytophthora* species isolated (Table 2, Figure 5a,b,d). The third mainly aquatic Clade 6 species, *Phytophthora thermophila*, was exclusively found in the Fiumefreddo river (Table 2, Figure 5a,b,d).

The Clade 2 species, *P. plurivora*, *P. multivora*, *P. citrophthora*, and *P. frigida*, were isolated from 36% of the rivers in five PNAs at lowland sites ranging from 6 to 220 m a.s.l. (Table 2, Figure 5a,b,d and Figure S3a,d,e). While *P. frigida* was only found in Ciane river, each of the other species of Clade 2 occurred in two rivers: *P. plurivora* in the Fiumara d'Agrò and Fiumefreddo rivers, *P. multivora* in the Fiumefreddo and Ciane rivers, and *P. citrophthora* in the Anapo and Fiumedinisi rivers (Table 2, Figure 5).

Phytophthora ×cambivora from Clade 7 and *P. polonica* from Clade 9 were both exclusively detected in the Della Saracena river in the Nebrodi RP (Table 2, Figure 5).

Two species from Clade 8 were found in five watercourses running through five PNAs (Table 2, Figure S3a,d,e). *Phytophthora pseudocryptogea* was widespread, occurring between 4 and 220 m a.s.l. in the Anapo, Ciane, Cassibile, Fiumefreddo, and Fiumara d'Agrò rivers, whereas *P.* sp. *kelmania* was exclusively isolated from the Ciane river (Table 2, Figure 5a,b,d).

Phytophthora hydropathica from Clade 9 was only found in the Cassibile river (Table 2, Figure 5a,b,d).

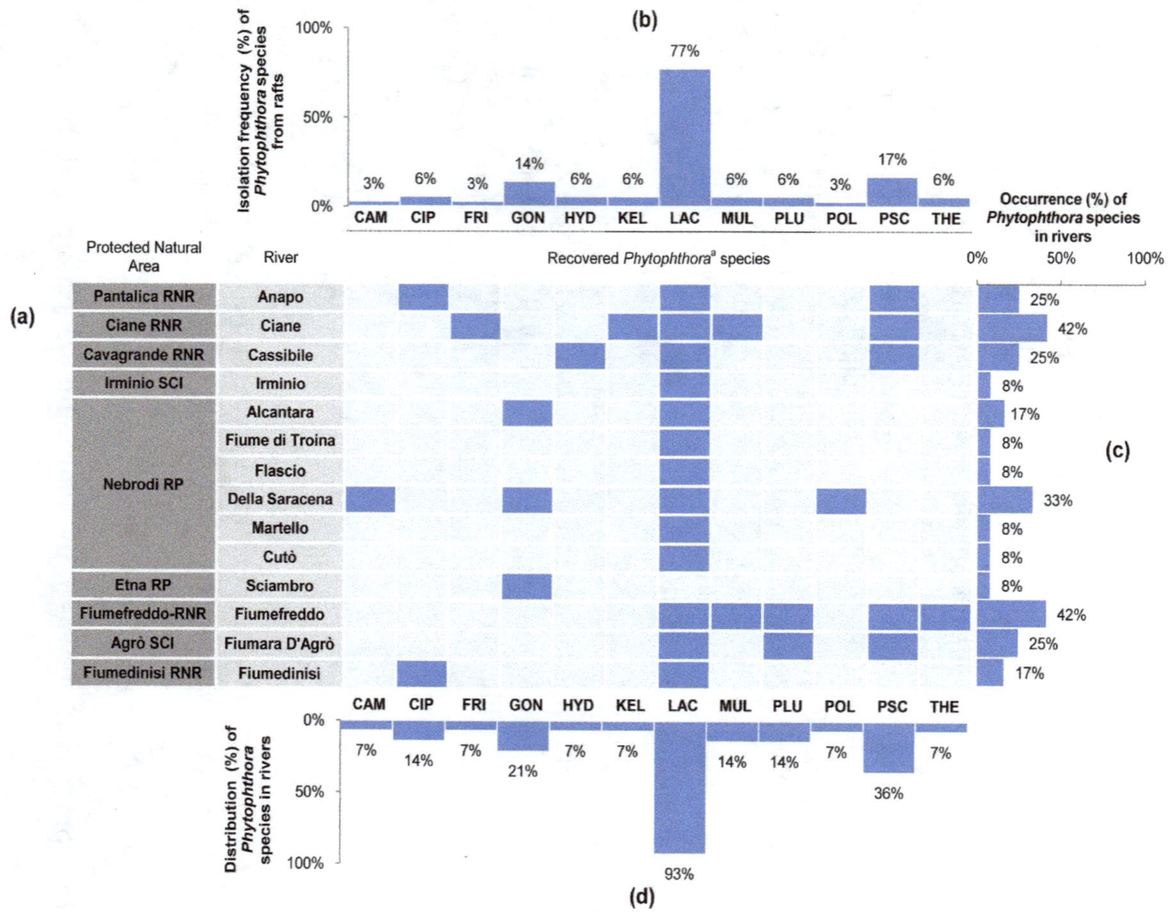

CAM = *Phytophthora* ×*cambivora*; CIP = *Phytophthora citrophthora*; FRI = *Phytophthora frigida*; GON = *Phytophthora gonapodyides*; HYD = *Phytophthora hydropatica*; KEL = *Phytophthora* sp. kelmania; LAC = *Phytophthora lacustris*; MUL = *Phytophthora multivora*; PLU = *Phytophthora plurivora*; POL = *Phytophthora polonica*; PSC = *Phytophthora pseudocryptogea*; THE = *Phytophthora thermophila*.

Figure 5. Distribution and diversity of *Phytophthora* species in sampled rivers from Protected Natural Areas in Sicily. (**a**) Etna RP = Etna Regional Park; Nebrodi RP = Nebrodi Regional Park; Madonie RP = Madonie Regional Park; Pantalica RNR = Pantalica, Valle dell'Anapo e Torrente Cavagrande Regional Natural Reserve (RNR); Ciane RNR = Fiume Ciane e Saline di Siracusa RNR; Cavagrande RNR = Cavagrande del Cassibile RNR; Fiumedinisi RNR = Fiume Fiumedinisi e Monte Scuderi RNR; Agrò SCI = ITA030019—Tratto Montano del Bacino della Fiumara di Agrò—Site of Community Importance (SCI); Irminio SCI = ITA080002—Alto corso del Fiume Irminio SCI, (**b**) isolation frequency (%) of *Phytophthora* species from *Phytophthora*-positive baiting rafts, (**c**) occurrence (%) of *Phytophthora* species in sampled rivers, (**d**) distribution (%) of *Phytophthora* species in the sampled rivers.

4. Discussion

This is the first study of *Phytophthora* diversity in Europe using conventional isolation methods and covering both a wide range of natural forest types and watercourses crossing these areas. Previously, the only surveys of *Phytophthora* diversity in both forests and rivers within the same region in Europe used only a metabarcoding approach which is based on DNA identification technologies and high-throughput DNA sequencing. In Spain, 13 and 35 *Phytophthora* phylotypes were detected in forest soils and streams, respectively [37]. Using a different molecular method, a survey in Scotland demonstrated the presence of 10 and 9 *Phytophthora* phylotypes in soil and water samples, respectively [47]. The present survey unveiled a rich community of 20 *Phytophthora* species in the Sicilian PNAs studied. With 17 different species from 8 of the 12 known phylogenetic clades, including the two newly described species *P. tyrrhenica* and *P. vulcanica* [6], *Phytophthora* diversity in 15 natural forest stands was higher than in previous broadleaved forest surveys in Europe using similar isolation methods. In oak forests across Italy, northeastern France, Austria, and Turkey, and in oak and beech forests in Bavaria 11, 8, 5, 7, and 13 *Phytophthora* species, respectively, were found [9,13,17,48,49]. However, the lower *Phytophthora* diversity in these surveys may partly be due to the limited number of tree species and forest types included. With nine *Phytophthora* species from four phylogenetic clades the diversity found in 14 rivers in Sicily was almost as high as in previous surveys in Australia, the USA, and South Africa which covered much larger areas and a higher numbers of rivers [40,50–52], but lower than in Taiwan where four described *Phytophthora* species and 14 previously unknown *Phytophthora* taxa were discovered in 19 rivers [21].

The high diversity of *Phytophthora* species in natural forests and rivers in Sicily is particularly impressive considering the relatively small area of less than 10,000 km^2 covered by this survey. This may be explained by the diversity of forest types and altitudinal zones surveyed and Sicily's long and changing history of human colonization and the introduction of non-native horticultural plants. Thirteen of the 20 *Phytophthora* species occurring in the sampled Sicilian ecosystems are considered introduced pathogens: *P. cactorum*, *P.* ×*cambivora*, *P. citricola* 12, *P. citrophthora*, *P. crassamura*, *P. frigida*, *P. hydropathica*, *P. multivora*, *P. plurivora*, *P. polonica*, *P. pseudocryptogea*, *P. thermophila*, and *P.* sp. kelmania [21,25,32,42,53,54]. In contrast, *P. psychrophila*, *P. quercina*, *P. tyrrhenica*, and *P. vulcanica* are considered endemic to Europe resulting from species radiation following adaptation to different Fagaceae species [6].

Amongst the 17 *Phytophthora* species obtained from forest stands, *P.* ×*cambivora*, *P. plurivora*, and *P. pseudocryptogea* were the most widespread whereas the other species had a more scattered or even punctual distribution. The allopolyploid hybrid pathogen *P.* ×*cambivora* was most common, occurring in the majority of meso- and supra-Mediterranean forest stands sampled in the Nebrodi, Etna, and Madonie Regional Parks. In a previous study, *P.* ×*cambivora* was also found in Corleone near Palermo [49]. Although the recovery from *I. aquifolium* extended the known host range of *P.* ×*cambivora*, this pathogen was mainly associated with known susceptible host species like *Quercus* spp. and *F. sylvatica* [9,13,42,49]. In most cases, oak and beech trees showed typical disease symptoms like thinning and dieback of crowns, fine root losses, and in some cases bleeding stem cankers, all indicative of *Phytophthora* infections. Due to the high aggressiveness of *P.* ×*cambivora* to oaks and beech [6,11,42,55] it seems likely that this pathogen is associated with the widespread decline and dieback of oak and beech stands recently reported in Sicily [5]. The results of this work confirm previous studies in Germany and Italy demonstrating that *P.* ×*cambivora* preferentially occurs in acidic and clayey soils [9,11,13,49,56]. Of note, *P.* ×*cambivora* was not isolated from riparian thermo-Mediterranean forests in Sicily. Compared to *P.* ×*cambivora*, *P. plurivora* showed an opposite distribution pattern, being the most common species in riparian thermo-Mediterranean forest stands dominated by willows, poplars, plane, and ash trees. However, it was only infrequently isolated from seasonally dry, meso- and supra-Mediterranean forests. This distribution is most likely caused by the thin oospore walls which make *P. plurivora* susceptible to droughts [53]. Although *P. plurivora* was already reported from more than 80 woody host species including *Castanea sativa* Mill., *F. sylvatica*, *Fraxinus* spp., *Quercus* spp., and *Salix* spp. [10,11,13,14,21,32,53,57–61], the recoveries from rhizosphere soil of *P. nigra*, *P. orientalis*, *Nerium oleander* L., and *Ostrya carpinifolia* Scop. in the present study constituted first-time records for this wide host range pathogen. Interestingly, *P. plurivora* showed a similar upper limit of vertical distribution as in the Bavarian Alps (ca. 870 m a.s.l.) [53]. Despite being an aggressive beech pathogen across Europe and in the USA [9,14,32,60,62,63], *P. plurivora* did not occur in the rhizosphere of *F. sylvatica* forests in Sicily, which at this southern latitude grow at altitudes above ca. 1400 m a.s.l. However, in contrast to Bavaria, in Sicily this vertical limit is most likely caused by extremely dry summers, causing desiccation of the thin-walled oospores [53], rather than by deep winter temperatures. In Taiwan, *P. plurivora* occurs at altitudes around 2000 m in regions with mild winters and humid summers [21]. *Phytophthora multivora*, the second species from the '*Phytophthora citricola* complex' found in this survey, was less common than *P. plurivora*, being isolated only from *Q. ilex* and *P. orientalis* in each one of the meso-Mediterranean evergreen oak and riparian thermo-Mediterranean forest stand, respectively, and in the Ciane and Fiumefreddo rivers. Due to its particularly thick oospore walls, *P. multivora* has adapted perfectly to severe summer droughts in Mediterranean regions such as Western Australia and South Africa, where it is widespread in both native vegetation and urban environments [51,64–66]. In Europe, *P. multivora* was recently introduced and is currently spreading through the nursery sector and in young plantings [32,67]. Prior to this study, it had only been occasionally recovered from the wider environment [60,68]. Hitherto, *P. frigida* from Clade 2 was only known from *Eucalyptus* plantations in South Africa and from rainforests in eastern Australia [54,69]. The number of known species of Clade 2 is rapidly increasing; besides *P. plurivora* and *P. multivora* it includes numerous other aggressive *Phytophthora* species. Many Clade 2 species pose serious threats to natural ecosystems across the world [10,43,53,70,71]. The findings of *P. frigida* and the new species *P. citricola* 12, and the widespread occurrence of *P. plurivora* and *P. multivora* in Sicilian PNAs are of serious concern.

In the present study, *P. pseudocryptogea* from Clade 8 was frequently recovered from six tree species in riparian thermo-Mediterranean forest stands and from five rivers. It is the first report of this species in Sicily. While *P. pseudocryptogea* was previously not reported from Sicily, its close relative *P. cryptogea* commonly causes damage to several non-native ornamentals in nurseries and tomato crops under plastic-houses [72–76]. *Phytophthora cryptogea* has a scattered, but widespread, distribution in periodically dry Mediterranean natural ecosystems [25,56,58]. In Europe, *P. cryptogea* is an established exotic pathogen, whereas *P. pseudocryptogea* and the phylogenetically close taxon *P.* sp. kelmania [46], are considered as recently introduced emerging pathogens [32].

In accord with previous studies in other areas of the world [38,39,41,50], Clade 6 species prevailed in rivers, indicating their adaptation to aquatic environments. Interestingly, two mainly aquatic opportunistic pathogens from Clade 6, *P. gonapodyides* and *P. lacustris*, which often co-occur in river systems in temperate regions of North America, Europe, and Asia [40,77–79], showed opposite distribution patterns in Sicily. *Phytophthora gonapodyides* occurred exclusively at altitudes above 620 m, where it was mainly associated with meso-Mediterranean oak stands on acidic non-calcareous soils and with rivers running through oak stands. In previous studies, *P. gonapodyides* was also often found in oak stands and on acidic sites [11,13,49,77]. In contrast, in this study, *P. lacustris* was only isolated below 850 m altitude from *Salix*-dominated riparian forests on both silica-rich acidic and calcareous alkaline sites, and from rivers running through these forests. Both *Phytophthora* species co-occurred only in three rivers in a transition zone between 624 and 811 m a.s.l. The different altitudinal preferences of both species reflect their different cardinal temperatures for growth [79]. In this survey, two other species, *P. hydropathica* and *P. thermophila* from Clade 9, were exclusively detected in rivers confirming their mainly aquatic lifestyle. Prior to the present study, *P. hydropathica* was found in rivers and irrigation reservoirs in the Eastern USA [80–82] and in rivers in Galicia, Spain [83]. In Italy, this species was only reported from ornamental plants in commercial nurseries [84]. *Phytophthora thermophila* was previously exclusively detected in streams and native forests of *Eucalyptus* and *Banksia* spp. in Australia [20] and, hence, the finding in the Fiumefreddo river constitutes the first-time report for

Europe. The presence of both a nursery and a young *Eucalyptus* plantation close to the Fiumefreddo River suggests an introduction via infested nursery plants.

Phytophthora quercina is commonly occurring across Europe, causing chronic fine root losses in different oak species which, in interaction with climatic extremes and secondary pests and pathogens, lead to decline, dieback, and mortality of oak forests [10,13,14,17,32,49,85–87]. The present findings in Sicily extend the known distribution of this pathogen to the southern oak stands of Europe.

Two previously unknown *Phytophthora* species, which have been recently described as *P. vulcanica* and *P. tyrrhenica*, were isolated from a beech stand on Mount Etna and a *Q. ilex* stand in the Madonie mountains, respectively. In a multigene phylogenetic study, both species were placed in Clade 7, closely related to *P. uliginosa*, a cryptic species which seems to be restricted to Europe [6,55]. *Phytophthora tyrrhenica* was also detected in oak stands in Sardinia [6] whereas *P. vulcanica* was recovered in Sicily for the first time. Since decline symptoms in the infested stands were only mild and both species showed limited aggressiveness to their respective host species in pathogenicity tests; they are considered as endemic species in Europe resulting from species radiation driven by adaptation to different Fagaceae hosts [6].

With 11 *Phytophthora* species from five phylogenetic clades, the four thermo-Mediterranean riparian forest stands located at altitudes between ca. 4 and 430 m a.s.l. showed the highest *Phytophthora* diversity. In contrast, despite the higher number of 11 sampled forest stands and the wide altitudinal range between ca. 700 and 1900 m a.s.l., the meso- and supra-Mediterranean forests contained only seven *Phytophthora* species from three clades. Interestingly, only three *Phytophthora* species, *P. megasperma*, *P. multivora*, and *P. plurivora*, occurred in both categories of forest stands. Similar to the La Maddalena archipelago in Sardinia [25], also in Sicily *Q. ilex* trees hosted with nine *Phytophthora* species the highest diversity of all tree species tested. The presence of a rich community of six *Phytophthora* species in the rhizosphere of *P. orientalis* trees was surprising and warrants further investigations of their potential involvement in the decline of Sicilian plane trees, in particular, in stands with the absence of the canker and wilt pathogen *Ceratocystis fimbriata* [88]. With five *Phytophthora* species, diversity in the rhizosphere of *F. oxycarpa* trees in Sicily was similar to *Fraxinus excelsior* forests in Denmark and Poland, where five *Phytophthora* species had also been recovered [26,61].

Analogous to the forest stands, altitude also had a strong influence on the diversity and composition of the *Phytophthora* populations in the rivers. While rivers below 400 m a.s.l. contained nine *Phytophthora* species from four phylogenetic clades, only two *Phytophthora* species, *P. gonapodyides*, and *P. lacustris*, from Clade 6 and, in one river, *P.* ×*cambivora* and *P. polonica* from Clades 7 and 9, respectively, could be recovered from rivers above 600 m altitude. Eight of the 12 *Phytophthora* species recovered from rivers, *P. citrophthora*, *P. gonapodyides*, *P. lacustris*, *P. multivora*, *P. plurivora*, *P. polonica*, *P. pseudocryptogea*, and *P.* sp. kelmania, were also found in rhizosphere soil of the thermo-Mediterranean riparian forest stands. In contrast, only four of the nine *Phytophthora* species found in non-flooded meso- and supra-Mediterranean forests, *P. gonapodyides*, *P. megasperma*, *P. multivora*, and *P. plurivora*, also occur in rivers. These results indicate that the mutual exchange of *Phytophthora* inoculum between river water and forest soils is largely dependent on seasonal or episodic flooding. The results also show that several typical forest *Phytophthora* species, in particular, *P. cactorum*, *P. crassamura*, *P. quercina*, and *P. psychrophila*, cannot establish in aquatic ecosystems. Similar results were found in forests and rivers in Taiwan [21].

5. Conclusions

This study demonstrated that in ecological and environmental studies the combined use of an efficient leaf baiting technique and a reliable molecular identification method is an efficient approach for studying the diversity and distribution of *Phytophthora* species in diverse protected natural ecosystems. Eleven of the 18 known *Phytophthora* species found in this survey, including *P. crassamura*, *P. frigida*, *P. hydropathica*, *P. polonica*, *P. pseudocryptogea*, *P. quercina*, *P. thermophila*, and *P.* sp. kelmania, were detected for the first time in Sicily. The findings of *P. frigida*, *P. thermophila* and the three new species *P. vulcanica*, *P. tyrrhenica*, and *P. citricola* 12 are first-time records for Europe. Another four species, *P. cactorum*, *P. citrophthora*, *P. megasperma*, and *P. multivora*, were previously only recorded in Sicily from nurseries or ornamental and horticultural plantings, but not from natural environments [32,67,75,89,90]. *Phytophthora cactorum*, *P. plurivora*, *P. multivora*, and *P.* ×*cambivora* are exotic, invasive wide-host-range pathogens with high aggressiveness to many native European tree species. Since their widespread occurrence in protected natural areas in Sicily poses a serious threat to the long-term stability of the infested ecosystems, management concepts are urgently required to prevent further spread of these pathogens to non-infested areas and to increase tree vigor and ecosystem stability.

Supplementary Materials: Table S1: Geographic location, geomorphological features, land area covered and ecological features of the 10 Protected Natural Areas included in the *Phytophthora* survey in Sicily; Table S2: Isolate details and GenBank accession numbers of *Phytophthora* isolates obtained during the Phytophthora survey of forest stands and river systems in 10 Protected Natural Areas in Sicily; Figure S1: Geographical location of Protected Natural Areas included in the *Phytophthora* survey of forest stands and river systems in Sicily, projected using the Universal Transverse Mercator (UTM); Figure S2: Distribution of phylogenetic *Phytophthora* Clades in sampled forest stands and protected natural areas in Sicily. Brown color represents the presence of a clade. (a) Etna RP = Etna Regional Park; Nebrodi RP = Nebrodi Regional Park; Madonie RP = Madonie Regional Park; Pantalica RNR = Pantalica, Valle dell'Anapo e Torrente Cavagrande Regional Natural Reserve (RNR); Ciane RNR = Fiume Ciane e Saline di Siracusa RNR; Cavagrande RNR = Cavagrande del Cassibile RNR; Irminio SCI = ITA080002—Alto corso del Fiume Irminio Site of Community Importance (SCI), (b) proportion (%) of forest stands in which individual *Phytophthora* Clades were present, (c) proportion (%) of *Phytophthora* Clades present in individual sampled forest stands, (d) proportion (%) of protected natural areas in which individual *Phytophthora* Clades were present, (e) proportion (%) of *Phytophthora* Clades present in individual protected natural areas. Figure S3: Distribution of phylogenetic *Phytophthora* Clades in baited river systems and protected natural areas in Sicily. (a) Etna RP = Etna Regional Park; Nebrodi RP = Nebrodi Regional Park; Madonie RP = Madonie Regional Park; Pantalica RNR = Pantalica, Valle dell'Anapo e Torrente Cavagrande Regional Natural Reserve (RNR); Ciane RNR = Fiume Ciane e Saline di Siracusa RNR; Cavagrande RNR = Cavagrande del Cassibile RNR; Fiumedinisi RNR = Fiume Fiumedinisi e Monte Scuderi RNR; Agrò SCI = ITA030019—Tratto Montano del Bacino della Fiumara di Agrò—Site of Community Importance (SCI); Irminio SCI = ITA080002—Alto corso del Fiume Irminio SCI, (b) isolation frequency (%) of phylogenetic *Phytophthora* Clades from baiting rafts, (c) proportion (%) of Phytophthora Clades present in individual baited rivers, (d) proportion (%) of rivers from which phylogenetic *Phytophthora* Clades were isolated, (e) proportion (%) of natural protected areas from which phylogenetic *Phytophthora* Clades were isolated, (f) proportion (%) of *Phytophthora* Clades present in rivers of individual protected natural areas.

Author Contributions: Conceptualization: T.J., S.O.C., G.M.d.S.L.; data curation: T.J., F.L.S., M.H.J., S.O.C., B.S., L.S.; formal analysis: T.J., S.O.C., M.H.J., F.L.S., I.P., L.S.; investigation: T.J., C.R., R.F., M.E., F.L.S, A.P., F.A.; methodology: S.O.C., T.J., A.P.; project administration: A.P.; resources: A.P., G.M.d.S.L., S.O.C., T.J.; Supervision: S.O.C.; writing—original draft: F.L.S., S.O.C., T.J.; writing—review and editing: S.O.C., T.J., B.S., F.L.S., G.M.d.S.L.

Acknowledgments: The authors are grateful to the Regional Forestry Board Agency (DRAFT—Dipartimento Regionale Azienda Foreste Demaniali), Luca Ferlito (Commissioner of Nebrodi Regional Park; formerly Head of the Provincial Operational Center) and Giovanni Granata for their invaluable professional assistance during surveys. The authors also warmly thank the Sicilian forest rangers for their qualified support and Ann Davies for the English revision.

References

1. Martino, A.D.; Raimondo, F.M. Biological and chorological survey of the Sicilian Flora. *Webbia* **1979**, *34*, 309–335. [CrossRef]
2. Schönfelder, I.; Schönfelder, P. *La Flora Mediterranea*; Istituto Geografico De Agostini: Novara, Italy, 1996.
3. Giardina, G.; Raimondo, F.M.; Spadaro, V. *A Catalogue of Plants Growing in Sicily*; Bocconea: Palermo, Italy, 2007; Volume 20, ISBN 9788879150224.
4. Conti, F.; Abbate, G.; Alessandrini, A.; Blasi, C.; Bonacquisti, S.; Scassellati, E. La flora vascolare italiana: Ricchezza e originalità a livello nazionale e regionale. In *Stato Delle Conoscenze Sulla Flora Vascolare D'italia*; Scoppola, A., Blasi, C., Eds.; Palombi Editori: Rome, Italy, 2005; pp. 18–22.

5. Rizza, C.; Scibetta, S.; Pane, A.; Maetzke, F.; La Mela Veca, D.S.; Culotta, S.; Granata, G.; La Spada, F.; Aloi, F.; Faedda, R.; et al. A new approach in the monitoring of the phytosanitary conditions of forests: The case of oak and beech stands in the Sicilian Regional Parks. *Italian J. Mycol.* **2016**, *45*, 29–46.
6. Jung, T.; Horta Jung, M.; Cacciola, S.O.; Cech, T.; Bakonyi, J.; Seress, D.; Mosca, S.; Schena, L.; Seddaiu, S.; Pane, A.; et al. Multiple new cryptic pathogenic Phytophthora species from Fagaceae forests in Austria, Italy and Portugal. *IMA Fungus* **2017**, *8*, 219–244. [CrossRef]
7. Erwin, D.C.; Ribeiro, O.K. *Phytophthora Diseases Worldwide*; American Phytopathological Society (APS Press): St. Paul, MN, USA, 1996.
8. Hansen, E.M.; Reeser, P.W.; Sutton, W. *Phytophthora* Beyond Agriculture. *Annu. Rev. Phytopathol.* **2012**, *50*, 359–378. [CrossRef] [PubMed]
9. Jung, T. Beech decline in Central Europe driven by the interaction between *Phytophthora* infections and climatic extremes. *For. Pathol.* **2009**, *39*, 73–94. [CrossRef]
10. Jung, T.; Pérez-Sierra, A.; Durán, A.; Jung, M.H.; Balci, Y.; Scanu, B. Canker and decline diseases caused by soil- and airborne *Phytophthora* species in forests and woodlands. *Persoonia Mol. Phylogeny Evolut. Fungi* **2018**, *40*, 182–220. [CrossRef] [PubMed]
11. Jung, T.; Blaschke, H.; Neumann, P. Isolation, identification and pathogenicity of *Phytophthora* species from declining oak stands. *Eur. J. For. Pathol.* **1996**, *26*, 253–272. [CrossRef]
12. Frisullo, S.; Lima, G.; Magnano di San Lio, G.; Camele, I.; Melissano, L.; Puglisi, I.; Pane, A.; Agosteo, G.E.; Prudente, L.; Cacciola, S.O. *Phytophthora cinnamomi* Involved in the Decline of Holm Oak (*Quercus ilex*) Stands in Southern Italy. *For.Sci.* **2018**, *64*, 290–298. [CrossRef]
13. Jung, T.; Blaschke, H.; Oßwald, W. Involvement of soilborne *Phytophthora* species in Central European oak decline and the effect of site factors on the disease. *Plant Pathol.* **2000**, *49*, 706–718. [CrossRef]
14. Jung, T.; Vettraino, A.M.; Cech, T.; Vannini, A. The impact of invasive *Phytophthora* species on European forests. In *Phytophthora: A Global Perspective*; Lamour, K., Ed.; CABI: Wallingford, UK, 2013; pp. 146–158.
15. Ristaino, J.B.; Gumpertz, M.L. New frontiers in the study of dispersal and spatial analysis of epidemics caused by species in the Genus *Phytophthora*. *Annu. Rev. Phytopathol.* **2000**, *38*, 541–576. [CrossRef]
16. Rizzo, D.M.; Garbelotto, M.; Hansen, E.M. *Phytophthora ramorum*: Integrative research and management of an emerging pathogen in California and Oregon forests. *Annu. Rev. Phytopathol.* **2005**, *43*, 309–335. [CrossRef]
17. Balcì, Y.; Halmschlager, E. *Phytophthora* species in oak ecosystems in Turkey and their association with declining oak trees. *Plant Pathol.* **2003**, *52*, 694–702. [CrossRef]
18. Balci, Y.; Balci, S.; Mac Donald, W.L.; Gottschalk, K.W. Pathogenicity of Phytophthora species isolated from rhizosphere soil in the eastern United States. In Proceedings of the Sudden Oak Death Third Science Symposium; Frankel, S.J., Kliejunas, J.T., Palmieri, K.M., Eds.; Department of Agriculture, Forest Service, Pacific Southwest Research Station: Albany, CA, USA, 2008; pp. 225–226.
19. Cooke, D.E.L.; Schena, L.; Cacciola, S.O. Tools to detect, identify and monitor *Phytophthora* species in natural ecosystems. *J. Plant Pathol.* **2007**, *89*, 13–28.
20. Jung, T.; Stukely, M.J.C.; Hardy, G.E.S.J.; White, D.; Paap, T.; Dunstan, W.A.; Burgess, T.I. Multiple new *Phytophthora* species from ITS Clade 6 associated with natural ecosystems in Australia: Evolutionary and ecological implications. *Persoonia Mol. Phylogeny Evolut. Fungi* **2011**, *26*, 13–39. [CrossRef]
21. Jung, T.; Chang, T.T.; Bakonyi, J.; Seress, D.; Pérez-Sierra, A.; Yang, X.; Hong, C.; Scanu, B.; Fu, C.H.; Hsueh, K.L.; et al. Diversity of *Phytophthora* species in natural ecosystems of Taiwan and association with disease symptoms. *Plant Pathol.* **2017**, *66*, 194–211. [CrossRef]
22. Pérez-Sierra, A.; López-García, C.; León, M.; García-Jiménez, J.; Abad-Campos, P.; Jung, T. Previously unrecorded low-temperature *Phytophthora* species associated with Quercus decline in a Mediterranean forest in eastern Spain. *For. Pathol.* **2013**, *43*, 331–339. [CrossRef]
23. Rea, A.J.; Burgess, T.I.; Hardy, G.E.S.J.; Stukely, M.J.C.; Jung, T. Two novel and potentially endemic species of Phytophthora associated with episodic dieback of Kwongan vegetation in the south-west of Western Australia. *Plant Pathol.* **2011**, *60*, 1055–1068. [CrossRef]
24. Ruano-Rosa, D.; Schena, L.; Agosteo, G.E.; Magnano di San Lio, G.; Cacciola, S.O. Phytophthora oleae sp. nov. causing fruit rot of olive in southern Italy. *Plant Pathol.* **2018**, *67*, 1362–1373. [CrossRef]
25. Scanu, B.; Linaldeddu, B.T.; Deidda, A.; Jung, T. Diversity of *Phytophthora* Species from Declining Mediterranean Maquis Vegetation, including Two New Species, *Phytophthora crassamura* and *P. ornamentata* sp. nov. *PLoS ONE* **2015**, *10*, e0143234. [CrossRef]

26. Tkaczyk, M.; Nowakowska, J.A.; Oszako, T. *Phytophthora* species isolated from ash stands in Białowieża Forest nature reserve. *Forest Pathol.* **2016**, *46*, 660–662. [CrossRef]
27. Linaldeddu, B.T.; Scanu, B.; Maddau, L.; Franceschini, A. *Diplodia corticola* and *Phytophthora cinnamomi*: The main pathogens involved in holm oak decline on Caprera Island (Italy). *For. Pathol.* **2014**, *44*, 191–200. [CrossRef]
28. Scanu, B.; Hunter, G.C.; Linaldeddu, B.T.; Franceschini, A.; Maddau, L.; Jung, T.; Denman, S. A taxonomic re-evaluation reveals that *Phytophthora cinnamomi* and *P. cinnamomi* var. parvispora are separate species. *For. Pathol.* **2014**, *44*, 1–20.
29. Abad, Z.G.; Abad, J.A.; Cacciola, S.O.; Pane, A.; Faedda, R.; Moralejo, E.; Pérez-Sierra, A.; Abad-Campos, P.; Alvarez-Bernaola, L.A.; Bakonyi, J.; et al. *Phytophthora niederhauserii* sp. nov., a polyphagous species associated with ornamentals, fruit trees and native plants in 13 countries. *Mycologia* **2014**, *106*, 431–447. [CrossRef] [PubMed]
30. Brasier, C.M. The biosecurity threat to the UK and global environment from international trade in plants. *Plant Pathol.* **2008**, *57*, 792–808. [CrossRef]
31. Jung, T.; Blaschke, M. *Phytophthora* root and collar rot of alders in Bavaria: Distribution, modes of spread and possible management strategies. *Plant Pathol.* **2004**, *53*, 197–208. [CrossRef]
32. Jung, T.; Orlikowski, L.; Henricot, B.; Abad-Campos, P.; Aday, A.G.; Aguín Casal, O.; Bakonyi, J.; Cacciola, S.O.; Cech, T.; Chavarriaga, D.; et al. Widespread *Phytophthora* infestations in European nurseries put forest, semi-natural and horticultural ecosystems at high risk of Phytophthora diseases. *For. Pathol.* **2016**, *46*, 134–163. [CrossRef]
33. Migliorini, D.; Ghelardini, L.; Tondini, E.; Luchi, N.; Santini, A. The potential of symptomless potted plants for carrying invasive soilborne plant pathogens. *Divers. Distrib.* **2015**, *21*, 1218–1229. [CrossRef]
34. Pérez-Sierra, A.; Jung, T. Phytophthora in woody ornamental nurseries. In *Phytophthora: A Global Perspective*; Lamour, K., Ed.; CABI: Wallingford, UK, 2013; pp. 166–177.
35. Prigigallo, M.I.; Mosca, S.; Cacciola, S.O.; Cooke, D.E.L.; Schena, L. Molecular analysis of *Phytophthora* diversity in nursery-grown ornamental and fruit plants. *Plant Pathol.* **2015**, *64*, 1308–1319. [CrossRef]
36. Prigigallo, M.I.; Abdelfattah, A.; Cacciola, S.O.; Faedda, R.; Sanzani, S.M.; Cooke, D.E.L.; Schena, L. Metabarcoding analysis of *Phytophthora* diversity using genus-specific primers and 454 pyrosequencing. *Phytopathology* **2016**, *106*, 305–313. [CrossRef]
37. Català, S.; Pérez-Sierra, A.; Abad-Campos, P. The use of genus-specific amplicon pyrosequencing to assess *Phytophthora* species diversity using eDNA from soil and water in northern spain. *PLoS ONE* **2015**, *10*, e0119311. [CrossRef]
38. Dunstan, W.A.; Howard, K.; Stj. Hardy, G.E.; Burgess, T.I. An overview of Australia's *Phytophthora* species assemblage in natural ecosystems recovered from a survey in Victoria. *IMA Fungus* **2016**, *7*, 47–58. [CrossRef] [PubMed]
39. Nagel, J.H.J.H.; Slippers, B.; Wingfield, M.J.M.J.J.; Gryzenhout, M. Multiple *Phytophthora* species associated with a single riparian ecosystem in South Africa. *Mycologia* **2015**, *107*, 915–925. [CrossRef]
40. Reeser, P.W.; Sutton, W.; Hansen, E.M.; Remigi, P.; Adams, G.C. *Phytophthora* species in forest streams in Oregon and Alaska. *Mycologia* **2011**, *103*, 22–35. [CrossRef]
41. Stamler, R.A.; Sanogo, S.; Goldberg, N.P.; Randall, J.J. *Phytophthora* species in rivers and streams of the Southwestern United sStates. *Appl. Environ. Microbiol.* **2016**, *82*, 4696–4704. [CrossRef]
42. Jung, T.; Horta Jung, M.; Scanu, B.; Seress, D.; Kovács, G.M.; Maia, C.; Pérez-Sierra, A.; Chang, T.T.; Chandelier, A.; Heungens, K.; et al. Six new Phytophthora species from ITS Clade 7a including two sexually functional heterothallic hybrid species detected in natural ecosystems in Taiwan. *Persoonia Mol. Phylogeny Evolut. Fungi* **2017**, *38*, 100–135. [CrossRef] [PubMed]
43. Yang, X.; Tyler, B.M.; Hong, C. An expanded phylogeny for the genus *Phytophthora. IMA Fungus* **2017**, *8*, 355–384. [CrossRef]
44. Cooke, D.E.; Drenth, A.; Duncan, J.M.; Wagels, G.; Brasier, C.M. A molecular phylogeny of *Phytophthora* and related oomycetes. *Fungal Genet. Biol.* **2000**, *30*, 17–32. [CrossRef]
45. White, T.J.; Bruns, T.; Lee, S.; Taylor, J.W. Amplification and direct sequencing of fungal ribosomal RNA genes for phylogenetics. In *PCR Protocols: A Guide to Methods and Applications*; Innis, M.A., Gelfand, D.H., Sninsky, J.J., White, T.J., Eds.; Academic Press, Inc.: San Diego, CA, USA, 1990; Volume 18, pp. 315–322.

46. Safaiefarahani, B.; Mostowfizadeh-Ghalamfarsa, R.; Hardy, G.E.S.J.; Burgess, T.I. Re-evaluation of the *Phytophthora cryptogea* species complex and the description of a new species, *Phytophthora pseudocryptogea* sp. nov. *Mycol. Prog.* **2015**, *14*, 1–12. [CrossRef]
47. Scibetta, S.; Schena, L.; Chimento, A.; Cacciola, S.O.; Cooke, D.E.L. A molecular method to assess *Phytophthora* diversity in environmental samples. *J. Microbiol. Methods* **2012**, *88*, 356–368. [CrossRef]
48. Hansen, E.; Delatour, C. *Phytophthora* species in oak forests of north-east France. *Ann. For. Sci.* **1999**, *56*, 539–547. [CrossRef]
49. Vettraino, A.M.; Barzanti, G.P.; Bianco, M.C.; Ragazzi, A.; Capretti, P.; Paoletti, E.; Luisi, N.; Anselmi, N.; Vannini, A. Occurrence of *Phytophthora* species in oak stands in Italy and their association with declining oak trees. *Forest Pathology* **2002**, *32*, 19–28. [CrossRef]
50. Hüberli, D.; Hardy, G.E.S.J.; White, D.; Williams, N.; Burgess, T.I. Fishing for *Phytophthora* from Western Australia's waterways: A distribution and diversity survey. *Australasian Plant Pathol.* **2013**, *42*, 251–260. [CrossRef]
51. Oh, E.; Gryzenhout, M.; Wingfield, B.D.; Wingfield, M.J.; Burgess, T.I. Surveys of soil and water reveal a goldmine of *Phytophthora* diversity in South African natural ecosystems. *IMA Fungus* **2013**, *4*, 123–131. [CrossRef]
52. Shrestha, S.K.; Zhou, Y.; Lamour, K. Oomycetes baited from streams in Tennessee 2010-2012. *Mycologia* **2013**, *105*, 1516–1523. [CrossRef] [PubMed]
53. Jung, T.; Burgess, T.I. Re-evaluation of *Phytophthora citricola* isolates from multiple woody hosts in Europe and North America reveals a new species, *Phytophthora plurivora* sp. nov. *Persoonia* **2009**, *22*, 95–110. [CrossRef] [PubMed]
54. Maseko, B.; Burgess, T.I.; Coutinho, T.A.; Wingfield, M.J. Two new Phytophthora species from South African Eucalyptus plantations. *Mycol. Res.* **2007**, *111*, 1321–1338. [CrossRef]
55. Jung, T.; Hansen, E.M.; Winton, L.; Oswald, W.; Delatour, C. Three new species of *Phytophthora* from European oak forests. *Mycol. Res.* **2002**, *106*, 397–411. [CrossRef]
56. Scanu, B.; Vannini, A.; Franceschini, A.; Vettraino, A.M.; Ginetti, B.; Moricca, S. *Phytophthora* spp. in Mediterranean forests. In Proceedings of the Second International Congress of Silviculture, Florence, Italy, 26–29 November 2014; Ciancio, O., Ed.; Accademia Italiana di Scienze Forestali: Florence, Italy, 2014; pp. 402–407.
57. Vettraino, A.M.; Natili, G.; Anselmi, N.; Vannini, A. Recovery and pathogenicity of *Phytophthora* species associated with a resurgence of ink disease in *Castanea sativa* in Italy. *Plant Pathol.* **2001**, *50*, 90–96. [CrossRef]
58. Vettraino, A.M.; Morel, O.; Perlerou, C.; Robin, C.; Diamandis, S.; Vannini, A. Occurrence and distribution of *Phytophthora* species in European chestnut stands, and their association with Ink Disease and crown decline. *Eur. J. Plant Pathol.* **2005**, *111*, 169. [CrossRef]
59. Jung, T.; Hudler, G.W.; Jensen-Tracy, S.L.; Griffiths, H.M.; Fleischmann, F.; Osswald, W. Involvement of Phytophthora species in the decline of European beech in Europe and the USA. *Mycologist* **2005**, *19*, 159–166. [CrossRef]
60. Mrázková, M.; Černý, K.; Tomšovský, M.; Strnadová, V.; Gregorová, B.; Holub, V.; Pánek, M.; Havrdová, L.; Hejná, M. Occurrence of *Phytophthora multivora* and *Phytophthora plurivora* in the Czech Republic. *Plant Protect. Sci.* **2013**, *49*, 155–164. [CrossRef]
61. Orlikowski, L.B.; Ptaszek, M.; Rodziewicz, A.; Nechwatal, J.; Thinggaard, K.; Jung, T. *Phytophthora* root and collar rot of mature *Fraxinus excelsior* in forest stands in Poland and Denmark. *Forest Pathol.* **2011**, *41*, 510–519. [CrossRef]
62. Jung, T.; Blaschke, H. *Phytophthora* root rot in declining forest trees. *Phyton (Austria)* **1996**, *36*, 95–102.
63. Weiland, J.E.; Nelson, A.H.; Hudler, G.W. Aggressiveness of *Phytophthora cactorum*, *P. citricola* I, and *P. plurivora* from European Beech. *Plant Dis.* **2010**, *94*, 1009–1014. [CrossRef]
64. Barber, P.A.; Paap, T.; Burgess, T.I.; Dunstan, W.; Hardy, G.E.S.J. A diverse range of *Phytophthora* species are associated with dying urban trees. *Urban For. Urban Green.* **2013**, *12*, 569–575. [CrossRef]
65. Burgess, T.I.; White, D.; McDougall, K.M.; Garnas, J.; Dunstan, W.A.; Català, S.; Carnegie, A.J.; Worboys, S.; Cahill, D.; Vettraino, A.M.; et al. Distribution and diversity of *Phytophthora* across Australia. *Pac. Conserv. Biol.* **2017**, *23*, 150–162. [CrossRef]

66. Scott, P.M.; Burgess, T.I.; Barber, P.A.; Shearer, B.L.; Stukely, M.J.C.; Hardy, G.E.S.J.; Jung, T. *Phytophthora multivora* sp. nov., a new species recovered from declining *Eucalyptus, Banksia, Agonis* and other plant species in Western Australia. *Persoonia* **2009**, *22*, 1–13. [CrossRef]
67. Pane, A.; Granata, G.; Cacciola, S.O.; Puglisi, I.; Evoli, M.; Aloi, F.; La Spada, F.; Magnano di San Lio, G.; Zambounis, A. First Report of Root Rot of White Mulberry Caused by Simultaneous Infections of *Phytophthora megasperma* and *P. multivora* in Italy. *Plant Dis.* **2017**, *101*, 260. [CrossRef]
68. Szabó, I.; Lakatos, F.; Sipos, G. Occurrence of soilborne *Phytophthora* species in declining broadleaf forests in Hungary. *Eur. J. Plant Pathol.* **2013**, *137*, 159–168. [CrossRef]
69. Scarlett, K.; Daniel, R.; Shuttleworth, L.A.; Roy, B.; Bishop, T.F.A.; Guest, D.I. *Phytophthora* in the Gondwana Rainforests of Australia World Heritage Area. *Australasian Plant Pathol.* **2015**, *44*, 335–348. [CrossRef]
70. Crous, P.W.; Wingfield, M.J.; Burgess, T.I.; Hardy, G.E.S.J.; Barber, P.A.; Alvarado, P.; Barnes, C.W.; Buchanan, P.K.; Heykoop, M.; Moreno, G.; et al. Fungal Planet description sheets. *Persoonia Mol. Phylogeny Evolut. Fungi* **2017**, 558–624. [CrossRef]
71. Puglisi, I.; De Patrizio, A.; Schena, L.; Jung, T.; Evoli, M.; Pane, A.; Van Hoa, N.; Van Tri, M.; Wright, S.; Ramstedt, M.; et al. Two previously unknown *Phytophthora* species associated with brown rot of Pomelo (*Citrus grandis*) fruits in Vietnam. *PLoS ONE* **2017**, *12*, e0172085. [CrossRef]
72. Cacciola, S.O.; Pane, A.; Raudino, F.; Davino, S. First Report of Root and Crown Rot of Sage Caused by *Phytophthora cryptogea* in Italy. *Plant Dis.* **2002**, *86*, 1176. [CrossRef]
73. Cacciola, S.O.; Chimento, A.; Pane, A.; Cooke, D.E.L.; Magnano di San Lio, G. Root and Foot Rot of Lantana Caused by *Phytophthora cryptogea*. *Plant Dis.* **2005**, *89*, 909. [CrossRef]
74. Pane, A.; Agosteo, G.E.; Cacciola, S.O. *Phytophthora* species causing crown and root rot of tomato in southern Italy. *EPPO Bull.* **2000**, *30*, 251–255. [CrossRef]
75. Pane, A.; Faedda, R.; Cacciola, S.O.; Rizza, C.; Scibetta, S.; Magnano di San Lio, G. Root and Basal Stem Rot of Mandevillas Caused by *Phytophthora* spp. in Eastern Sicily. *Plant Dis.* **2010**, *94*, 1374. [CrossRef]
76. Pane, A.; Faedda, R.; Granata, G.; Puglisi, I.; Aloi, F.; La Spada, F.; Evoli, M.; Stracquadanio, C.; Cacciola, S.O. First Report of Root and Basal Stem Rot Caused by *Phytophthora cryptogea* and *P. inundata* on Dwarf Banana in Italy. *Plant Dis.* **2018**, *102*, 684. [CrossRef]
77. Brasier, C.M.; Cooke, D.E.L.; Duncan, J.M.; Hansen, E.M. Multiple new phenotypic taxa from trees and riparian ecosystems in Phytophthora gonapodyides-P. megasperma ITS Clade 6, which tend to be high-temperature tolerant and either inbreeding or sterile. *Mycol. Res.* **2003**, *107*, 277–290. [CrossRef]
78. Huai, W.X.; Tian, G.; Hansen, E.M.; Zhao, W.X.; Goheen, E.M.; Grünwald, N.J.; Cheng, C. Identification of Phytophthora species baited and isolated from forest soil and streams in northwestern Yunnan province, China. *Forest Pathol.* **2013**, *43*, 87–103. [CrossRef]
79. Nechwatal, J.; Bakonyi, J.; Cacciola, S.O.; Cooke, D.E.L.; Jung, T.; Nagy, Z.Á.; Vannini, A.; Vettraino, A.M.; Brasier, C.M. The morphology, behaviour and molecular phylogeny of *Phytophthora* taxon Salixsoil and its redesignation as *Phytophthora lacustris* sp. nov. *Plant Pathol.* **2013**, *62*, 355–369. [CrossRef]
80. Copes, W.E.; Yang, X.; Hong, C. *Phytophthora* Species Recovered From Irrigation Reservoirs in Mississippi and Alabama Nurseries and Pathogenicity of Three New Species. *Plant Dis.* **2015**, *99*, 1390–1395. [CrossRef] [PubMed]
81. Hong, C.X.; Gallegly, M.E.; Richardson, P.A.; Kong, P.; Moorman, G.W.; Lea-Cox, J.D.; Ross, D.S. *Phytophthora hydropathica*, a new pathogen identified from irrigation water, *Rhododendron catawbiense* and *Kalmia latifolia*. *Plant Pathol.* **2010**, *59*, 913–921. [CrossRef]
82. Hulvey, J.; Gobena, D.; Finley, L.; Lamour, K. Co-occurrence and genotypic distribution of *Phytophthora* species recovered from watersheds and plant nurseries of eastern Tennessee. *Mycologia* **2010**, *102*, 1127–1133. [CrossRef]
83. Pintos, C.; Rial, C.; Aguín, O.; Ferreiroa, V.; Mansilla, J.P. First report of *Phytophthora hydropathica* in river water associated with riparian alder in Spain. *New Dis. Rep.* **2016**, *33*, 25.
84. Vitale, S.; Luongo, L.; Galli, M.; Belisario, A. First Report of *Phytophthora hydropathica* Causing Wilting and Shoot Dieback on *Viburnum* in Italy. *Plant Dis.* **2014**, *98*, 1582. [CrossRef] [PubMed]
85. Jönsson, U.; Lundberg, L.; Sonesson, K.; Jung, T. First records of soilborne *Phytophthora* species in Swedish oak forests. *Forest Pathol.* **2003**, *33*, 175–179. [CrossRef]
86. Jonsson, U.; Jung, T.; Sonesson, K.; Rosengren, U. Relationships between health of *Quercus robur*, occurrence of *Phytophthora* species and site conditions in southern Sweden. *Plant Pathol.* **2005**, *54*, 502–511. [CrossRef]

87. Jung, T.; Cooke, D.E.L.; Blaschke, H.; Duncan, J.M.; Oßwald, W. *Phytophthora quercina* sp. nov., causing root rot of European oaks. *Mycol. Res.* **1999**, *103*, 785–798. [CrossRef]
88. Granata, G.; Pennisi, A.M. Estese morie di platani orientali in forestazioni naturali causate da *Ceratocystis fimbriata* (Ell. Et. Halst.) Davidson f. *platani* Walter. *Inf. Fitopatol.* **1989**, *12*, 59–61.
89. Cacciola, S.O.; Pane, A.; Cooke, D.E.L.; Raudino, F.; Magnano di San Lio, G. First Report of Brown Rot and Wilt of Fennel Caused by *Phytophthora megasperma* in Italy. *Plant Dis.* **2006**, *90*, 110. [CrossRef]
90. Cacciola, S.O.; Scibetta, S.; Pane, A.; Faedda, R.; Rizza, C. *Callistemon citrinus* and *Cistus salvifolius*, Two New Hosts of *Phytophthora* taxon *niederhauserii* in Italy. *Plant Dis.* **2009**, *93*, 1075. [CrossRef] [PubMed]

6

Decline of European Beech in Austria: Involvement of *Phytophthora* spp. and Contributing Biotic and Abiotic Factors

Tamara Corcobado [1,2], Thomas L. Cech [2], Martin Brandstetter [2], Andreas Daxer [2], Christine Hüttler [2], Tomáš Kudláček [1], Marília Horta Jung [1,3] and Thomas Jung [1,3,*]

[1] Phytophthora Research Centre, Mendel University in Brno, Zemědělská 1, 61300 Brno, Czech Republic; tamara.sanchez@mendelu.cz (T.C.); kudlak@seznam.cz (T.K.); marilia.jung@mendelu.cz (M.H.J.)

[2] Federal Research and Training Centre for Forests, Natural Hazards and Landscape, Unit of Phytopathology, Department of Forest Protection, Seckendorff-Gudent-Weg 8, 1131 Vienna, Austria; thomas.cech@bfw.gv.at (T.L.C.); martin.brandstetter@bfw.gv.at (M.B.); andreas.daxer@bfw.gv.at (A.D.); christine.huettler@bfw.gv.at (C.H.)

[3] Phytophthora Research and Consultancy, Am Rain 9, 83131 Nussdorf, Germany

* Correspondence: thomas.jung@mendelu.cz or dr.t.jung@gmail.com.

Abstract: A severe decline and dieback of European beech (*Fagus sylvatica* L.) stands have been observed in Austria in recent decades. From 2008 to 2010, the distribution and diversity of *Phytophthora* species and pathogenic fungi and pests were surveyed in 34 beech forest stands in Lower Austria, and analyses performed to assess the relationships between *Phytophthora* presence and various parameters, i.e. root condition, crown damage, ectomycorrhizal abundance and site conditions. In total, 6464 trees were surveyed, and *Phytophthora*-associated collar rot and aerial bark cankers were detected on 133 trees (2.1%) in 25 stands (73.5%). Isolations tests were performed from 103 trees in 27 stands and seven *Phytophthora* species were isolated from bleeding bark cankers and/or from the rhizosphere soil of 49 trees (47.6%) in 25 stands (92.6%). The most common species were *P. ×cambivora* (16 stands) followed by *P. plurivora* (eight stands) and *P. cactorum* (four stands), while *P. gonapodyides*, *P. syringae*, *P. psychrophila* and *P. tubulina* were each found in only one stand. Geological substrate had a significant effect on the distribution of *P. ×cambivora* and *P. plurivora* while *P. cactorum* showed no site preferences. In addition, 21 fungal species were identified on beech bark, of which 19 and five species were associated with collar rot and aerial bark cankers, respectively. Four tested fine root parameters showed differences between declining and non-declining beech trees in both *Phytophthora*-infested and *Phytophthora*-free stands. In both stand categories, ectomycorrhizal frequency of fine root tips was significantly higher in non-declining than in declining trees. This study confirmed the involvement of *Phytophthora* species in European beech decline and underlines the need of more research on the root condition of beech stands and other biotic and abiotic factors interacting with *Phytophthora* infections or causing beech decline in absence of *Phytophthora*.

Keywords: *Phytophthora ×cambivora*; *Phytophthora plurivora*; root rot; bark canker; ectomycorrhiza

---

1. Introduction

European beech (*Fagus sylvatica* L.) is a dominant broadleaved tree species of temperate forests with a distribution of 17 million hectares across Europe [1–3]. Populations of beech are exposed to biotic and abiotic stressors, which are predicted to increase in intensity and frequency. Predicted climatic trends with increasing temperature and drought will affect the habitat suitability of European beech [2–4]. These climatic extremes are expected to restrict its xeric boundaries in Central and Southern

Europe [5,6]. During the last century, beech forests have suffered from repeated outbreaks of pests such as the limantrid moth (*Dasychira pudibunda* L.; syn. *Calliteara pudibunda* L.) and the beech scale insect (*Cryptococcus fagisuga*) [7,8], which occurred particularly often in Central and Northern Europe. Cankers caused by *Neonectria ditissima* and *N. coccinea* do not pose a serious risk to beech trees, but in combination with beech scale infestations they can trigger the complex "Beech Bark Disease" resulting in severe mortality [9–12]. The main biotic threat to European beech is posed by pathogens from the oomycete genus *Phytophthora*. First outbreaks of severe decline and mortality of beech forests caused by *Phytophthora* spp. were reported during the 1930s from the UK [13,14]. In recent decades, numerous studies have reported decline and dieback of beech stands caused by a diverse range of *Phytophthora* spp. in 16 European countries and in the USA [15–38]. Currently, 17 *Phytophthora* species are known to be associated with European beech forests [20,39]. In Central Europe, after the drought and heatwave of summer 2003, a dramatic increase in declining beech trees, showing crown defoliation, fine root destructions, collar rot and aerial cankers and eventually mortality, was observed in *Phytophthora* infested areas [19,30,40]. The predicted climatic changes and the increasing spread of non-native invasive *Phytophthora* spp. with infested nursery stock into the wider environment are expected to exacerbate the situation and further destabilize the European forests of beech, oaks and other tree species [20–22,30,41–47].

Site conditions have a strong effect on water availability, and hence, the physiological and health status of trees, and influence the spread and infections of *Phytophthora* species via sporangia and zoospores. In general, *Phytophthora* species have a preference for a soil pH higher than 3.5 and sandy-loamy to clayey soil texture, but the ecological amplitudes of individual *Phytophthora* species can differ substantially from each other [20,22,30,45,48]. Therefore, more field research is needed to establish significant associations between site conditions and specific *Phytophthora* species.

Most trees depend on their mutualistic associations with ectomycorrhizal fungi to acquire nutrients such as nitrogen or phosphorus and, in exchange, provide the fungal partners with carbohydrates [49]. Long-term infected declining trees have decreased photosynthetic rates and carbon storage, negatively affecting ectomycorrhizal abundance [50,51]. On the other hand, ectomycorrhizal symbiosis can play an important role in the interaction between trees and *Phytophthora* root pathogens by suppressing *Phytophthora* infections [52].

Since two decades, in the forests of Eastern Austria, increasing numbers of declining beech trees with *Phytophthora*-typical bleeding bark cankers and root rot symptoms, secondary attacks by insects and root, bark and wood infecting fungi have been observed [20,24]. In the present study, in 34 beech forest stands in Lower Austria, the presence of bleeding bark cankers, the occurrence of *Phytophthora* and fungal pathogens in the rhizosphere and the bark of beech trees and site conditions were assessed. In addition, in 10 of these stands, crown transparency, fine root parameters and ectomycorrhizal abundance were analyzed in order to develop a better understanding of the possible relationships between the different factors.

2. Material and Methods

2.1. Study Sites and Field Assessments

The selected 34 forest sites cover the full spectrum of beech forest types and sites occurring across Lower Austria, comprising the hilly areas of the "Leithagebirge" and "Rosaliengebirge" at the border to the province of Burgenland, the Vienna Forest, the mountainous Lower Austrian Prealps and the adjacent hilly areas of the "Waldviertel" and "Weinviertel", both regions bordering the Czech Republic, as well as the region "Bucklige Welt" along the border to the Austrian province of Styria (Figure 1; Tables 1 and 2). Selection of sites was supported by local forest authorities and private forest owners reporting observations of crown thinning and/or *Phytophthora*-typical bleeding bark cankers. Selected sites varied in size from 1 to 8.6 ha and comprised only mature beech stands (>60 years old) with distinct crown thinning of varying degree.

Ten of the 34 sites were selected for detailed investigations, hereafter named intensive survey, while the remaining 24 sites were subjected to a more extensive survey. From 2008 to 2010, in total, 6464 beech trees in the 34 forest sites (50–500 trees per site) were visually assessed for the presence of *Phytophthora*-typical bleeding bark lesions on stems, collars or surface roots, for the presence of fungal fruiting bodies and symptoms caused by pathogenic fungi and for the presence and symptoms of pests, in particular the exit holes and breeding galleries of bark beetles (Figures 2 and 3). Since fungal species were only macroscopically identified several fungi could only be identified to genus level, i.e., *Neonectria*, *Armillaria* and *Hypoxylon*. A differential diagnosis was performed for the trees with bark lesions to clarify whether the lesions were caused by biotic or abiotic agents or mechanical injuries. For each site, general data like geographical coordinates, altitude and geological substrate, were collected (Tables 1 and 2, Table S1 (Supplementary Materials)).

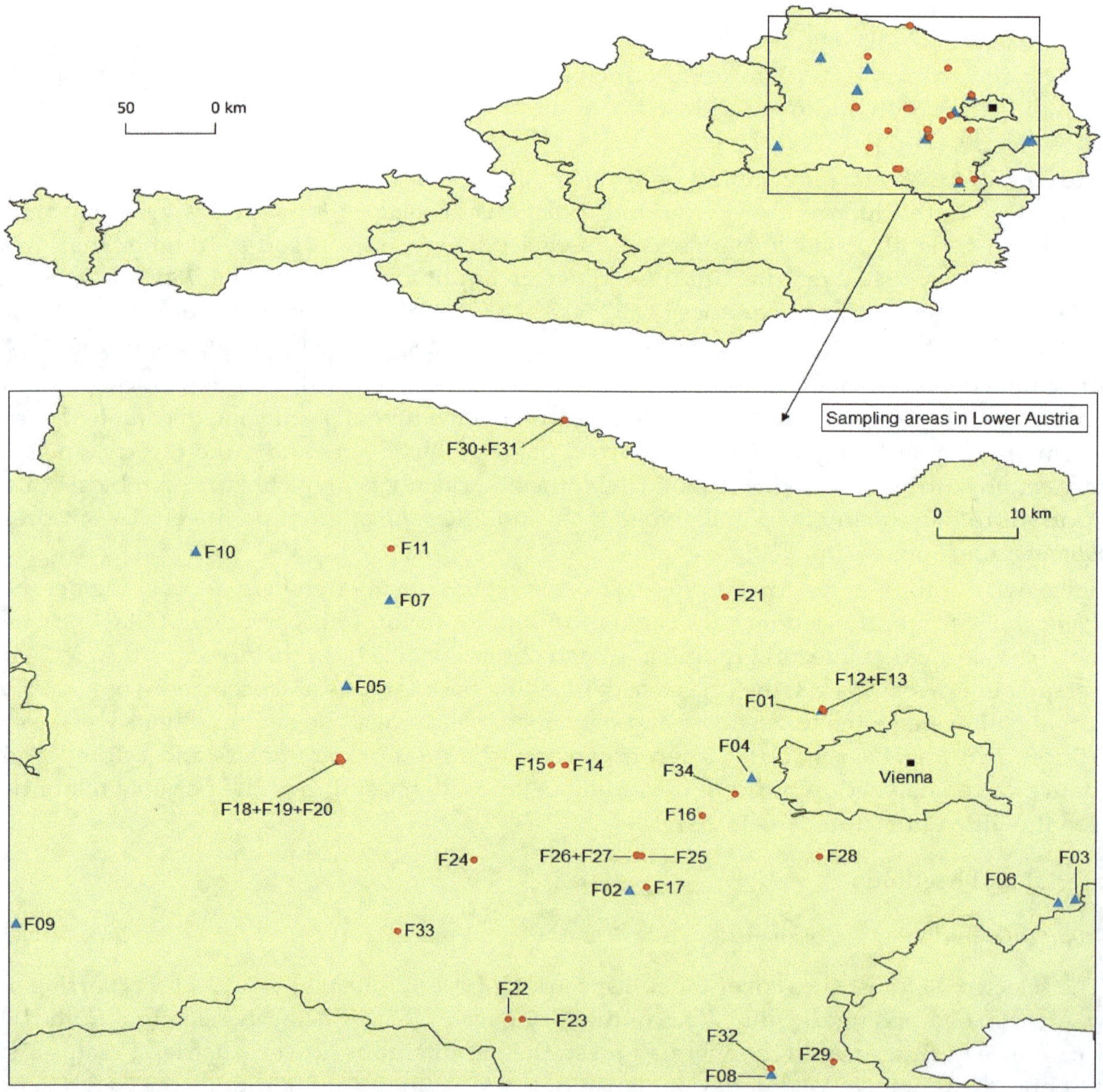

Figure 1. Location of Lower Austria within Austria (small rectangle) and the detailed location of the 10 intensively surveyed (F01–F10; blue triangles) and the 24 extensively surveyed (F11–F34; red dots) beech forest stands included in the *Phytophthora* survey in Lower Austria. For GPS coordinates, see Tables 1 and 2.

Table 1. Altitude, geological substrate, soil texture, mean pH and concentrations ($g \times Kg^{-1}$) of phosphorous, nitrogen and organic carbon in 10 intensively surveyed beech stands in Lower Austria, and *Phytophthora* species isolated from rhizosphere soil and bleeding bark cankers of declining and non-declining mature beech trees (n = 6–9 per stand).

Forest Stand		Altitude (m. a.s.l.)	Geological Substrate	Soil Texture	pH ($CaCl_2$)	P	N	Corg	*Phytophthora* spp. Isolated from	
no.	**Name**								**Rhizosphere Soil (no. of Trees)** [a]	**Collar Rot and Aerial Cankers (no. of Trees)** [a]
F01	Hadersfeld 1	351	Sandstone	Sandy loam	3.9	0.3	1.1	13.0	CAM (1) [b,c]	
F02	Hocheck 1	992	Limestone	Sandy loam	7.4	1.7	9.1	113.0		
F03	Kaiserstein-bruch	237	Limestone	Clay	7.2	0.3	2.8	39.5	PLU (1), SYR (1) [b]	PLU (1) [d]
F04	Purkersdorf	406	Claystone and sandstone	Loamy sand	4.2	0.3	1	13.8	CAM (2) [b]	
F05	Rossatz	614	Gneiss	Clayey sand	3.7	0.3	1.1	22.8	PLU (1) [b]	
F06	Sommerein	308	Schist	Loamy silt	4.4	0.2	1.3	20.0	CAC (2), PSY (1) [bc]	CAC (1) [e]
F07	Stackelberg	527	Gneiss	Clayey sand	3.8	1.0	1.0	17.0	CAM (1) [b,c]	
F08	Thernberg 2	560	Limestone	Sandy silt	7.5	0.6	3.6	57.5	PLU (1)	PLU (1) [e]
F09	Ybbsitz	1061	Limestone	Clayey loam	7.2	0.9	5.9	82.3	[b]	
F10	Zwettl	519	Granodiorite	Clayey sand	4.2	2.7	3.2	36.0	CAM (2) [b]	CAM (1) [e]

[a] CAC = *P. cactorum*, CAM = *P. ×cambivora*, PLU = *P. plurivora*, PSY = *P. psychrophila*, SYR = *P. syringae*. [b] *Globisporangium* sp. also isolated. [c] *Saprolegnia* sp. also isolated. [d] Aerial bleeding canker. [e] Bleeding collar rot lesion.

Table 2. Isolation of *Phytophthora* species from rhizosphere soil samples and bleeding bark cankers of declining and non-declining mature beech trees in 24 extensively surveyed forest stands in Lower Austria.

Forest Stand		Altitude (m a.s.l.)	Geological Substrate	*Phytophthora* spp. Isolated from	
no.	Name			Rhizosphere Soil (no. of Trees) [a]	Collar rot and Aerial Cankers (no. of Trees) [a]
F11	Fuglau	420	Schist	n.a.	CAM (1) [b]
F12	Hadersfeld 2	380	Sandstone	n.a.	CAM (4) [b]
F13	Hadersfeld 3	313	Sandstone	n.a.	n.a.
F14	Haspelwald 1	363	Claystone	CAM (1)	n.a.
F15	Haspelwald 2	353	Claystone	n.a.	CAM (1) [b]
F16	Hengstlberg	530	Claystone	n.a.	CAM (6) [b,c]
F17	Hocheck 2	630	Limestone	n.a.	PLU (2) [b,d], negative (1) [b]
F18	Hohenegg 1	390	Gneiss	n.a.	CAM (2) [b]
F19	Hohenegg 2	397	Gneiss	CAM (1), GON (1), PLU (1), TUB (1)	n.a.
F20	Hohenegg 3	481	Gneiss	n.a.	n.a.
F21	Hollabrunn	309	Alluvial deposits	n.a.	n.a.
F22	Höllental-Schwarza	562	Limestone	CAC (1) [e]	n.a.
F23	Höllental-Weichtalklamm	968	Limestone	n.a.	n.a.
F24	Kerschenbach	434	Claystone and sandstone	CAM (1) [e]	negative (1) [b]
F25	Kleinmariazell 1	531	Claystone	n.a.	n.a.
F26	Kleinmariazell 2	581	Claystone	n.a.	n.a.
F27	Kleinmariazell 3	637	Claystone	n.a.	CAM (1), negative (1) [b]
F28	Mödling Richardhof	457	Limestone	PLU (2) [f]	n.a.
F29	Rosalia		Gneiss	n.a.	CAM (1) [b]
F30	Thayatal 1	360	Limestone	CAC (1), PLU (1) [e]	n.a.
F31	Thayatal 2	330	Gneiss	CAM (1)	n.a.
F32	Thernberg 1	394	Schist	n.a.	PLU (1) [b]
F33	Türnitz	1255	Limestone	n.a.	n.a.
F34	Wienerwaldsee	338	Claystone	CAC (1)	CAM (1)

[a] CAC = *P. cactorum*, CAM = *P.* ×*cambivora*, GON = *P. gonapodyides*, PLU = *P. plurivora*, TUB = *P. tubulina*. [b] Bleeding collar rot lesions. [c] Bleeding bark lesions on surface roots. [d] Aerial bleeding canker. [e] *Globisporangium* sp. also isolated. [f] Both trees with inactive aerial cankers. n.a. = isolation test not attempted.

Declining beech trees show deteriorations of the crown structure characterized by long-term stunted growth of shoots, clustering of lateral branches around the major branches and at the ends of branches leading to brush- and claw-like structures, excessive losses of lateral twigs and small branches and as a result, crown transparency >25% (Jung 2009). In 2008, in each of the 10 intensively surveyed stands, three declining beech trees showing the aforementioned crown deteriorations and crown transparency ≥30% and three healthy trees with crown transparency ≤20% were selected. The 60 sample trees had no bleeding bark cankers. The crown transparency of the 60 trees was recorded in summer 2008 according to [53].

2.2. Sampling and Isolation Procedures, Species Identification

The assessment of *Phytophthora* presence was performed in 27 of the 34 beech stands between 2008 and 2010 by sampling rhizosphere soil and bleeding bark lesions according to [30,44].

2.2.1. Soil Sampling and Baiting

At the 10 sites selected for the intensive survey, rhizosphere soil samples were taken in 2008 from each three healthy and declining beech trees per site, and in 2010, in stands F05 and F10 from three additional declining trees with collar rot lesions (in total 63 trees). In eight of the other 24 stands, soil samples were taken between 2008 and 2010 from 13 trees with inactive, dry dark-brown *Phytophthora*-typical bark lesions, from which direct isolation attempts are usually not promising [30]. Three monoliths (ca. 30 × 30 × 30 cm) of rhizosphere soil were excavated around each sample tree at 50–100 cm distance from the stem. After the removal of the upper organic soil layer (ca. 5–10 cm) which is usually not infested by *Phytophthora* [21], mineral soil was taken from all the monoliths and mixed into a bulked sample of ca. 2 l per tree. For the baiting tests, each soil sample was carefully mixed and flooded with distilled water so that the soil was covered by ca. 3 cm of water. Young soft leaves from European beech as well as from cork oak (*Quercus suber*), pedunculate oak (*Q. robur*) and Turkey oak (*Q. cerris*) were used as baits, floating on the water surface for 3–7 days [21,44]. Leaves developing necrotic areas were checked under the light microscope at ×80 magnification for *Phytophthora* sporangia and then plated onto selective PARPNH-agar [44] and incubated at 20 °C in the dark.

2.2.2. Isolations from Bleeding Collar Rot and Aerial Bark Lesions

Isolations were performed from bleeding bark lesions of 27 beech trees in 15 stands (Tables 1 and 2). Approximately 10–15 cm long bark samples, including phloem and cambium, were taken from the upper lesion margins using a hammer and a chisel, immediately placed in distilled water and taken to the laboratory. As a first attempt to detect the presence of *Phytophthora*, a quick test consisting of a lateral flow-device (Pocket Diagnostic Kit *Phytophthora*, CSL now FERA, Sand Hutton, York YO41 1LZ) was performed using a small amount of the inner bark lesions. In case of a positive result, the remaining bark sample was incubated in distilled water at 16–18 °C for 2–3 days. The distilled water was replaced three times per day in order to remove the polyphenols released by the bark. Then, in the case of active lesions with an orange-brown, flamed appearance of the inner bark, 3–5 mm pieces of inner bark were blotted dry on filter paper and plated onto selective PARPNH and incubated at 20 °C in the dark [30,54]. Inactive, dry dark-brown lesions were shredded and the small pieces flooded with distilled water and baited with oak and beech leaflets (Figure 3d). The water was replaced daily in order to remove excess polyphenols and decrease bacterial populations [30,54]. Isolations from necrotic baiting leaves were performed as described for soil baitings. With one inactive bark lesion in stand F17 both direct plating and baiting were used in parallel.

Beginning at 12h after plating, all PARPNH plates with plated baiting leaves or bark tissues were regularly checked under the stereomicroscope at ×20 for developing *Phytophthora* colonies which were transferred onto V8 juice agar (V8A) [44] and after 3 weeks growth at 20 °C, they were stored at 8 °C in the fridge.

Phytophthora species were identified by examining the morphological characters of sporangia, oogonia, antheridia, chlamydospores and hyphal swellings under the light microscope at ×320 magnification, and comparing them together with colony growth patterns on V8A and carrot agar (CA) [55] and cardinal temperatures of growth on V8A with descriptions in the Literature [48,56–59]. Sporangia were produced by cutting discs (15 mm diam) from the growing edge of a 5–7 days old culture grown on V8A at 20 °C in the dark, and immersing them in non-sterile soil extract water [44]. In case the classical species identification was ambiguous, the isolates were identified using ITS DNA sequence analysis according to [60,61]. ITS sequences from representative isolates of all *Phytophthora* species and *cox1* sequences of representative isolates of new species obtained in this study were deposited at GenBank and accession numbers are given in Supplementary Table S2 (Supplementary Materials).

2.3. Root Biometry

Biometrical analyses of roots were performed in the 10 intensively surveyed sites following the method described by [45]. At each site, in 2008, four of the six selected trees (two declining and two healthy trees) were sampled. All the roots with diameters ≤5 mm were collected from the three soil–root monoliths sampled per tree for the soil baitings (see Section 2.2.1) and mixed to one sample per tree. The root samples were transported in sealed plastic zip bags to the laboratory where they were immersed in water for 12–18 h. Then, the roots were washed thoroughly in running tap water and subsequently cleaned of adhering soil particles in an ultrasonic cleaning box for 10 min. After the removal of roots from other plant species, the roots were divided into five sub-samples per beech tree which were then immersed in a water tub and scanned in a root scanner (Epson Transparency Unit, model EU-22, Meerbusch, Germany).

The scans were analyzed using the program WINRHIZO 2004a (Regent Instruments INC., Ch Saint-foy, QC, Canada). The roots were separated into fine roots (≤2 mm diameter) and coarse roots (2–5 mm diameter) and, according to [45], the following parameters were measured: total fine root length (FRL); total coarse root length (CRL); total fine root surface (FRS); total coarse root surface (CRS); total number of fine root tips (FRT). These parameters were used to calculate the root ratio parameters FRL/CRL, FRS/CRL, FRT/CRL and FRT/CRS. In order to allow the comparisons between different stands, the influence of different site and climatic conditions was reduced by calculating relative root ratio parameters (FRL/CRL_{rel}, FRS/CRL_{rel}, FRT/CRL_{rel}, FRT/FRL_{rel}, FRT/CRS_{rel}) by using the site-specific mean values of the non-declining sample trees from the respective sites as reference values (100%) [45].

2.4. Ectomycorrhizal Frequency

Following the root scanning, the ectomycorrhizal frequency of fine root tips in the 40 declining and non-declining sample trees of the 10 intensively surveyed stands was assessed. The water tubs with the immersed roots of a subsample were underlaid by a grid of 2 × 2 cm, and 502 × 2 cm units per subsample were randomly selected. In each unit, the number of ectomycorrhized root tips (MT) with a turgid mantle and a shiny appearance and non-mycorrhized root tips (NMT) [62] was assessed under the stereomicroscope at ×40, and the ratio MT/NMT was calculated for each tree. Then, analogous to the root ratio parameters (see Section 2.3), the relative MT/NMT_{rel} ratios were calculated using the mean MT/NMT ratio of the non-declining trees from the respective site as a site-specific reference value (100%).

2.5. Climate, Geological Substrate and Soil Analysis

For the 10 intensively surveyed stands, the long-term mean precipitation and temperature values for 30 years from the meteorological stations closest to the surveyed sites were provided by the Hydrographischer Dienst Österreich, Abteilung I/3-Wasserhaushalt (HZB; Vienna, Austria). For eight stands, the mean annual precipitation (522–779 mm) and mean monthly precipitation during the vegetation period April–September (60.8–76.7 mm) was comparable to each other whereas the two high altitude sites F02 Hocheck 1 (1650.5 and 163.9 mm, respectively) and F09 Ybbsitz (1450 and 139.2 mm, respectively) received considerably higher precipitation (Table S4 (Supplementary Materials)). Mean annual temperatures and mean temperature values for the vegetation period of the 10 stands ranged from 6.4 to 9.6 °C and from 12.5 to 16.4 °C, respectively, with site F02 Hocheck 1 being the coldest and sites F03 Kaisersteinbruch 1 and F06 Sommerein being the warmest sites (Table S3 (Supplementary Materials)).

Data on the geological substrates of the 34 surveyed beech forests were obtained from detailed geological maps (Geological Maps 1:50,000 and 1:200,000, Geological Survey of Austria, Geologische Bundesanstalt, Vienna, Austria). The sites showed a wide range of geological substrates representative for the distribution of beech forests in Lower Austria (Tables 1 and 2).

A detailed analysis of soil physical and chemical properties was performed for the 10 intensively surveyed stands F01–F10 (Table 1). Soil parameters included soil texture, i.e. percentages of sand, clay and silt, pH (measured in calcium chloride = pH_{CaCl2}), and the total contents of N, P and organic carbon (C_{org}). The soil samples were homogenized, sieved (2 mm mesh) and air-dried. Soil dry weight was determined by thermogravimetry (Moisture Analyzer HR73, Mettler Toledo, Columbus, OH, USA). Total soil carbon (C_t) and total nitrogen (N_t) contents were determined by dry combustion using a LECO C/N TruMAC Analyzer (LECO, Saint Joseph, MI, USA). Carbonate ($CaCO_3$) content was measured using a Scheibler calcimeter. Total contents of C_{org} were calculated as the difference between the contents of C_t and $CaCO_3$. Total P was determined using an acid-assisted (HNO_3) microwave leaching plus ICP-OES analysis (Perkin Elmar Optima 8300 (Perkin Elmar Inc., Waltham, MA, USA).

2.6. Statistical Analysis

Statistical analyses were performed in R software statistical package 3.6.2 (R Core Team, 2019; https://www.r-project.org/) and CANOCO 5 [63]. For all tests, the significance level $\alpha = 0.05$ was used.

The relationship between the geological substrate and *Phytophthora* spp. assemblages was studied by multivariate statistical methods. As the gradient length tested by detrended correspondence analysis (DCA) showed the unimodal nature of the data, the canonical correspondence analysis (CCA) was used to assess the links between the individual *Phytophthora* species and the geological substrate. To eliminate the distorting effect of rare species, which is an issue for all unimodal methods, rare species were down weighted using the preset function of the software.

The effect of the geological substrate on *Phytophthora* spp. occurrence was tested by two different tests: ANOSIM (analysis of similarities) and PERMANOVA (permutational multivariate analysis of variance), both implemented in the R package vegan. Statistical significance was based on 9999 permutations. *Phytophthora* species contributing most to the dissimilarity between individual geological substrates, i.e. being most responsible for the differences in *Phytophthora* species assemblages between geological substrates, were identified by similarity percentage breakdown (SIMPER).

The relationship between tree decline, root parameters and ectomycorrhizal frequency was analyzed by means of generalized linear mixed models (GLMM) with Gamma distribution and inverse link function implemented in the R package lme4 and follow-up type III likelihood-ratio test. Multiple comparisons were carried out using the Tuckey's post hoc test. Correlations between the root parameters, ectomycorrhizal frequency and crown transparency were assessed by Kendall's Tau.

3. Results

3.1. *Phytophthora*-Related Disease Symptoms

In beech stands infested by *Phytophthora* species, in particular *P.* ×*cambivora* and *P. plurivora*, individual trees and groups of trees showed thinning and dieback of crowns (Figure 2a). Bleeding bark cankers on surface roots (Figure 2b) were found on individual trees while open callusing lesions on coarse roots and small woody roots (Figure 2c) and extensive losses of lateral roots and fine roots (Figure 2d) were widespread.

In total, 6464 trees were surveyed in the 34 beech stands. *Phytophthora*-typical collar rot (Figure 2e) and aerial bleeding bark cankers (Figure 2f) were detected on 133 trees (2.1%) in 25 stands (73.5%) (Table S1 (Supplementary Materials)). Collar rot cankers were observed in 87 trees (1.3%) in 24 stands (70.6%) including, seven of the 10 intensively surveyed stands and 17 of the 24 extensively surveyed stands. Aerial cankers were found in 46 trees (0.7%) in 13 stands (38.2%) including four and nine intensively and extensively surveyed stands, respectively (Table S1).

Collar rots and aerial bleeding cankers were characterized by tongue-shaped necrosis of the inner bark and the cambium, with orange-brown discoloration in active cankers and dark-brown discoloration in inactive cankers, and tarry or rusty spots on the surface of the bark (Figure 2e,f). Collar rots usually extended 1–2 m from the stem base (Figure 2e), but in some cases, could reach up to 5 m. Aerial bleeding cankers were observed along the stems up into the canopy, had no particular orientation and were often located above collar rots and below stem forks (Figure 2f). In three beech stands on temporarily waterlogged sites with high clay contents (F04 Purkersdorf, F16 Hengstlberg, F26 Kleinmariazell 2) and in stand F08 Thernberg 2, individual trees and groups of trees were recently wind-thrown (Figure 2g).

Most of these trees showed severe destructions of the root system with extensive losses of lateral and fine roots and numerous lesions on woody roots (Figure 2c,d). Often beech trees with collar rots were concentrated along forest roads (Figure 3a,b). In several stands, the breeding of bark beetles, in particular *Taphrorychus bicolor*, in active *Phytophthora* lesions could be observed (Figure 3c).

Figure 2. Decline and dieback symptoms caused by *Phytophthora* spp. on *Fagus sylvatica* in Lower Austria; (**a**) severe dieback, chlorosis and mortality due to root and collar rot caused by *Phytophthora ×cambivora* in stand F01 Hadersfeld 1; (**b**) bleeding bark lesions caused by *P. ×cambivora* on surface roots in stand F16 Hengstlberg; (**c**) coarse roots of wind-thrown beech trees in stand F16 Hengstlberg, with severe losses of lateral roots and open callusing lesions caused by *P. ×cambivora*; (**d**) small woody roots with extensive losses of lateral roots and fine roots caused by *P. plurivora* and *P. syringae* in stand F03 Kaisersteinbruch; (**e**) bleeding collar rot lesion caused by *P. ×cambivora* in stand F34 Wienerwaldsee; (**f**) aerial bleeding canker caused by *P. plurivora* in stand F19 Hohenegg 2; (**g**) wind-thrown beech trees in stand F16 Hengstlberg due to severe root losses and root lesions caused by *P. ×cambivora*.

Figure 3. (**a**,**b**) Bleeding collar rot lesions caused by *Phytophthora* ×*cambivora* on *Fagus sylvatica* trees along forest roads in stand F15 Haspelwald 2; (**c**) surface root of a beech tree in stand F16 Hengstlberg with exit holes of the bark beetle *Taphrorychus bicolor* being mostly connected to the bleeding bark lesions caused by *P.* ×*cambivora*; (**d**) baiting test from an inactive aerial canker of a beech tree in stand F17 Hocheck 2 with lesions on *F. sylvatica* baiting leaves caused by *P. plurivora*.

3.2. Distribution of Phytophthora Species and Their Association with Disease Symptoms

Phytophthora isolation tests from bark cankers and rhizosphere soil using baiting and direct isolation methods were performed with 103 beech trees in 27 stands, and *Phytophthora* species were isolated from 49 trees (47.6%) in 25 stands (92.6%) (Tables 1 and 2). The most common species was *P.* ×*cambivora*, which was obtained from 28 trees in 16 stands, followed by *P. plurivora* (12 trees in eight stands) and *P. cactorum* (six trees in four stands). *Phytophthora gonapodyides*, *P. syringae*, *P. psychrophila* and *P. tubulina* were each isolated from only one stand (Tables 1 and 2).

From bleeding collar rot and aerial bark lesions, isolations were attempted from 27 trees in 15 stands (Tables 1 and 2) and three *Phytophthora* species were isolated from 24 trees (88.9%) in 14 stands (93.3%). *Phytophthora* ×*cambivora* was the most common species isolated from 19 trees in nine stands, whereas *P. plurivora* and *P. cactorum* were recovered from four trees in four stands and one tree in one stand, respectively. In one stand (F17 Hocheck 2), *P. plurivora* could be baited from an inactive aerial bark lesion (Figure 3d) while the direct plating of necrotic canker tissue on PARPNH-agar failed.

Rhizosphere soil was collected from 63 beech trees in the 10 intensively surveyed stands and from 13 beech trees with inactive dry bark cankers and/or severe crown dieback in eight of the 24 extensively surveyed stands. Phytophthoras were present in the rhizosphere soil from 25 trees (32.9%) in 16 stands (88.9%; each eight intensively and extensively surveyed stands; Tables 1 and 2, Table S4 (Supplementary Materials)). The most common species were *P.* ×*cambivora* with 10 trees in eight stands and *P. plurivora* with seven trees in six stands. *Phytophthora cactorum* was found in five trees in four stands while *P. gonapodyides*, *P. syringae*, *P. psychrophila* and *P. tubulina* were each isolated only once.

3.3. Influence of Geological Substrate, Soil Texture and Soil pH on Phytophthora Distribution

The 34 beech stands were growing on six main geological substrates, including limestone (10 stands), claystone (nine stands), gneiss/granodiorite (eight stands), sandstone (five stands), schist (three stands) and alluvial deposits (one stand) (Tables 1 and 2). In the 10 intensively surveyed stands, *P.* ×*cambivora* was present at four sites with acidic soil reaction (pH_{CaCl2} 3.8–4.2) and clayey and sandy texture of the soils which were derived from claystones, gneiss, granodiorite and sandstones (Table 1). Additionally, in the 24 extensively surveyed stands, *P.* ×*cambivora* was only found on geological substrates which form acidic soils with high contents of clay and sand, including claystones, gneiss, sandstones and schist (12 stands; Table 2). *Phytophthora* ×*cambivora* was not recovered from any of the 10 limestone sites. *Phytophthora plurivora* showed a different ecological amplitude occurring in three intensively surveyed stands with soils of different textures (clayey sand, clay and sandy silt) and pH_{CaCl2} values of 3.7, 7.2 and 7.5, respectively (Table 1). Over all 34 stands, the eight sites from which *P. plurivora* was recovered were situated on gneiss (two sites), limestone (five sites) and schist (Tables 1 and 2). The four stands with presence of *P. cactorum* were located on claystone, limestone (two sites) and schist. Since *P. gonapodyides*, *P. psychrophila*, *P. syringae* and *P. tubulina* were each detected in only one stand, their association with site conditions remains unclear.

The CCA revealed that geological substrate explained 29.9% of the total variability in *Phytophthora* distribution. There is a clear association between *P.* ×*cambivora* and sandstone and between *P. plurivora* and limestone, while *P. cactorum* shows no evident substrate preference (Figure 4). The results of ANOSIM (R = 0.3316; $p \leq 0.001$) and PERMANOVA (F = 7.505; $p \leq 0.001$) confirm that the geological substrate has a significant effect on *Phytophthora* species distribution. Pairwise PERMANOVA revealed significant differences in *Phytophthora* species occurrence between claystone and limestone (F = 34.138; $R^2 = 0.7563$; $p = 0.007$), gneiss/granodiorite and limestone (F = 12.208; $R^2 = 0.5260$; $p = 0.02$) and sandstone and limestone (F = 22.788; $R^2 = 0.7341$; $p = 0.02$), respectively. SIMPER results further specify that for all these substrate pairs, the differences in *Phytophthora* distribution are given by the presence/absence of *P.* ×*cambivora* and *P. plurivora*, i.e. these two species are responsible for 84.5%, 83.7% and 86.1%, respectively, of the pairwise differences in *Phytophthora* assemblages between these substrates.

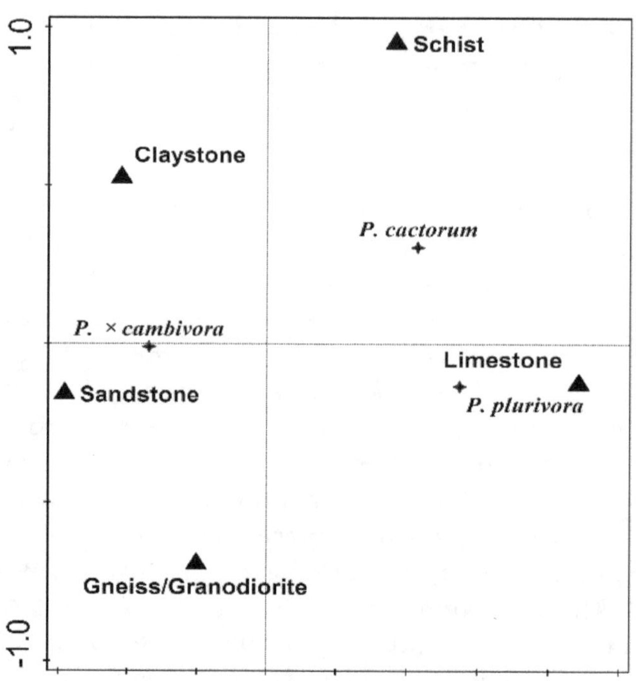

Figure 4. Canonical correspondence analysis (CCA) of *Phytophthora* community structure (CAC = *P. cactorum*, CAM = *P.* ×*cambivora*, PLU = *P. plurivora*) according to the geological substrate in 34 beech stands in Lower Austria.

3.4. Fungal Pathogens and Pests

In 21 of the 34 beech stands, a total of 21 fungal species were recorded on various types of bark damages and on dead wood (Table S5 (Supplementary Materials)). In 19 stands, 18 fungal species were associated with collar rot cankers, in many cases as secondary invaders of lesions where a *Phytophthora* species was isolated from the active lesion margins (Figure 5). With 12 stands, *Armillaria* spp. which were not identified to the species level, were most common, followed by *Fomes fomentarius* (10 stands), *Trametes* spp. and *Hypoxylon* sp. (each eight stands), *Schizophyllum commune* (six stands) and *Ustulina deusta* (four stands). Other aggressive secondary bark pathogens like *Bjerkandera adusta* and *Neonectria* sp./*Cylindrocarpon* sp. were more rare (Table S5 (Supplementary Materials); Figure 5). Eight of the 19 fungal invaders of primary *Phytophthora* collar rot lesions were also colonizing mechanical injuries on buttroots and stems (Table S5 (Supplementary Materials)). In three stands, *F. fomentarius*, *Inonotus nidus-pici*, *Oudemansiella mucida* and *Polyporus squamosus* were secondary invaders of aerial *Phytophthora* bark cankers. *Neonectria ditissima* was locally widespread in the stands in Hohenegg, causing in twigs, branches and in rare cases, also stems, typical perennial cankers which were not related to primary *Phytophthora* infections. The wood decay fungus *S. commune* was typically associated with sunburn injuries in five stands.

Figure 5. Secondary fungal bark infections on *Fagus sylvatica* trees infected by *Phytophthora* spp. in Lower Austria: (**a**) bleeding collar rot lesion (white arrows) caused by *P.* ×*cambivora* in stand F16 Hengstlberg with secondary infection by *Cylindrocarpon* sp./*Neonectria* sp. (red arrows); (**b**) white pycnidia of *C. candidum* (red arrows) and red perithecia of *N. coccinea* in stand F16 Hengstlberg; (**c**) bleeding collar rot lesion (white arrow) caused by *P.* ×*cambivora* in stand F12 Hadersfeld 2 with fruiting bodies of *Bjerkandera adusta*; (**d**,**e**) aerial bleeding cankers (white arrows) caused by *P. cactorum* in stand F06 Sommerein with fruiting bodies of the secondary invaders *Oudemansiella mucida* (red arrows) and *Fomes fomentarius* (yellow arrows); (**f**) fruiting bodies of the secondary invader *Fomitopsis pinicola* above a collar rot lesion caused by *P.* ×*cambivora* (not shown) on a beech tree in stand F04 Purkersdorf.

In five stands (F01 Hadersfeld 1, F04 Purkersdorf, F07 Stackelberg, F12 Hadersfeld 2, F16 Hengstlberg), in 26 of the 33 trees with *Phytopthora*-collar rot lesions, secondary attacks by the opportunistic bark beetle *Taphrorychus bicolor* were detected. In the stands F04 Purkersdorf and F16 Hengstlberg, in three trees, numerous exit holes of *T. bicolor* were detected in actively bleeding collar rot lesions from which *P.* ×*cambivora* could be isolated (Figure 3c).

3.5. Relationship between Root Parameters, Ectomycorrhizal Frequency, Crown Transparency and Phytophthora Infestation in Ten Intensively Surveyed Beech Stands

In the eight *Phytophthora*-infested sites, declining sample trees had 38% higher crown transparency than non-declining sample trees. In the two sites without *Phytophthora* infestation, the difference was 23.4%. In both stand categories, this difference was statistically significant ($p \leq 0.001$; Table 3).

In the eight *Phytophthora*-infested stands, the fine root status of the non-declining trees was better than in the declining trees. The relative root ratio parameters FRL/CRL_{rel}, FRS/CRS_{rel}, FRT/CRL_{rel}, FRT/FRL_{rel}, FRT/CRS_{rel} of the 16 non-declining trees were 7.8–13.4% higher compared to the 16 declining trees. For the parameters FRL/CRL_{rel} and FRS/CRS_{rel}, the differences between the declining and non-declining trees were statistically significant ($p \leq 0.05$) and non-significant ($p \leq 0.1$), respectively. However, due to low sample numbers and relatively high variation for the parameters FRT/CRL_{rel} and FRT/CRS_{rel}, the differences between declining and non-declining trees showed no statistical significance (Table 3). In the two *Phytophthora*-free high-altitude stands, the differences of all relative root ratio parameters between the four non-declining and the four declining trees were even higher (21.4–27.1%) than in the *Phytophthora*-infested stands (Table 3).

In both the *Phytophthora*-infested and *Phytophthora*-free stands, the non-declining trees showed 57.3% and 64.3% higher ectomycorrhizal frequency of fine root tips than the declining trees. However, in both stand categories, the differences in the relative ratio mycorrhized/non-mycorrhized fine root tips (MT/NMT_{rel}) between non-declining and declining trees were statistically not significant (Table 3). In both stand categories, the ectomycorrhizal frequency (MT/NMT) was not correlated with any of the root parameters.

Table 3. Crown transparency and relative values of root parameters and ectomycorrhizal frequency of healthy and declining beech trees in eight *Phytophthora*-infested and two non-infested forest stands in Lower Austria, and the significance of differences (Type III likelihood-ratio test).

Forest Stands	*Phytophthora* spp. [a]	Crown Transparency (%)		FRL/CRL_{rel} [b]		FRS/CRS_{rel} [b]		FRT/CRL_{rel} [b]		FRT/CRS_{rel} [b]		MT/NMT_{rel} [b]	
8 *Phytophthora*-Infested Stands (16 Healthy and 16 Declining Beech Trees)													
		H	D	H	D	H	D	H	D	H	D	H	D
F01 Hadersfeld 1	CAM	10.0	37.5	100.0	118.1	100.0	142.9	100.0	111.7	100.0	129.0	100.0	26.0
F03 Kaisersteinbruch 1	PLU, SYR	10.0	50.0	100.0	112.1	100.0	97.5	100.0	153.1	100.0	147.0	100.0	19.7
F04 Purkersdorf	CAM	10.0	42.5	100.0	91.6	100.0	98.4	100.0	70.0	100.0	69.7	100.0	43.5
F05 Rossatz	PLU	10.0	62.5	100.0	116.1	100.0	132.2	100.0	113.3	100.0	119.8	100.0	52.5
F06 Sommerein	CAC, PSY	12.5	47.5	100.0	96.4	100.0	88.1	100.0	134.6	100.0	122.1	100.0	79.1
F07 Stackelberg	CAM	10.0	30.0	100.0	30.5	100.0	35.4	100.0	30.2	100.0	28.1	100.0	58.6
F08 Thernberg 2	PLU	10.0	37.5	100.0	89.7	100.0	94.4	100.0	89.3	100.0	88.2	100.0	24.8
F10 Zwettl	CAM	12.5	32.5	100.0	38.7	100.0	36.2	100.0	35.3	100.0	32.2	100.0	37.7
Mean of 8 infested stands		**10.6**	**34.0**	**100.0**	**86.6**	**100.0**	**90.6**	**100.0**	**92.2**	**100.0**	**92.0**	**100.0**	**42.7**
Significance [c]		***		*				n.s.		n.s.		n.s.	
2 *Phytophthora*-Free Stands (4 Healthy and 4 Declining Beech Trees)													
		H	D	H	D	H	D	H	D	H	D	H	D
F02 Hocheck 1	-	10.0	55.0	100.0	78.6	100.0	71.3	100.0	85.6	100.0	82.3	100.0	23.2
F09 Ybbsitz	-	5.0	37.5	100.0	77.8	100.0	82.3	100.0	71.6	100.0	63.6	100.0	48.1
Mean of 2 non-infested stands		**7.5**	**46.3**	**100.0**	**78.2**	**100.0**	**76.8**	**100.0**	**78.6**	**100.0**	**72.9**	**100.0**	**35.7**
Significance [c]		***		n.s.		n.s.		n.s.		n.s.		n.s.	

[a] CAC = Phytophthora cactorum, CAM = P. ×cambivora, PLU = P. plurivora, PSY = P. psychrophila, SYR = P. syringae. [b] FRL = fine root length, CRL = coarse root length, FRS = fine root surface, CRS = coarse root surface, FRT = number of fine root tips, MT = mycorrhized root tips, NMT = non-mycorrhized root tips; for detailed explanations of root parameters, see Material and Methods. [c] $p \leq 0.1$, * = $p \leq 0.05$, *** = $p \leq 0.001$, n.s. = non-significant.

4. Discussion

This study in 34 beech forests across Lower Austria demonstrated the widespread occurrence of *Phytophthora*-typical disease symptoms and soilborne *Phytophthora* species. In total, seven *Phytophthora* species were isolated from rhizosphere soil and from bleeding bark lesions of 48% of sample trees in 25 of the 27 stands from which isolation tests were performed. Considering the size of Lower Austria of 19.186 km^2, *Phytophthora* diversity in the 34 surveyed beech stands was relatively high in comparison to surveys in beech stands of other European countries. In Bavaria, southern Germany, where 134 beech stands were surveyed between 2003 and 2007, across an area three times the size of Lower Austria, nine *Phytophthora* species were recorded from 81% of the sample trees in 93% of the stands [20,30]. In several separate surveys in Italy, 10 *Phytophthora* species have been detected in 13 beech stands [17,20,23,28,36]. In contrast, in 13 other European countries, only between one and six *Phytophthora* species were recorded from beech stands [16,18–22,25–27,29,31,32,35,37,38]. Across Europe, 17 *Phytophthora* species are currently known to occur in European beech forests [20–23,36,39]. As in other European countries, in Lower Austria, *P.* ×*cambivora*, *P. plurivora* and *P. cactorum*, all considered in Europe as introduced invasive pathogens [20–22,39,59], were the most common *Phytophthora* species in beech forests, whereas *P. syringae* and the putatively native *P. gonapodyides*, *P. psychrophila* and *P. tubulina* [21,39] showed only rare occurrence. A population genetic study demonstrated that the genetic diversity of *P. plurivora* is higher in Europe than in North America and a European origin of this pathogen was proposed [64]. However, in this study, no Asian isolates were included and recent findings of *P. plurivora* in remote healthy forests in Nepal, Yunnan and Taiwan [61,65,66] and the ubiquitous distribution of *P. plurivora* in healthy forest ecosystems and streams across Japan (T. Jung, C.M. Brasier and K. Kageyama, unpublished results) clearly indicate an Asian origin for this pathogen. This is also supported by the fact that the diversity of both known and yet undescribed species from phylogenetic *Phytophthora* Clade 2c, to which also *P. plurivora* belongs, is extremely high in Asia [61,67]. In addition, the high aggressiveness of *P. plurivora* to major European forest tree species like beech, *Tilia cordata*, *Acer* spp. and *Quercus* spp., and the widespread association of *P. plurivora* with the decline, dieback and mortality of forests across Europe [20–23,30,33,35,44,45,54,56,57,59] demonstrate a lack of long-term co-evolution, whereas in Asia, *P. plurivora* is associated with healthy ecosystems [61,65,66]. Interestingly, *P. tubulina*, a new species closely related to *P. quercina*, which is one of the main drivers of chronic European oak decline [16,20,22,39,45,68–70], has never been found anywhere else before. *Phytophthora pseudosyringae*, commonly associated with beech and oak forests in Germany and Italy, [17,19,20,22,25,30,57] was not isolated from beech forests in Lower Austria. Results from recent surveys in Taiwan and Vietnam suggest that in the absence of invasive *Phytophthora* species, native *Phytophthora* species are widespread and common in natural and semi-natural forests [61,67]. It appears, hence, that multiple invasive *Phytophthora* species are outcompeting native *Phytophthora* species in the soils of European beech stands, possibly due to their higher aggressiveness to native tree species and their higher cardinal temperatures, which provide them with a selective advantage over native low-temperature *Phytophthora* species in times of rising annual and winter temperatures. This was also recently suggested for the invasive aggressive *P. cinnamomi* and potentially native *Phytophthora* species in the Valdivian rainforests of Chile [71]. However, more surveys and long-term field studies are needed to confirm this hypothesis.

This study demonstrated the site preferences of *P.* ×*cambivora* and *P. plurivora* which were in agreement with the results from previous surveys in beech and oak forests in other countries [19,20,22, 25,30–32,37,45,59]. *Phytophthora* ×*cambivora* was exclusively found on geological substrates forming acidic soils with high contents of clay and sand and a tendency for temporary waterlogging, including claystones, gneiss, granodiorite, sandstones and schist. In contrast, *P. plurivora* was mainly, though not exclusively, found in limestone sites. Like in Bavarian beech stands, *P. cactorum* did not show clear site preferences [30]. The geological substrate alone explains 30% of the *Phytophthora* species distribution in the 25 *Phytophthora*-infested beech stands in Lower Austria. In comparison to other ecological studies, this is a remarkably high value for a single ecological factor [72]. The upper altitudinal limits of *P. ×cambivora*, *P. plurivora* and *P. cactorum* in Lower Austria were slightly lower than in Bavaria (637, 614 and 562 m a.s.l. versus 750, 870 and 600 m a.s.l. [30]).

The question arises how exotic invasive *Phytophthora* species were introduced to the mature beech forests in Lower Austria. Almost ubiquitous infestations of nursery fields and young planting sites of beech and numerous other tree species across Europe, including Austria, with *P. ×cambivora*, *P. plurivora*, *P. cactorum*, and more than 50 other *Phytophthora* species demonstrated, beyond any reasonable doubt, that the planting of infested nursery stock is the major pathway of non-native *Phytophthora* pathogens into forest ecosystems [21]. In addition, in several forest stands in Lower Austria beech trees with bleeding cankers caused by *P. ×cambivora* or *P. plurivora* were concentrated along forest roads, suggesting the relatively recent introduction and spread of these invasive pathogens with infested road-building materials and/or infested soil particles adhering to vehicles and hikers boots, as previously shown for *P. cinnamomi* in Western Australia and *P. lateralis* in Oregon [73,74].

Phytophthora ×cambivora, *P. plurivora* and *P. cactorum* are the most common *Phytophthora* pathogens associated with bark cankers and the decline of beech stands in other European countries [16,19–23, 25–32,35,37,59]. Across Europe, all three pathogens are also involved in the widespread decline of oak forests, the ink disease of *Castanea sativa* and the declines of numerous other broadleaved tree species including *Aesculus hippocastanum*, *Acer* spp., *Fraxinus excelsior*, *Populus* spp. and *Tilia* spp. [15,16,20–22, 29,44,45,59,65,74–78]. In pathogenicity trials, *P. ×cambivora*, *P. plurivora* and *P. cactorum* demonstrated high aggressiveness to roots and the bark of *F. sylvatica*, several European oak species and *C. sativa* while *P. gonapodyides*, *P. psychrophila* and *P. tubulina* were only moderately aggressive [15,28,39,44,56–58,66,79–81]. *Phytophthora ×cambivora*, *P. plurivora* and *P. cactorum* also showed high aggressiveness to several poplar clones commonly used in riparian plantations in Europe [78]. In addition, in pathogenicity trials, *P. plurivora* and *P. cactorum* caused considerable bark lesions on *F. excelsior* and *Alnus glutinosa* while *P. ×cambivora* was highly aggressive to *Prunus laurocerasus*, a rare understorey species in Southern European beech forests [82–84].

In Lower Austria, *Phytophthora*-typical bleeding bark cankers were found in 74% of the surveyed 34 beech stands with 65% of these cankers in 71% of the stands located at the collar while 35% of the cankers in 38% of stands had no connection to the roots (aerial cankers). Similar results were reported from a survey in 134 beech stands in Bavaria with collar rot and aerial bleeding cankers occurring in 74 and 36% of surveyed stands, respectively [30]. It is noteworthy that no *Phytophthora*-typical bleeding cankers were observed in the four high-altitude stands (F02, F09, F23, F33) supporting the altitudinal limits of *Phytophthora* spp. in Lower Austria suggested by the isolation records. As in previous surveys in Bavaria, Lower Saxony (Northern Germany), Belgium, Norway and Sicily [20,21,23,25,30,31,37], *P. ×cambivora* was the most common *Phytophthora* species isolated from necrotic beech bark in Lower Austria. However, while *P. ×cambivora* was only detected in collar rot lesions, *P. plurivora* could be isolated from both collar rot and aerial cankers. Also in Bavaria, *P. plurivora* was the most common species recovered from aerial cankers in beech trees, but other *Phytophthora* species like *P. ×cambivora*, *P. cactorum* and *P. gonapodyides* were infrequently isolated, too [30]. *Phytophthora plurivora*, like most other soilborne *Phytophthora* species, lacks caducous sporangia for aerial spread and its movement to higher stem heights was demonstrated to be achieved via the passive transport inside non-symptomatic xylem vessels of beech trees resulting in isolated aerial bark cankers along the stems [26]. Snails sucking on the exudates of bleeding cankers were also suggested to act as *Phytophthora* vectors along beech stems [30].

Like in other European countries and the USA [19,20,22,23,25,30–33,37,59], in Lower Austria, the beech trees affected by *Phytophthora* cankers were usually showing a severe decline and dieback. However, apart from the stands F12, F27 and F31 with 14–20% canker incidence, the proportions of beech trees affected by bark cankers were in most of the 25 *Phytophthora*-infested beech stands too low to explain the observed decline of beech trees in groups or at stand level. Also in Bavaria, in 87% of the surveyed 134 beech stands, the distribution of beech trees with bleeding bark lesions was scattered, and similar to oak decline, the *Phytophthora*-related destruction of the root system was suggested as main driver of beech decline [19,20,22,30,44,45]. In the present study, the fine root conditions of each two healthy and declining beech trees were assessed in 10 beech stands in Lower Austria. In the eight *Phytophthora*-infested stands, the four relative root ratio parameters, FRL/CRL_{rel}, FRS/CRS_{rel}, FRT/CRL_{rel}, FRT/FRL_{rel}, FRT/CRS_{rel} were between 7.8 and 13.4% higher in the 16 non-declining compared to the 16 declining trees. However, due to the low sample numbers, these differences were statistically only significant or almost significant for two root parameters (FRL/CRL_{rel} and FRS/CRS_{rel}). In a comparable study of 19 *Phytophthora*-infested oak stands in Bavaria, the parameters FRL/CRL_{rel} and FRT/CRL_{rel} for the 59 healthy oak trees were on average 21.6 and 35.4% higher than in the 65 declining trees [45]. Several factors might explain the smaller differences in the root parameters between healthy and declining trees in beech as compared to oaks. Trees are in a functional equilibrium between the water absorbing fine root system and the transpiring and producing leaf area [85,86]. The threshold for fine root losses leading to the onset of crown thinning and eventually decline is most likely considerably lower in the drought-sensitive *F. sylvatica* than in drought-tolerant oak species [87–89]. Furthermore, oak decline in temperate Central Europe is mainly driven by *P. quercina*, an oak-specific specialized fine root pathogen [16,20,22,44,45,64–66], whereas the most common species involved in beech decline, *P.* ×*cambivora*, *P. cactorum* and *P. plurivora*, are both fine root and bark pathogens causing bleeding lesions on stems and woody roots. Like in Bavaria [30], the root systems of wind-thrown beech trees in the three beech stands on temporarily waterlogged sites with high clay contents (F04 Purkersdorf, F16 Hengstlberg, F26 Kleinmariazell 2) and in the limestone stand F08 Thernberg 2 showed, besides extensive losses of lateral and fine roots, also numerous *Phytophthora*-like lesions on woody roots. These lesions certainly reduce both the supply of distant parts of the root system with carbohydrates and the transport of water to the crowns. As demonstrated in the present study, bark lesions on roots and stems can be colonized by a range of secondary fungal pathogens like *Armillaria* spp., *Ustulina deusta*, *Neonectria coccinea*, *Bjerkandera adusta*, *Fomes fomentarius*, *Oudemansiella mucida* etc., exacerbating the primary bark damages and colonizing the underlying xylem leading to a reduced water supply. Furthermore, this study showed that *Phytophthora* bark lesions are weakening affected beech trees and predisposing them to attacks by opportunistic bark beetles. The observation of active breeding galleries in fresh collar lesions caused by *P.* ×*cambivora* raises the question whether, similar to ant species [90], bark beetles might act as *Phytophthora* vectors. In Hungary, *T. bicolor,* together with other bark beetles, was involved in the mass mortality of beech forests [91]. Unfortunately, in Hungary, a possible involvement of *Phytophthora* species has not been investigated. Additional studies of declining and non-declining beech trees are needed to assess the extent of fine root losses and of bark lesions on woody roots, their effect on the health status of beech trees and the synergistic interaction between primary *Phytophthora* infections, secondary fungal infections and bark beetle colonization of beech bark.

The ectomycorrhizal frequency of fine root tips (MT/NMT_{rel}) was considerably higher, though not statistically significant, in non-declining versus declining beech trees and seemed to be a good indicator of beech vitality. In the 10 intensively studied beech stands in Lower Austria, no consistent ecological factor was found which would explain the selective reduction of the ectomycorrhizal frequency of individual trees within a relatively homogeneous forest stand. Ectomycorrhizal symbiosis relies on a solid bidirectional exchange of carbohydrates and nutrients between plant and fungal partners [49]. The reduced mycorrhizal frequency of declining beech trees might have been caused by a reduced supply of the fungal partners with carbohydrates due to reduced tree vitality. The reduced mycorrhization of fine roots results in reduced nutrient uptake and in turn further decreases tree vitality. *Phytophthora* pathogens are known to alter ectomycorrhizal symbioses [52]. In the eight *Phytophthora*-infested beech stands in Lower Austria, the observed losses of fine roots and necrotic lesions on the woody roots of declining trees could certainly have negative impacts on the ectomycorrhiza by reducing both the available number of fine root tips and the fine root length for mycorrhizal colonization, and also the transport of carbohydrates to the fungal partners. Also in *Phytophthora*-infested oak forests in Europe, declining trees were associated with a decrease in ectomycorrhizal frequency [50,92]. In contrast, in a chestnut stand in Italy severely affected by ink disease caused by *P.* ×*cambivora* ectomycorrhizal frequency, and species richness, were both higher compared to a healthy chestnut forest [93]. Studies of Swiss needle cast of Douglas-fir caused by *Phaeocryptopus gaeumannii* in Oregon demonstrated that ectomycorrhizal frequency did not vary between sites with different disease incidences. However, ectomycorrhizal density and species richness were positively correlated with needle retention [94]. Unfortunately, in the present work, ectomycorrhizal species richness had not been examined. Additional research in beech stands is needed on the relationship between *Phytophthora* presence, root parameters, tree vitality and ectomycorrhizal frequency and species richness.

In Bavaria, *Phytophthora*-related decline and dieback of beech forests reached an epidemic extent during summer and autumn 2003 and in 2004 as a consequence of the exceptionally wet year 2002, followed by the severe drought during the summer of 2003 [30,40]. In Lower Austria, according to the data from the meteorological stations of the HZB, similar successions of abnormally wet and dry periods occurred which might have triggered beech decline. In almost all 10 intensively surveyed beech stands, from 2004 to 2006, precipitation in late winter and spring exceeded the long-term average, while summer 2004 and autumn 2006 were drier than the long-term average (data not shown). Climatic models predict for Europe continuously rising annual temperatures and a significant increase in the frequency and duration of both heavy summer rains and prolonged summer droughts [6,46,95]. These climatic changes will most likely trigger more severe *Phytophthora* disease epidemics in forest ecosystems including beech forests, and favor the spread and establishment of invasive high-temperature *Phytophthora* species like *P.* ×*cambivora*, *P. cactorum*, *P. cinnamomi*, *P. niederhauserii*, *P. multivora* and *P. plurivora* and provide them with a selective advantage over European native low-temperature species like *P. castanetorum*, *P. psychrophila*, *P. pseudosyringae*, *P. tubulina* and *P. vulcanica* [20,22,30,41,42,44,45,47].

5. Conclusions

Our study demonstrated the widespread occurrence of soilborne *Phytophthora* species, in particular the introduced invasive *P.* ×*cambivora*, *P. cactorum* and *P. plurivora*, in beech forests of Lower Austria, and their site preferences and their involvement in beech decline. Primary *Phytophthora* collar rot and aerial bark cankers on beech stems were found to be widespread, but in the majority of stands, their distribution was scattered. The analyses of several root parameters showed that the fine root status of declining beech trees was reduced compared to healthy trees. The reduced water uptake due to the fine root losses in combination with reduced water transport due to *Phytophthora* lesions on woody roots most likely triggered crown thinning, eventually leading to decline. However, additional studies of declining and non-declining beech trees in a statistically representative number of both *Phytophthora*-free and *Phytophthora*-infested stands are needed to quantify fine root losses, ectomycorrhizal frequency and the extent of bark lesions on woody roots, and assess their effect on the health status of beech trees and the synergistic interaction between primary *Phytophthora* infections and secondary fungal infections and bark beetle colonization of beech bark.

Supplementary Materials: Table S1: Location and altitude of 34 forest stands of *Fagus sylvatica* in Lower Austria, the areas and numbers of trees surveyed, and the numbers of trees with *Phytophthora*-typical collar rot and aerial bark cankers, Table S2: GenBank accession numbers of ITS and partial *cox1* sequences generated in this study for representative *Phytophthora* isolates from Lower Austrian beech forests, Table S3: Long-term (1961–1991) climatic data of climatic stations closest and most comparable in altitude to ten intensively surveyed beech forest stands in Lower Austria, provided by the Hydrographischer Dienst Österreich (HZB; Vienna, Austria), Table S4: Crown condition of 60 beech trees in the 10 intensively surveyed mature beech stands in 2008 and the occurrence of *Phytophthora* spp, Table S5: Diversity of fungi associated with different symptoms and injuries in 21 beech stands in Lower Austria.

Author Contributions: Conceptualization: T.J., T.L.C.; data curation: T.J., T.L.C., T.K., T.C., M.H.J.; formal analysis: T.J., T.L.C., T.C., T.K., M.H.J.; investigation: T.J., T.L.C., T.C.; methodology: T.J., T.L.C., T.K., M.H.J.; technical support: A.D., C.H., M.B.; writing—original draft: T.J., T.C.; writing—review and editing: T.J., T.C. All authors have read and agreed to the published version of the manuscript.

Acknowledgments: The authors are very grateful to Erich Fischer (HZB, Vienna, Austria) for providing climatic data, Rainer Reiter (Institute of Forest Ecology and Soil, BFW, Vienna, Austria) for the soil analyses, and Reinhard Hagen from the Forest Service of Lower Austria and private forest owners for providing data for the selection of sampling sites and for their permission to perform this study and take samples in their forests.

References

1. Petit, R.J.; Hampe, A. Some evolutionary consequences of being a tree. *Annu. Rev. Ecol. Evol. Syst.* **2006**, *37*, 187–214. [CrossRef]
2. Fritz, P. (Ed.) *Ökologischer Waldumbau in Deutschland (Ecological Reconstruction of Forests in Germany)*; Oekom Verlag: Munich, Germany, 2006; p. 351, ISBN 978-3-86581-001-4.
3. Walentowski, H.; Ewald, J.; Fischer, A.; Kölling, C.; Türk, W. *Handbuch der Natürlichen Waldgesellschaften Bayerns (Handbook of the Natural Forest Types of Bavaria)*, 2nd ed.; Geobotanica: Freising, Germany, 2006; p. 441, ISBN 3-930560-04-6.
4. Del Río, S.; Álvarez-Esteban, R.; Cano, E.; Pinto-Gomes, C.; Penas, Á. Potential impacts of climate change on habitat suitability of *Fagus sylvatica* L. forests in Spain. *Plant Biosyst.* **2018**, *152*, 1205–1213. [CrossRef]
5. Jump, A.S.; Hunt, J.M.; Penuelas, J. Rapid climate change-related growth decline at the southern range edge of *Fagus sylvatica*. *Glob. Chang. Biol.* **2006**, *12*, 2163–2174. [CrossRef]
6. Czúcz, B.; Galhidy, L.; Matyas, C. Present and forecasted xeric climatic limits of beech and sessile oak distribution at low altitudes in Central Europe. *Ann. For. Sci.* **2011**, *68*, 99–108. [CrossRef]
7. Gora, V.; König, J.; Lunderstädt, J. Population dynamics of beech scale (*Cryptococcus fagisuga*) (Coccina, Pseudococcidae) related to physiological defence reactions of attacked beech trees (*Fagus sylvatica*). *Chemoecology* **1996**, *7*, 112–120. [CrossRef]
8. Mazzoglio, P.J.; Paoletta, M.; Patetta, A.; Currado, I. *Calliteara pudibunda* (Lepidoptera, Lymantriidae) in Northwest Italy. *Bull. Insectology* **2005**, *58*, 25–34.
9. Lonsdale, D. Nectria infection of beech bark: Variations in disease in relation to predisposing factors. *Ann. Sci. For.* **1980**, *37*, 307–317. [CrossRef]
10. Lonsdale, D.; Wainhouse, D. Beech bark disease. In *Forestry Commission Bulletin*; HMSO: London, UK, 1987; p. 15, ISBN 0-11-710207-5.
11. Metzler, B.; Meierjohann, E.; Kublin, E.; Von Wuehlisch, G. Spatial dispersal of *Nectria ditissima* canker of beech in an international provenance trial. *For. Pathol.* **2002**, *32*, 137–144. [CrossRef]
12. Houston, D.R. Beech bark disease: 1934 to 2004: What's new since Ehrlich? In *Beech Bark Disease, Proceedings of the Beech Bark Disease Symposium, Saranac Lake, NY, USA, 16–18 June 2004*; Gen. Tech. Rep. NE-331; US. Department of Agriculture Forest Service, Northern Research Station: Newtown Square, PA, USA, 2005; pp. 2–13. Available online: https://www.nrs.fs.fed.us/pubs/7401 (accessed on 18 July 2020).
13. Day, W.R. Root-rot of sweet chestnut and beech caused by species of *Phytophthora*. I. Cause and symptoms of disease: Its relation to soil conditions. *Forestry* **1938**, *12*, 101–116. [CrossRef]
14. Day, W.R. Root-rot of sweet chestnut and beech caused by species of *Phytophthora*. II. Inoculation experiments and methods of control. *Forestry* **1939**, *13*, 46–58. [CrossRef]
15. Jung, T.; Blaschke, H. *Phytophthora* root rot in declining forest trees. *Phyton (Horn, Austria)* **1996**, *36*, 95–102.
16. Balci, Y.; Halmschlager, E. *Phytophthora* species in oak ecosystems in Turkey and their association with declining oak trees. *Plant Pathol.* **2003**, *52*, 694–702. [CrossRef]
17. Motta, E.; Annesi, T.; Pane, A.; Cooke, D.E.L.; Cacciola, S.O. A new *Phytophthora* sp. causing a basal canker on beech in Italy. *Plant Dis.* **2003**, *87*, 1005. [CrossRef] [PubMed]
18. Brasier, C.M.; Beales, P.A.; Denman, S.; Rose, J. *Phytophthora kernoviae* sp. nov., an invasive pathogen causing bleeding stem lesions on forest trees and foliar necrosis of ornamentals in the UK. *Mycol. Res.* **2005**, *109*, 853–859. [CrossRef] [PubMed]
19. Jung, T.; Hudler, G.W.; Jensen-Tracy, S.L.; Griffiths, H.M.; Fleischmann, F.; Oßwald, W. Involvement of *Phytophthora* spp. in the decline of European beech in Europe and the USA. *Mycologist* **2005**, *19*, 159–166. [CrossRef]
20. Jung, T.; Vettraino, A.M.; Cech, T.L.; Vannini, A. The impact of invasive *Phytophthora* species on European forests. In *Phytophthora: A Global Perspective*; Lamour, K., Ed.; CABI: Wallingford, UK, 2013; pp. 146–158, ISBN 978-1-78064-093-8.

21. Jung, T.; Orlikowski, L.; Henricot, B.; Abad-Campos, P.; Aday, A.G.; Aguín Casal, O.; Bakonyi, J.; Cacciola, S.O.; Cech, T.; Chavarriaga, D.; et al. Widespread *Phytophthora* infestations in European nurseries put forest, semi-natural and horticultural ecosystems at high risk of *Phytophthora* diseases. *For. Pathol.* **2016**, *46*, 134–163. [CrossRef]
22. Jung, T.; Pérez–Sierra, A.; Durán, A.; Horta Jung, M.; Balci, Y.; Scanu, B. Canker and decline diseases caused by soil- and airborne *Phytophthora* species in forests and woodlands. *Persoonia* **2018**, *40*, 182–220. [CrossRef]
23. Jung, T.; La Spada, F.; Pane, A.; Aloi, F.; Evoli, M.; Horta Jung, M.; Scanu, B.; Faedda, R.; Rizza, C.; Puglisi, I.; et al. Diversity and distribution of *Phytophthora* species in protected natural areas in Sicily. *Forests* **2019**, *10*, 259. [CrossRef]
24. Cech, T.L.; Jung, T. *Phytophthora*—Wurzelhalsfäulen an Buchen nehmen auch in Österreich zu (*Phytophthora* root rot of beech is also increasing in Austria). *Forstsch. Aktuell* **2005**, *34*, 7.
25. Hartmann, G.; Blank, R.; Kunca, A. Collar rot of *Fagus sylvatica* caused by *Phytophthora cambivora*: Damage, site relations and susceptibility of broadleaf hosts. In *Progress in Research on Phytophthora Diseases of Forest Trees, Proceedings of the 3rd International IUFRO Working Party 7.02.09 Meeting, Freising, Germany, 11–17 September 2004*; Brasier, C., Jung, T., Osswald, W., Eds.; Forest Research: Farnham, UK, 2006; pp. 135–138, ISBN 0-85538-721-1.
26. Brown, A.V.; Brasier, C.M. Colonization of tree xylem by *Phytophthora ramorum*, *P. kernoviae* and other *Phytophthora* species. *Plant Pathol.* **2007**, *56*, 227–241. [CrossRef]
27. Munda, A.; Zerjav, M.; Schroers, H.J. First report of *Phytophthora citricola* occurring on *Fagus sylvatica* in Slovenia. *Plant Dis.* **2007**, *91*, 907. [CrossRef] [PubMed]
28. Vettraino, A.M.; Jung, T.; Vannini, A. First report of *Phytophthora cactorum* associated with beech decline in Italy. *Plant Dis.* **2008**, *92*, 1708. [CrossRef] [PubMed]
29. Cerny, K.; Strnadova, V.; Gregorova, B.; Holub, V.; Tomsovsky, M.; Mrazkova, M.; Gabrielova, S. *Phytophthora cactorum* causing bleeding canker of common beech, horse chestnut, and white poplar in the Czech Republic. *Plant Pathol.* **2009**, *58*, 394. [CrossRef]
30. Jung, T. Beech decline in Central Europe driven by the interaction between *Phytophthora* infections and climatic extremes. *For. Pathol.* **2009**, *39*, 73–94. [CrossRef]
31. Schmitz, S.; Zini, J.; Chandelier, A. Involvement of *Phytophthora* species in the decline of beech (*Fagus sylvatica*) in the southern part of Belgium. In *Phytophthoras in Forests and Natural Ecosystems: Fourth Meeting of the International Union of Forest Research Organizations (IUFRO) Working Party S07.02.09*; General Technical Report PSW-GTR-221; Goheen, E.M., Frankel, S.J., Eds.; USDA Forest Service Pacific Southwest Research Station: Albany, CA, USA, 2009; pp. 320–323.
32. Stępniewska, H.; Dłuszyński, J. Incidence of *Phytophthora cambivora* in bleeding lesions on beech stems in selected forest stands in South-eastern Poland. *Phytopathologia* **2010**, *56*, 39–51.
33. Weiland, G.E.; Nelson, A.H.; Hudler, G.W. Aggressiveness of *Phytophthora cactorum*, *P. citricola* I and *P. plurivora* from European beech. *Plant Dis.* **2010**, *94*, 1009–1014. [CrossRef]
34. Nechwatal, J.; Hahn, J.; Schönborn, A.; Schmitz, G. A twig blight of understorey European beech (*Fagus sylvatica*) caused by soilborne *Phytophthora* spp. *For. Pathol.* **2011**, *41*, 493–500. [CrossRef]
35. Milenković, I.; Keča, N.; Karadžić, D.; Nowakowska, J.A.; Borys, M.; Sikora, K.; Oszako, T. Incidence of *Phytophthora* species in beech stands in Serbia. *Folia For. Pol.* **2012**, *54*, 223–232.
36. Cacciola, S.O.; Motta, E.; Raudino, F.; Chimento, A.; Pane, A.; Magnano di San Lio, G. *Phytophthora pseudosyringae* the causal agent of bleeding cankers of beech in central Italy. *J. Plant Pathol.* **2005**, *87*, 289.
37. Telfer, K.H.; Brurberg, M.B.; Herrero, M.L.; Stensvand, A.; Talgø, V. *Phytophthora cambivora* found on beech in Norway. *For. Pathol.* **2015**, *45*, 415–425. [CrossRef]
38. Cleary, M.; Blomquist, M.; Ghasemkhani, M.; Witzell, J. First report of *Phytophthora gonapodyides* causing stem canker on European beech (*Fagus sylvatica*) in southern Sweden. *Plant Dis.* **2016**, *100*, 2174. [CrossRef]
39. Jung, T.; Horta Jung, M.; Cacciola, S.O.; Cech, T.; Bakonyi, J.; Seress, D.; Mosca, S.; Schena, L.; Seddaiu, S.; Pane, A.; et al. Multiple new cryptic pathogenic *Phytophthora* species from Fagaceae forests in Austria, Italy and Portugal. *IMA Fungus* **2017**, *8*, 219–244. [CrossRef] [PubMed]
40. Fink, A.H.; Brücher, T.; Krüger, A.; Leckebusch, G.C.; Pinto, J.G.; Ulbrich, U. The 2003 European summer heatwaves and drought–synoptic diagnosis and impacts. *Weather* **2004**, *59*, 209–216. [CrossRef]
41. Brasier, C.M.; Scott, J.K. European oak declines and global warming: A theoretical assessment with special reference to the activity of *Phytophthora cinnamomi*. *EPPO Bull.* **1994**, *24*, 221–234. [CrossRef]

42. Brasier, C.M. *Phytophthora cinnamomi* and oak decline in southern Europe. Environmental constraints including climate change. *Ann. Sci. For.* **1996**, *53*, 347–358. [CrossRef]
43. Brasier, C.M. The biosecurity threat to the UK and global environment from international trade in plants. *Plant Pathol.* **2008**, *57*, 792–808. [CrossRef]
44. Jung, T.; Blaschke, H.; Neumann, P. Isolation, identification and pathogenicity of *Phytophthora* species from declining oak stands. *Eur. J. For. Pathol.* **1996**, *26*, 253–272. [CrossRef]
45. Jung, T.; Blaschke, H.; Oßwald, W. Involvement of soilborne *Phytophthora* species in Central European oak decline and the effect of site factors on the disease. *Plant Pathol.* **2000**, *49*, 706–718. [CrossRef]
46. Anonymous. Climate Change 2014—Synthesis Report: Impacts, Adaptation, and Vulnerability. In Proceedings of the Intergovernmental Panel on Climate Change (IPCC), Geneva, Switzerland; 2015. Available online: https://www.ipcc.ch/report/ar4/wg2/ (accessed on 18 July 2020).
47. Burgess, T.I.; Scott, J.K.; McDougall, K.L.; Stukely, M.J.C.; Crane, C.; Dunstan, W.A.; Brigg, F.; Andjic, V.; White, D.; Rudman, T.; et al. Current and projected global distribution of *Phytophthora cinnamomi*, one of the world's worst plant pathogens. *Glob. Chang. Biol.* **2017**, *23*, 1661–1674. [CrossRef]
48. Erwin, D.C.; Ribeiro, O.K. *Phytophthora Diseases Worldwide*; APS Press, American Phytopathological Society: St. Paul, MI, USA, 1996; p. 592, ISBN 0-89054-212-0.
49. Fields, K.J.; Pressel, S. Unity in diversity: Structural and functional insights into the ancient partnerships between plants and fungi. *New Phytol.* **2018**, *220*, 996–1006. [CrossRef]
50. Corcobado, T.; Vivas, M.; Moreno, G.; Solla, A. Ectomycorrhizal symbiosis in declining and non-declining *Quercus ilex* trees infected with or free of *Phytophthora cinnamomi*. *For. Ecol. Manag.* **2014**, *324*, 72–80. [CrossRef]
51. Sapsford, S.J.; Paap, T.; Hardy, G.E.S.J.; Burgess, T.I. The 'chicken or the egg': Which comes first, forest tree decline or loss of mycorrhizae? *Plant Ecol.* **2017**, *218*, 1093–1106. [CrossRef]
52. Marx, D.H. Ectomycorrhizae as biological deterrents to pathogenic root infections. *Annu. Rev. Phytopathol.* **1972**, *10*, 429–454. [CrossRef] [PubMed]
53. Anonymous. Forest Condition in Europe. Results of the 1993 Survey. Convention on Long-Range Transboundary Air Pollution. In *International Co-Operative Programme on Assessment and Monitoring of Air Pollution Effects on Forests*; EC-UN/ECE: Brussels, Geneva, 1994.
54. Jung, T.; Blaschke, M. *Phytophthora* root and collar rot of alders in Bavaria: Distribution, modes of spread and possible management strategies. *Plant Pathol.* **2004**, *53*, 197–208. [CrossRef]
55. Scanu, B.; Hunter, G.C.; Linaldeddu, B.T.; Franceschini, A.; Maddau, L.; Jung, T.; Denman, S. A taxonomic re-evaluation reveals that *Phytophthora cinnamomi* and *P. cinnamomi* var. *parvispora* are separate species. *For. Pathol.* **2014**, *44*, 1–20. [CrossRef]
56. Jung, T.; Hansen, E.M.; Winton, L.; Oßwald, W.; Delatour, C. Three new species of *Phytophthora* from European oak forests. *Mycol. Res.* **2002**, *106*, 397–411. [CrossRef]
57. Jung, T.; Nechwatal, J.; Cooke, D.E.L.; Hartmann, G.; Blaschke, M.; Oßwald, W.F.; Duncan, J.M.; Delatour, C. *Phytophthora pseudosyringae* sp. nov., a new species causing root and collar rot of deciduous tree species in Europe. *Mycol. Res.* **2003**, *107*, 772–789. [CrossRef]
58. Jung, T.; Horta Jung, M.; Scanu, B.; Seress, D.; Kovács, D.M.; Maia, C.; Pérez-Sierra, A.; Chang, T.T.; Chandelier, A.; Heungens, A.; et al. Six new *Phytophthora* species from ITS Clade 7a including two sexually functional heterothallic hybrid species detected in natural ecosystems in Taiwan. *Persoonia* **2017**, *38*, 100–135. [CrossRef]
59. Jung, T.; Burgess, T.I. Re-evaluation of *Phytophthora citricola* isolates from multiple woody hosts in Europe and North America reveals a new species, *Phytophthora plurivora* sp. nov. *Persoonia* **2009**, *22*, 95–110. [CrossRef]
60. Cooke, D.E.L.; Drenth, A.; Duncan, J.M.; Wagels, G.; Brasier, C.M. A molecular phylogeny of *Phytophthora* and related oomycetes. *Fungal Genet. Biol.* **2000**, *30*, 17–32. [CrossRef]
61. Jung, T.; Chang, T.T.; Bakonyi, J.; Seress, D.; Pérez-Sierra, A.; Yang, X.; Hong, C.; Scanu, B.; Fu, C.H.; Hsueh, K.L.; et al. Diversity of *Phytophthora* species in natural ecosystems of Taiwan and association with disease symptoms. *Plant Pathol.* **2017**, *66*, 194–211. [CrossRef]
62. Agerer, R. (Ed.) *Colour Atlas of Ectomycorrhizae*; Einhorn Verlag: Schwäbisch Gmünd, Germany, 1987; Volume 1, p. 58, ISBN 0-89054-212-0.
63. Ter Braak, C.J.F.; Šmilauer, P. *Canoco Reference Manual and User's Guide: Software for Ordination*, 5th ed.; Microcomputer Power: Ithaca, NY, USA, 2012; p. 496. Available online: http://www.canoco5.com/ (accessed on 18 July 2020).

64. Schoebel, C.N.; Stewart, J.; Gruenwald, N.J.; Rigling, D.; Prospero, S. Population history and pathways of spread of the plant pathogen *Phytophthora plurivora*. *PLoS ONE* **2014**, *9*, e85368. [CrossRef] [PubMed]
65. Vettraino, A.M.; Brasier, C.M.; Brown, A.V.; Vannini, A. *Phytophthora himalsilva* sp. nov. an unusually phenotypically variable species from a remote forest in Nepal. *Fungal Biol.* **2011**, *115*, 275–287. [CrossRef] [PubMed]
66. Huai, W.X.; Tian, G.; Hansen, E.M.; Zhao, W.X.; Goheen, E.M.; Grünwald, N.J.; Cheng, C. Identification of *Phytophthora* species baited and isolated from forest soil and streams in northwestern Yunnan province, China. *For. Pathol.* **2013**, *43*, 87–103. [CrossRef]
67. Jung, T.; Scanu, B.; Brasier, C.M.; Webber, J.; Milenković, I.; Corcobado, T.; Tomšovský, T.; Pánek, M.; Bakonyi, J.; Maia, C.; et al. A survey in natural forest ecosystems of Vietnam reveals high diversity of both new and described *Phytophthora* taxa including *P. ramorum*. *Forests* **2020**, *11*, 93. [CrossRef]
68. Jung, T.; Cooke, D.E.L.; Blaschke, H.; Duncan, J.M.; Oßwald, W. *Phytophthora quercina* sp. nov., causing root rot of European oaks. *Mycol. Res.* **1999**, *103*, 785–798. [CrossRef]
69. Balci, Y.; Halmschlager, E. Incidence of *Phytophthora* species in oak forests in Austria and their possible involvement in oak decline. *For. Pathol.* **2003**, *33*, 157–174. [CrossRef]
70. Pérez–Sierra, A.; López–García, C.; León, M.; García-Jiménez, J.; Abad-Campos, P.; Jung, T. Previously unrecorded low temperature *Phytophthora* species associated with *Quercus* decline in a Mediterranean forest in Eastern Spain. *For. Pathol.* **2013**, *43*, 331–339. [CrossRef]
71. Jung, T.; Durán, A.; Sanfuentes von Stowasser, E.; Schena, L.; Mosca, S.; Fajardo, S.; González, M.; Navarro Ortega, A.D.; Bakonyi, J.; Seress, D.; et al. Diversity of *Phytophthora* species in Valdivian rainforests and association with severe dieback symptoms. *For. Pathol.* **2018**, *48*, e12443. [CrossRef]
72. Lotter, A.F.; Birks, H.J.B.; Hofmann, W.; Marcetto, A. Modern diatom, cladocera, chironomid, and chrysophyte cyst assemblages as quantitative indicators for the reconstruction of past environmental conditions in the Alps. I. Climate. *J. Paleolimnol.* **1997**, *18*, 395–420. [CrossRef]
73. Shearer, B.L.; Tippett, J.T. *Jarrah Dieback: The Dynamics and Management of Phytophthora Cinnamomi in the Jarrah (Eucalyptus Marginata) Forests of South-Western Australia*; Department of Conservation and Land Management: Perth, Australia, 1989; p. 76. ISSN 1032-8106.
74. Hansen, E.M.; Goheen, D.J.; Jules, E.S.; Ullian, B. Managing Port–Orford–Cedar and the introduced pathogen *Phytophthora lateralis*. *Plant Dis.* **2000**, *84*, 4–14. [CrossRef] [PubMed]
75. Vettraino, A.M.; Barzanti, G.P.; Bianco, M.C.; Ragazzi, A.; Capretti, P.; Paoletti, E.; Luisi, N.; Anselmi, N.; Vannini, A. Occurrence of *Phytophthora* species in oak stands in Italy and their association with declining oak trees. *For. Pathol.* **2002**, *32*, 19–28. [CrossRef]
76. Vettraino, A.M.; Morel, O.; Perlerou, C.; Robin, C.; Diamandis, S.; Vannini, A. Occurrence and distribution of *Phytophthora* species associated with ink disease of chestnut in Europe. *Eur. J. Plant Pathol.* **2005**, *111*, 169–180. [CrossRef]
77. Brasier, C.M.; Jung, T. Recent developments in *Phytophthora* diseases of trees and natural ecosystems in Europe. In *Progress in Research on Phytophthora Diseases of Forest Trees, Proceedings of the 3rd International IUFRO Working Party 7.02.09 Meeting, Freising, Germany, 11–17 September 2004*; Brasier, C., Jung, T., Osswald, W., Eds.; Forest Research: Farnham, Surrey, UK, 2006; pp. 5–16, ISBN 0-85538-721-1.
78. Milenković, I.; Keča, N.; Karadžić, D.; Radulović, Z.; Nowakowska, J.A.; Oszako, T.; Sikora, K.; Corcobado, T.; Jung, T. Isolation and pathogenicity of *Phytophthora* species from poplar plantations in Serbia. *Forests* **2018**, *9*, 330. [CrossRef]
79. Fleischmann, F.; Schneider, D.; Matyssek, R.; Oßwald, W.F. Investigations on Net CO_2 assimilation, transpiration and root growth of *Fagus sylvatica* infested with four different *Phytophthora* species. *Plant Biol.* **2002**, *4*, 144–152. [CrossRef]
80. Brasier, C.M.; Jung, T. Progress in understanding *Phytophthora* diseases of trees in Europe. In *Phytophthora in Forests and Natural Ecosystems, Proceedings of the 2nd International IUFRO Working Party 7.02.09 Meeting, Albany, Western Australia, 30 September–5 October 2001*; McComb, J.A., Hardy, G.E.S.J., Eds.; Murdoch University Print: Perth, Australia, 2003; pp. 4–18, ISBN 0-86905-825-5.
81. Cleary, M.; Blomquist, M.; Vetukuri, R.R.; Böhlenius, H.; Witzell, J. Susceptibility of common tree species in Sweden to *Phytophthora cambivora*, *P. plurivora* and *P. cactorum*. *For. Pathol.* **2017**, *47*, e12329. [CrossRef]
82. Jung, T.; Nechwatal, J. *Phytophthora gallica* sp. nov., a new species from rhizosphere soil of declining oak and reed stands in France and Germany. *Mycol. Res.* **2008**, *112*, 1195–1205. [CrossRef]

83. Orlikowski, L.B.; Ptaszek, M.; Rodziewicz, A.; Nechwatal, J.; Thinggaard, K.; Jung, T. *Phytophthora* root and collar rot of mature *Fraxinus excelsior* in forest stands in Poland and Denmark. *For. Pathol.* **2011**, *41*, 510–519. [CrossRef]
84. Milenković, I.; Keča, N.; Karadžić, D.; Radulović, Z.; Tomšovský, M.; Jung, T. Occurrence and pathogenicity of *Phytophthora* ×*cambivora* on *Prunus laurocerasus* in Serbia. *For. Pathol.* **2018**, *48*, e12436. [CrossRef]
85. Eamus, D.; Chen, X.; Kelley, G.; Hutley, L.B. Root biomass and root fractal analyses of an open *Eucalyptus* forest in a savanna of north Australia. *Aust. J. Bot.* **2002**, *50*, 31–41. [CrossRef]
86. Meng, S.; Jia, Q.; Zhou, G.; Zhou, H.; Liu, Q.; Yu, J. Fine Root Biomass and Its Relationship with aboveground traits of *Larix gmeliniit* trees in Northeastern China. *Forests* **2018**, *9*, 35. [CrossRef]
87. Ellenberg, H. *Vegetation Mitteleuropas Mit den Alpen (Vegetation of Central Europe and the Alps)*, 4th ed.; Eugen Ulmer: Stuttgart, Germany, 1986; p. 989, ISBN 3800134306.
88. Friedrichs, D.A.; Trouet, V.; Büntgen, U.; Frank, D.C.; Esper, J.; Neuwirth, B.; Löffler, J. Species-specific climate sensitivity of tree growth in Central-West Germany. *Trees* **2009**, *23*, 729–739. [CrossRef]
89. Scharnweber, T.; Manthey, M.; Criegee, C.; Bauwe, A.; Schröder, C.; Wilmking, M. Drought matters—Declining precipitation influences growth of *Fagus sylvatica* L. and *Quercus robur* L. in north-eastern Germany. *For. Ecol. Manag.* **2011**, *262*, 947–961. [CrossRef]
90. El-Hamalawi, Z.A.; Menge, J.A. The role of snails and ants in transmitting the avocado stem canker pathogen, *Phytophthora citricola*. *J. Am. Soc. Hortic. Sci.* **1996**, *121*, 973–977. [CrossRef]
91. Lakatos, F.; Molnár, M. Mass mortality of beech (*Fagus sylvatica* L.) in South-West Hungary. *Acta Silv. Lignaria Hung.* **2009**, *5*, 75–82.
92. Montecchio, L.; Causin, R.; Rossi, S.; Mutto Accordi, S. Changes in ectomycorrhizal diversity in a declining *Quercus ilex* coastal forest. *Phytopathol. Mediterr.* **2004**, *43*, 26–34. [CrossRef]
93. Bloom, J.M.; Vannini, A.; Vettraino, A.M.; Hale, M.D.; Godbold, D.L. Ectomycorrhizal community structure in a healthy and a *Phytophthora*-infected chestnut (*Castanea sativa* Mill.) stand in central Italy. *Mycorrhiza* **2008**, *20*, 25–38. [CrossRef]
94. Luoma, D.L.; Eberhard, J.C. Relationships between Swiss needle cast and ectomycorrhizal fungus diversity. *Mycologia* **2014**, *106*, 666–675. [CrossRef]
95. Spinoni, J.; Vogt, J.V.; Barbosa, P.; Dosio, A.; McCormick, N.; Bigano, A.; Füssel, H.-M. Changes of heating and cooling degree-days in Europe from 1981 to 2100. *Int. J. Climatol.* **2018**, *38*, e191–e208. [CrossRef]

7

Phytophthora ramorum and *Phytophthora gonapodyides* Differently Colonize and Contribute to the Decomposition of Green and Senesced *Umbellularia californica* Leaves in a Simulated Stream Environment

Kamyar Aram * and David M. Rizzo

Department of Plant Pathology, University of California, One Shields Drive, Davis, CA 95616, USA; dmrizzo@ucdavis.edu

* Correspondence: kamaram@ucdavis.edu

Abstract: Plant pathogenic as well as saprotrophic *Phytophthora* species are now known to inhabit forest streams and other surface waters. How they survive and function in aquatic ecosystems, however, remains largely uninvestigated. *Phytophthora ramorum*, an invasive pathogen in California forests, regularly occurs in forest streams, where it can colonize green leaves shed in the stream but is quickly and largely succeeded by saprotrophically competent clade 6 *Phytophthora* species, such as *Phytophthora gonapodyides*. We investigated, using controlled environment experiments, whether leaf litter quality, based on senescence, affects how *P. ramorum* and *P. gonapodyides* compete in leaf colonization and to what extent each species can contribute to leaf decomposition. We found that both *Phytophthora* species effectively colonized and persisted on green or yellow (senescing) bay leaves, but only *P. gonapodyides* could also colonize and persist on brown (fully senesced and dried) leaves. Both *Phytophthora* species similarly accelerated the decomposition of green leaves and yellow leaves compared with non-inoculated controls, but colonization of brown leaves by *P. gonapodyides* did not affect their decomposition rate.

Keywords: leaf decay; oomycetes; invasive species; aquatic fungi; trophic specialization; saprotroph; pathogen; parasite

1. Introduction

The ecology of *Phytophthora*, a genus of fungal-like oomycetes historically erected and known for plant pathogenic species primarily associated with destructive diseases in agriculture [1], has undergone substantial reconsideration in recent years [2]. The recent emergence of a number of *Phytophthora*-caused plant epidemics in forests and other non-agricultural ecosystems has clearly shown that many members of the genus have potential as invasive species that can threaten natural ecosystems [2–4]. As a consequence of research in non-agricultural environments, a surprising diversity and abundance of *Phytophthora* species have been discovered, many previously undescribed [2,4]. Incidental to this research has been the discovery that many species of *Phytophthora* are abundant in natural surface waters, especially in streams. Many such species are so widespread and regularly encountered that they are now considered resident, if not endemic, and characteristic of such environments [5–22]. Nevertheless, isolates of well-known plant pathogenic species or species complexes are also regularly recovered, often without discernible symptoms or signs of disease on the vegetation [7,9,15,16,21,23–26].

Though the prevalence of *Phytophthora* in surface waters is now well established, the ecology underpinning this phenomenon is largely speculative. Because these organisms are known primarily as causes of often devastating plant diseases, the nature of their presence in these environments and its implications for the persistence and spread of pathogenic species are important considerations for disease prevention and management. There is also a growing interest to understand the role of *Phytophthora*, among other Peronosporales, in decomposition of vegetative matter in aquatic environments [27]. The biology of *Phytophthora*, a genus of well adapted plant pathogens with a necrotrophic phase [1,28], suggests that their ecological role in leaf decomposition should be early colonization and breakdown of relatively fresh, live vegetative tissue. As they colonize leaves newly exposed in streams, they can open the integral tissues for colonization by saprotrophic organisms less able to penetrate the leaf cuticle, in a process analogous to 'conditioning' of leaf litter for palatability to shredder organisms [29,30]. The co-occurrence of both known plant pathogens and primarily stream-associated *Phytophthora* in aquatic environments also raises the question of whether these taxa have similar or divergent modes of life and whether they compete for resources in these environments.

In streams, vegetative litter is the primary source of nutrients for microorganisms [29,31,32], but the quality of vegetative tissues available varies with respect to senescence and degree of decomposition. Coastal forests of northern California largely consist of evergreen trees and shrubs [33] and so green leaves are a regular component of leaf litter introduced into streams, especially in winter and spring when, based on the region's climate, most rainstorms occur. Nevertheless, much vegetative litter is in the form of senesced leaves [34]. California bay (*Umbellularia californica* (Hook. and Arn.) Nutt.) is a common, broadleaf evergreen component of northern California's coastal forests and a frequently occurring tree species in riparian zones [35,36]. It is also a primary source of *P. ramorum* inoculum in California forests affected by sudden oak death, epidemic mortality of certain species in the beech family (Fagaceae) resulting from *P. ramorum* infection of the vascular cambium of the main trunk [37,38]. California bay leaves are highly conducive to sporulation by *P. ramorum* which, despite causing localized necrotic lesions and spots on leaves, nevertheless causes little damage to the tree species itself [39–41]. Additionally, bay leaves are sclerophyllous, as is typical for broadleaf evergreen plants in this Mediterranean climate, and so they decompose slowly [33]. Bay leaves are therefore both very common as leaf litter in northern California forest streams and a highly suitable substrate for *P. ramorum*.

Leaf senescence in California bay increases in the hot and dry summer months, peaking in late summer [34,37]. Thus, though green leaves often enter streams during winter and spring storms, as summer progresses, most of the bay leaves shed into streams are either dropped directly upon senescence from trees or are blown in from accumulated litter on the forest floor, nearby (as described by [29,42]). In general, fully senesced leaves have as much as 75% reduced protein content compared with green leaves, primarily from the dismantling of chloroplasts, and though yellow, senescing leaves still have live cells with active mitochondria, leaves that have turned brown as a result of drying no longer contain biologically active cells [30,43,44]. Therefore, green, senescing and fully senesced bay leaves are substrates that likely vary in their suitability for colonization by *P. ramorum* and stream-resident clade 6 *Phytophthora* species, taxa that commonly occur at high inoculum levels in northern California coastal forest streams [45,46].

We have shown that there is a difference in trophic specialization between the saprotrophically competent, clade 6 *Phytophthora* species, such as *P. gonapodyides*, and *P. ramorum* [45], an aggressive pathogen on many plant species [38,47,48]. In that study, green California bay leaves were rapidly colonized by *P. ramorum* in streams but were succeeded nearly completely within three weeks by clade 6 *Phytophthora* species [45]. It remains uncertain, however, whether *P. ramorum* was displaced by more competent saprotrophs or receded from an inability to persist in tissues that it had colonized as they progressively decomposed. Additionally, as most leaf litter consists of senesced leaves, it is important to know how these differently adapted taxa can compete for and persist on biologically inactive leaf tissue. Finally, though stream resident *Phytophthora* species are assumed to contribute to leaf decay given their regular recovery from streams and frequent association with decomposing vegetation [2,27], experimental evidence for the kind and extent of this contribution is lacking. Moreover, it is unknown how the introduction of an exotic and plant pathogenic species, like *P. ramorum*, into a stream ecosystem might affect the decomposition of leaf litter by other organisms, such as resident *Phytophthora* species. Therefore we undertook a laboratory study to determine: (1) How well *P. ramorum* and *P. gonapodyides* could use senesced leaves as a substrate in comparison to green, live leaves, (2) whether colonization by and persistence of *P. ramorum* on leaves was affected by competition with *P. gonapodyides*, and (3) how much each of these *Phytophthora* species contribute to the decay of each leaf type.

2. Materials and Methods

2.1. Experiment Overview

To test the capacity of *P. ramorum* and *P. gonapodyides* to colonize green and senesced bay leaves, we conducted controlled environment experiments exposing leaves to an inoculum of each species alone and in combination in microcosms designed to simulate an aquatic environment (Supplemental Figure S1). The experiment consisted of a randomized complete block design with treatments representing a complete factorial of bay leaf type (green/live or brown/senesced), stream water addition (autoclaved or not), and *Phytophthora* inoculation (none, *P. ramorum*, *P. gonapodyides*, or combined *P. ramorum* and *P. gonapodyides*). These 16 treatment combinations were replicated in five blocks arranged in three growth chambers (model PGR-15, Conviron Controlled Environment Ltd). The experimental unit was a mesh packet of five leaves which were sampled at intervals over 16 weeks from microcosms. One treatment packet per sampling served for decomposition as percent biomass loss and another for colonization based on isolations on a selective medium. We repeated the experiment once, with leaf types maintained in the same microcosm in the first and in separate microcosms in the second experiment. We conducted a separate experiment with yellow, senescing leaves collected while still attached to trees and with the cuticle intact, with *P. ramorum*-only and combined *P. ramorum*/*P. gonapodyides* treatments as well as non-inoculated controls, in a completely randomized design with four reps in a single growth chamber.

2.2. Experiment Preparation

2.2.1. Leaves

We collected leaves from two sites where our previous field experiments were conducted [45]. One was a canyon through which Graham creek runs at Jack London State Park (38°21′2″ N, 122°33′16″ W) which consists of redwood forest with California bay as a dominant riparian tree, along with redwood (*Sequoia sempervirens* [Lamb. ex D. Don] Endl.), Douglas fir (*Pseudotsuga menziesii* [Mirb.] Franco), tanoak (*Notholithocarpus densiflorus* [Hook. & Arn.] P.S. Manos, C.H. Cannon, & S.H. Oh), bigleaf maple (*Acer macrophyllum* Pursh), and less frequently, madrone (*Arbutus menziesii* Pursh) [49,50]. The second included canyons around Copeland Creek at Sonoma State University's Fairfield Osborn Preserve (38°20′37″ N, 122°35′41″ W) which is characterized by mixed evergreen forest with a prevalence of California bay, white alder (*Alnus rhombifolia* Nutt.), big leaf maple, and occasionally, tanoak, madrone and coast live oak (*Quercus agrifolia* Née) [39,49]. At each site, we collected green, symptom-free bay leaves with a mature cuticle from trees and brown, recently shed bay leaves from beneath trees in the manner of Wood et al. [51]. Brown leaves were collected from both sites in September 2014, allowed to air dry in the laboratory, and then stored in sealed plastic bags at room temperature until used in experiments. Green leaves were collected on 12 December 2014 from the redwood forest site and on 5 August 2015 from the mixed evergreen forest site. Yellow leaves were collected directly from trees at the mixed evergreen forest site on 7 September 2015. Green leaves were stored at 4 °C for up to three weeks prior to use in experiments and yellow leaves were likewise stored but deployed in experiments within one week of collection. We collected leaves primarily from riparian areas around the described creeks, though, at the mixed evergreen forest site, we had to seek symptomless leaves to some extent from plateaus above the canyons. Leaves collected from each forest type were used in separate experiments. Leaf treatments were in the same microcosm for the first experiment (40 containers) using leaves from the redwood forest and separate in the second experiment (80 containers) with leaves from the mixed evergreen forest. Yellow leaves were collected from the mixed evergreen forest site only. We tested a subsample of 50 of each leaf type for both sites—through isolations attempted on a selective medium as described below—to verify that there were no pre-existing *Phytophthora* infections. Brown leaves were soaked in sterile deionized water at 4 °C for two days prior to these test isolations. Leaf packets were prepared for each leaf type by packing five leaves into a flat envelope of 1 mm plastic mesh approximately 20 × 20 cm so that the leaf surfaces were in minimal contact with one another and each packet was sealed by folding over the open lip and securing it with two common metal staples.

2.2.2. Microcosms and Water

We assembled microcosms simulating an aquatic decomposition environment similar to the approach described by Medeiros et al. [52] (Supplemental Figure S1). White plastic buckets (2 gal., 21 × 24 cm, dia. × ht., Argee Corp, Santee, CA, USA) were used in the first (leaf types together) and yellow leaf experiments and opaque plastic containers (8 qt., 19.4 × 27.3 cm, dia. × ht., Continental Carlisle, Oklahoma City, OK, USA) in the second experiment (leaf types separate). Each container was aerated through a tube terminating in an aeration stone (3 cm dia., Uxcell®, Hong Kong, China) fed by an air pump (Commercial Air 1, EcoPlus®, 18W, 793 GPH, 12/Cs, Hawthorne Gardening Co, Vancouver, WA, USA) that was turned on for 30 minutes twice daily using an electric timer (Intermatic TIME-ALL®, TN311, Spring Grove, IL, USA). Aeration intensity was moderated with the addition of adjustable valves inserted in the tubing. A dilute nutrient solution was used as the base for the water mixtures in microcosms in order to avoid osmotic stress on spores. This was achieved by adding Hoagland's #2 salts (Caisson Laboratories, Inc., Smithfield, UT, USA) to autoclaved Millipore®filtered water for a final concentration of 0.01× the standard concentration (1.63 g/L). To test for any effect of natural stream microbiota on *Phytophthora* colonization or leaf decomposition, we included an addition of autoclaved or non-sterilized stream water as a treatment factor. The final composition of water in microcosms consisted of 4 L nutrient solution and 2 L stream water in the first experiment, and 4 L nutrient solution and 1 L stream water in the second. We collected water from streams in a bucket, pouring it through several layers of cotton mesh ("cheesecloth") into 4 L plastic bladders that we consolidated into larger plastic containers or used directly to transport water out of the field. Once brought to the laboratory, stream water was stored in plastic containers in a growth chamber at 12 °C and 12 h photoperiod (≈ 1800 lux) for 20 and 23 days prior to deployment in the first and second experiments, respectively. After storing the water for seven days, we submerged symptomless California bay leaves collected at each site as baits in each container for two days to confirm that *Phytophthora* zoospores were not present. We tested baits for infection using the isolation technique described below. No *Phytophthora* infections were detected from baits at this point. In the experiment with yellow leaves, we used only 4 L of a nutrient solution without stream water addition.

We measured stream pH, electrical conductivity (EC) and temperature on site at the time of stream water collection and subsequently in each microcosm throughout the experiments with a portable sensor (Combo pH and EC tester, model 98129N, Hanna Instruments, Woodsocket, RI, USA). Stream pH, EC, and water temperature were 8.55, 208 μS/cm, and 13.5 °C, respectively, for the redwood forest stream on the 9 December 2014 collection date, and 8.08, 363 μS/cm, and 17.7 °C, respectively, for the mixed evergreen forest stream on the 5 Aug 2015 collection date. To approximate natural stream pH in microcosms, we amended the mix of dilute nutrient solution and stream water in each microcosm with potassium carbonate buffer ("pH UP", General Hydroponics, Santa Rosa, CA, USA) at approximately 10 mg/L and adjusted it with KOH and HCl for a target of pH 8.3. The average pH (± SD) measured in microcosms periodically over the course of experiments was 7.99 (±0.27), 8.33 (±0.38) and 8.24 (±0.27) in the first, second and yellow leaf experiments, respectively. The average EC (±SD) was 208 (±63), 141 (±18), and 98 (±15) in the first, second, and yellow leaf experiments, respectively.

In the first experiment, where green and brown leaves were maintained together in treatment microcosms, the water darkened from leaf leachates shortly after experiment initiation. After 52 days, we removed two liters of water from each microcosm using an auto-siphon—sanitized with a 10% bleach solution in between each treatment—and added a fresh sterile nutrient solution to bring the volume back up to six liters. For the second experiment, we leached leaves prior to deployment in the experiment in approximately 300 mL autoclaved Millipore®-filtered water per 10 leaves and the water did not darken to the extent observed in the first experiment.

During all experiments, we periodically topped off the microcosms with autoclaved Millipore®-filtered water to 6, 5 or 4 L in the first, second, and yellow leaf experiments, respectively.

2.2.3. *Phytophthora* Inoculum

Phytophthora inoculum consisted of three isolates per species grown for three weeks at 20 °C in 10 mL 10% clarified V8 juice liquid culture (V8®original vegetable juice (Campbells Soup Co., Camden, NJ, USA) neutralized with 15 g/L $CaCO_3$, clarified by centrifuging at 7000 RPM for 10 minutes and diluted with deionized water) for the first experiment and in 5 mL of the same liquid culture for the second and yellow leaf experiments. Inoculum was introduced as mycelial mats to each container to initiate experiments with leaf packets already present for 24 h. Each container received six total inoculum doses: Those receiving only *P. ramorum* or *P. gonapodyides* receiving two doses of each isolate and the combined inoculation treatments receiving one dose of each isolate of each *Phytophthora* species. We used the same isolates in both experiments, all collected from the stream at the redwood forest site described above. *Phytophthora ramorum* isolates Pr-1906, Pr-1907, Pr-1908 and *P. gonapodyides* isolates P-1903, P-1904 and P-1905 are maintained in D.M. Rizzo's laboratory. Isolates of both *Phytophthora* species were originally identified by morphology, and the identity of *P. gonapodyides* isolates was confirmed through ITS sequence BLAST matches in GenBank (GenBank accessions: MK908979, MK908980, MK908981).

2.2.4. Experiment Conditions and Sampling

Experiments were maintained with 12 h photoperiod (≈1800 lux) and 18/14 °C light/dark temperatures, respectively, to reflect typical average stream temperatures and also to provide a temperature differential that would potentially encourage *Phytophthora* zoospore release. Temperatures were monitored hourly in each block using iButton®loggers (Maxim Integrated, Inc., San Jose, CA, USA) to verify chamber settings. At 4, 8 and 16 weeks after inoculating microcosms, we sampled one leaf packet for evaluating leaf decomposition as biomass loss and another for evaluating *Phytophthora* colonization. For experiments with green and brown leaves, we included an additional sampling for *Phytophthora* colonization at two weeks. Therefore, in the first experiment each microcosm contained seven packets of each leaf type for a total of 14, and, as leaves were maintained in separate microcosms in the second experiment, each contained a total of seven packets containing either green or brown leaves. In the yellow leaf experiment, microcosms contained six packets each.

2.3. *Data Collection*

To determine the rate of decomposition measured as leaf biomass loss [53,54], we weighed leaves to the hundredth decimal of a gram with an analytical balance (model EP612C, Ohaus Corporation, Pine Brook, NJ, USA) prior to packing and we labeled the packets with aluminum tree tags secured with a plastic tie for future identification. We estimated the original dry mass of both leaf types from the average dry weight (determined after oven-drying at 55–60 °C for 48 h) of a subsample of 50 fresh or air dried leaves. The average percent dry weight (±SD) for green and brown leaves, respectively, was 40.9 (±2.3) and 94.3 (±0.3) for the first experiment and 53.5 (±0.4) and 92.6 (±0.3) for the second experiment. The average percent dry weight for yellow leaves was 55.0 (±0.3). The average estimated weight in grams (±SD) for five green leaves was 0.84 (±0.08) and 1.13 (±0.06), and that for five brown leaves, 0.91 (±0.08) and 0.85 (±0.06) for the first and second experiments, respectively. For five yellow leaves, the estimated average weight in grams was 0.98 (±0.06). At each sampling, leaves were retrieved from tagged leaf packets, rinsed gently with deionized tap water to remove adhering debris, oven-dried in a paper envelope or an open aluminum foil envelope at 55–60 °C for 48 h, and weighed as described above. The fraction of original biomass was calculated for all leaves in a packet by dividing the weight at the time of sampling by the estimated original dry biomass.

To determine the level of *Phytophthora* colonization of leaves, at each sampling we collected a packet for each leaf type from each container to evaluate by culturing on *Phytophthora*-selective PARP-H medium (corn meal agar 1.7% w/v, pimaricin 5 ppm, ampicillin 250 ppm, rifampicin 10 ppm, PCNB (pentachloronitrobenzene) 50 ppm and hymexazol 25 ppm, [1]). Upon retrieval, leaves were submerged and gently rubbed free of biofilm in 1% household bleach solution (≈65 ppm hypochlorite), surface sterilized in fresh bleach solution for three to seven minutes, rinsed with deionized tap water, and then laid out on paper towels and the excess water allowed to evaporate. Finally, leaves were wrapped in a paper towel and stored at 4 °C until isolations by culturing could be performed. Isolations were attempted from all leaves belonging to treatment (a single packet) using a 'mosaic' sampling approach whereby the leaf discs are removed from the petiole, midrib and flanking lobes of the leaf at approximately 1 cm distance from one another in order to collect a representative sample from the entire leaf [45,55]. For experiments with green and brown leaves, isolations were initiated immediately after collection, with most samples (75%) processed within 29 days. All isolations were completed by 46 days after collection. Storage period did not alter results when included as a covariate in models for these experiments and was excluded from the final analyses. Isolations from leaves of the yellow leaf experiment were completed within nine days after collection, and all isolations from a single collection week were completed in one day. The presence of *P. ramorum* and *P. gonapodyides* was determined by microscopic examination of isolate morphologies directly from the isolation plates after four to five days and checked again periodically for three weeks [45].

To test for active sporulation from colonized leaves in the microcosms, periodically a California bay leaf disc (12 mm dia.) was floated as bait—either naked or in a roughly 35 mm^2 mesh envelope—on the surface of the water in each microcosm for three to seven days, after which it was surface sterilized and isolations attempted from it on selective PARP-H medium. We conducted these tests of sporulation four times during the first and yellow leaf experiments, and three times during the second experiment. Additionally, we tested for sporulation periodically for up to eight weeks after all leaves had been removed from microcosms to determine if *Phytophthora* spores could persist in the absence of a substrate.

The first experiment was initiated on 29 December 2014, but we delayed the first collection at two weeks by two additional weeks because zoospores were not detected in the microcosms until two weeks after inoculation, most likely due to excessive aeration of the water during the first week. All subsequent collection dates were shifted forward by two weeks accordingly. Collections are reported according to the originally planned intervals of 2, 4, 8 and 16 weeks, with time zero being two weeks after inoculation. The final collection for the first experiment was on 1 May 2015 (126 days). Subsequent experiments proceeded as expected and the collection week reflects the period elapsed since introducing inoculum. The second experiment was initiated 28 August 2015 and concluded with the last sampling on 18 December 2015 (112 days). The yellow leaf experiment was initiated on 13 September 2015 and the final collection made on 7 January 2016 (116 days).

2.4. *Analysis*

Due to the differences in how each experiment was set up, we analyzed results separately for each.

2.4.1. *Phytophthora* Colonization

To evaluate the colonization of leaves by each *Phytophthora* species in each treatment, we recorded the total number of pieces yielding *P. ramorum* or *P. gonapodyides* out of the total number of pieces sampled for each leaf. The average proportion of leaves colonized by either species was calculated for each packet from this ratio. This average leaf fraction colonized per packet was *logit* transformed to normalize variances, with a +0.005 correction applied to values of zero and −0.005 to values of one before transformation [56]. The transformed average proportions colonized were analyzed in linear mixed models (*lme* function) with the *nlme* package [57] in R statistical software, version 3.3.1 [58]. Replication block and microcosm were set as random variables with microcosm nested in a block. Because *Phytophthora* recovery followed a non-linear trend with respect to time, we treated the collection week as a categorical variable. As one *Phytophthora* species occurred almost exclusively in each treatment (see Results below)—*P. ramorum* and *P. gonapodyides* in the treatments where they were inoculated solely and *P. gonapodyides* in the combined inoculations—we simplified the analysis by comparing leaf colonization by the dominant species across treatments. That is, the response variable in the model was the average fraction of leaf discs colonized by *P. ramorum* in *P. ramorum*-only treatments, and by *P. gonapodyides* in *P. gonapodyides*-only and combined *Phytophthora* inoculum treatments. Therefore, the main independent variables for *Phytophthora* leaf colonization analyses were the inoculation treatment—with non-inoculated treatments excluded—and collection week. Leaf type (green or brown) and stream water type (autoclaved or not) were included as independent variables in the model for the experiments where the distinctions applied. The full set of interactions were included in the models for each experiment (see supplemental Tables S1–S4 and S6). We verified adherence to model assumptions by the Shapiro–Wilk and Levene's tests. We obtained *P*-values using the *anova* function in R with the sum of squares set to type III ("marginal"), and least square means comparisons with the *lsmeans* package [59]. Significance for means comparisons was determined with the default Tukey's HSD.

2.4.2. Leaf Decomposition

For leaf decomposition, we estimated a decay constant (k) for each treatment combination in each block based on the fraction of estimated original leaf mass remaining at each collection interval [53,54]. For this, we used the exponential decay equation $M_t = M_0 \cdot e^{-kt}$ where t is time as the number of incubation days, M_t is the fraction of leaf mass remaining at each collection interval, and M_0, fraction at time zero, is set to one [53,54]. Values for k were estimated using the *nls* function in R statistical program. The decay constants for each treatment combination were then analyzed in a mixed model using the *lme* function of the *nlme* package with inoculum, leaf and water type as independent variables and block as a random factor. For the yellow leaf experiment, only inoculum was used as an independent variable, and since replications were not blocked, an analysis of variance was performed using the *aov* function in R. For all experiments, we included treatments not inoculated with either *Phytophthora* species in the analysis to evaluate the effect of *Phytophthora* colonization on leaf decay. Two non-inoculated microcosms in the first experiment were contaminated with both *Phytophthora* species, and one non-inoculated microcosm in the second experiment became contaminated with *P. ramorum*, likely from a rare, undetected leaf infection. We excluded the results from these microcosms from the analysis.

3. Results

3.1. Phytophthora Leaf Colonization

When *P. ramorum* was inoculated alone, it rapidly colonized most of the green leaf area and persisted at this level throughout the 16 weeks of incubation (Figure 1). It did not effectively colonize brown, senesced leaves, though it could occasionally be recovered from a few pieces of some leaves. In contrast, *P. gonapodyides* colonized most of the area of both green and brown leaves in microcosms where it was inoculated (Figure 1). However, *P. gonapodyides* colonized brown leaves to a significantly lesser degree than green leaves when the leaves were exposed to inoculum in separate microcosms, while there was no difference between the colonization of green and brown leaves when they were maintained in the same microcosm (Figure 1). In combined inoculations of both *Phytophthora* species, *P. ramorum* was unexpectedly suppressed on both leaf types and the recovery of *P. gonapodyides* from this treatment was identical to that of *P. gonapodyides*-only treatments (Figure 1). Reflecting these results, in both experiments with green and brown leaves, the interaction of inoculation and leaf type was highly significant ($p < 0.0001$, Tables S1 and S2).

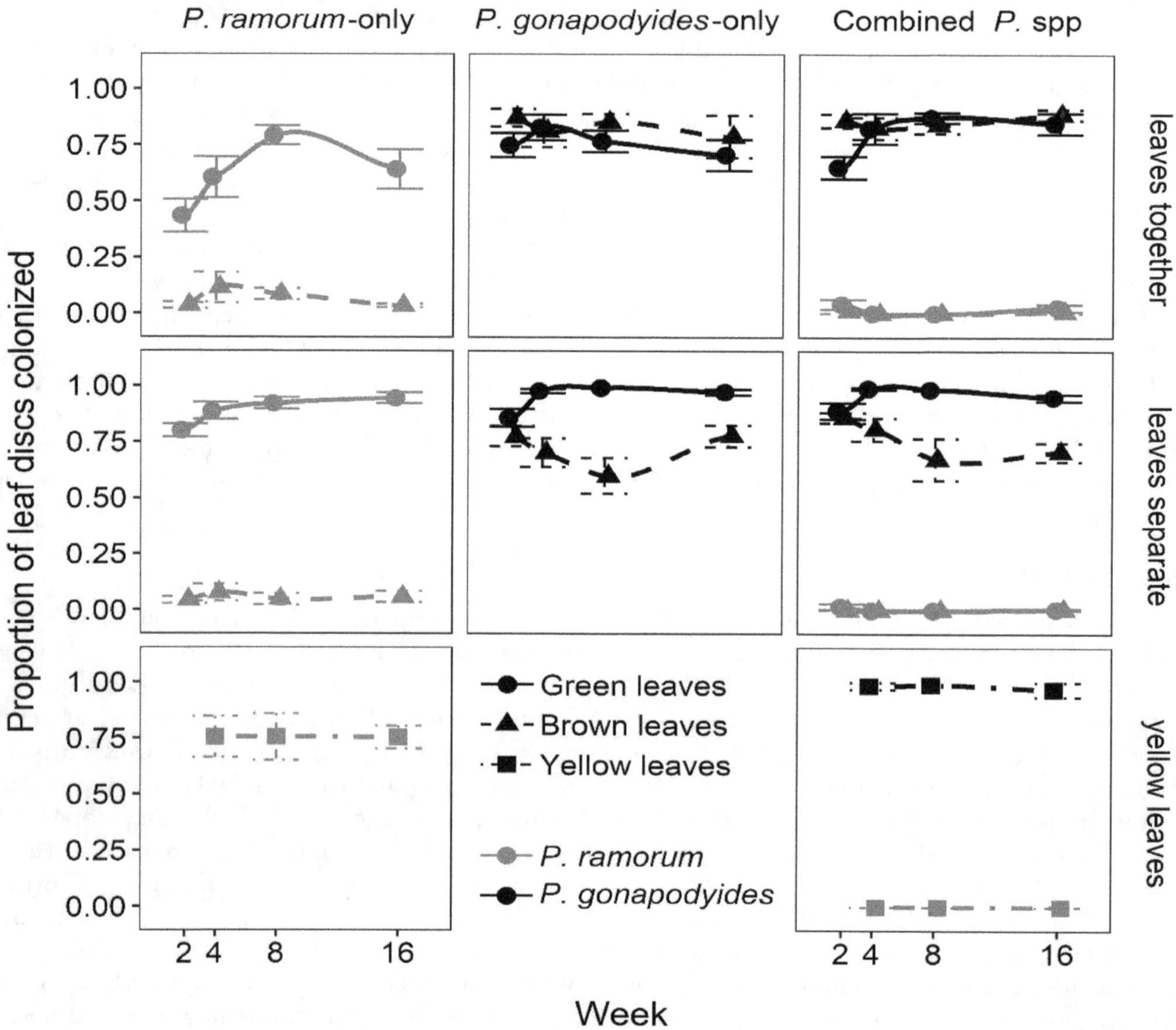

Figure 1. Proportion of green, brown and yellow leaves (designated by element shape and line type) colonized by *P. ramorum* or *P. gonapodyides* (designated by element and line shade)—determined as the proportion of leaf pieces colonized out of the total number sampled in "mosaic" isolations—for three different inoculum treatments (horizontal panels) at sampling intervals over 16 weeks of incubation in three different experiments (vertical panels). Two experiments included green and brown leaves, the first with both leaf types in the same microcosm and the second with each leaf type in different microcosms. One experiment included yellow leaves only with only *P. ramorum* and combined *Phytophthora* species inoculation treatments. Non-inoculated treatments are not shown, and results are averaged over stream water treatments which did not have a significant effect, except for the experiment with yellow leaves which used only sterile nutrient solution. Bars represent ± standard error, $n = 10$.

Though the difference was not significant in the overall model, in *P. ramorum*-only inoculated treatments where brown leaves were maintained separately from green leaves, *P. ramorum* colonized brown leaves at consistently higher levels in autoclaved water treatments compared with treatments with non-sterilized stream water added (0.094 and 0.030 mean fraction of leaf discs colonized, respectively). This difference was less apparent with green leaves (Supplemental Figure S2).

Nevertheless, there were no statistically significant differences between treatments based on stream water additions (Tables S1 and S2), and therefore, results are presented averaged over this factor.

In the experiment with yellow leaves, *P. ramorum*, when inoculated alone, colonized most of the leaf area and persisted at this level throughout the experiment, similar to the result with green leaves in other experiments (Figure 1). In combined *P. ramorum* and *P. gonapodyides* inoculations, *P. ramorum* was once again completely suppressed and *P. gonapodyides* colonized yellow leaves almost completely, at levels similar to its colonization of green leaves in both other experiments (Figure 1). The colonization of yellow leaves by *P. gonapodyides* in combined *Phytophthora* inoculum treatments was significantly higher than that by *P. ramorum* in *P. ramorum*-only inoculated treatments in this experiment, though both species colonized more than 70% of the leaf area.

Thus, in the experiment with yellow leaves, only the effect of *Phytophthora* inoculation was significant (p = 0.0159, Table S3). In all experiments, both *Phytophthora* species colonized leaves rapidly, in most cases reaching maximum levels by four weeks, and persisted at these levels throughout the 16 weeks experimental duration. A slight increase in the level of colonization by both *Phytophthora* species was apparent in many cases from two to four weeks, though for brown leaves maintained in separate microcosms in the second experiment, levels appeared to actually decline after the second week. This contrast is reflected in the significant interaction of leaf type and collection week for this experiment (p = 0.0135, Table S2).

3.2. Sporulation

The isolation of *P. ramorum* or *P. gonapodyides* from California bay leaf disc baits deployed on the water surface in microcosms indicated the presence of zoospores. The sum of successful bait isolations for each treatment across the five replication blocks in the green and brown leaf experiments, and across four replications in the yellow leaf experiment, are presented in Tables 1–3. The recovery of *P. gonapodyides* from *P. gonapodyides*-only and combined *Phytophthora* inoculation treatments was from nearly 100% of baits throughout the duration of all experiments.

The recovery of *P. ramorum* was more erratic, ranging from 40% to 90% of baits during the experiments with green and brown leaves. However, *P. ramorum* recovery from baits in microcosms that included green leaves and sterile rather than non-sterilized water was closer to 100%, excepting the second baiting of the second experiment, when *P. ramorum* was not recovered from most microcosms. *Phytophthora ramorum* was also rarely recovered by baiting from microcosms in the second experiment with only brown leaves, especially when excluding the first baiting, which was done a few days after inoculation.

Consistent with this, brown leaves were colonized at very low levels by *P. ramorum*. Nevertheless, at 14 weeks, *P. ramorum* could still be recovered from several of these microcosms (Table 2). *Phytophthora ramorum* was recovered somewhat more frequently from sterile than non-sterile stream water treatment. Such an effect was not apparent for *P. gonapodyides*. Both *Phytophthora* species were recovered at nearly 100% from baits throughout the yellow leaf experiment which used sterile dilute nutrient solution only (Table 3).

Additionally, we baited microcosms for weeks after all leaf packets had been collected to see how long spores may persist in the absence of leaves. *Phytophthora ramorum* could be recovered from a few microcosms up to six weeks after all leaves were removed, but its frequency generally diminished rapidly. In contrast, *P. gonapodyides* could be recovered for up to 12 weeks after all leaves had been removed from microcosms, and was relatively frequent even six weeks after leaves were removed in the second experiment.

Table 1. Count of *P. ramorum* (*Pr*) and *P. gonapodyides* (*Pg*) recovery from single leaf disc baits deployed for three to seven days in a total of five microcosms per treatment (i.e., out of five possible colonization events per sampling, 35 total. Dash indicates not inoculated and not recovered) in the first experiment where green and brown leaves were maintained together in microcosms with either sterile (st) or non-sterile (nst) stream water added. Grey shading indicates results from after all leaves had been removed from the microcosms (126 days).

		wk	day	wk	day	wk	day	wk	day	wk	day	wk	day	wk	day	Total	
		7	52	10	71	16	111	18	123	19	133	20	141	22	151		
Water	Inoculum	*Pr*	*Pg*	*Pr*	*Pg*	*Pr*	*Pg*	*Pr*	*Pg*	*Pr*	*Pg*	*Pr*	*Pg*	*Pr*	*Pg*	*Pr*	*Pg*
nst	*Pr*	3	-	3	-	4	-	3	-	0	-	1	-	1	-	15	-
st		5	-	5	-	5	-	3	-	3	-	2	-	0	-	23	-
nst	*Pg*	-	5	-	5	-	5	-	4	-	0	-	0	-	1	-	20
st		-	3	-	5	-	5	-	4	-	3	-	2	-	2	-	24
nst	*Pr* + *Pg*	0	5	0	5	0	5	0	5	1	3	0	0	0	0	1	23
st		0	5	0	5	0	5	1	3	0	2	0	0	0	1	1	21

Table 2. Count of *P. ramorum* (*Pr*) and *P. gonapodyides* (*Pg*) recovery from single leaf disc baits deployed for three to seven days in a total of five microcosms per treatment (i.e., out of five possible colonization events per sampling, 35 total. Dash indicates not inoculated and not recovered) in the second experiment where green and brown leaves were maintained in separate microcosms with either sterile (st) or non-sterile (nst) stream water added.. Grey shading indicates results from after all leaves had been removed from the microcosms (112 days).

			wk	day	wk	Day	wk	day	wk	day	wk	day	wk	day	wk	day	Total	
			0	2	10	69	14	100	17	117	19	134	22	153	28	193		
Water	Leaf	Inoculum	*Pr*	*Pg*	*Pr*	*Pg*	*Pr*	*Pg*	*Pr*	*Pg*	*Pr*	*Pg*	*Pr*	*Pg*	*Pr*	*Pg*	*Pr*	*Pg*
nst	Green	*Pr*	2	-	2	-	5	-	1	-	0	-	0	-	0	-	10[a]	-
st			5	-	0	-	5	-	3	-	5	-	2	-	0	-	20	-
nst	Brown		5	-	0	1	2	-	0	-	1	-	0	-	0	-	8	1
st			5	-	0	-	1	-	0	-	0	-	0	-	0	-	6	-
nst	Green	*Pg*	-	5	-	5	-	5	-	4	-	4	-	5	-	2	-	30 a
st			-	5	-	5	-	5	-	5	-	4	-	5	-	1	-	30
nst	Brown		-	5	-	5	-	5	-	3	-	2	-	4	-	2	-	26
st			-	5	-	5	-	5	-	4	-	3	-	4	-	1	-	27
nst	Green	*Pr* + *Pg*	0	4	0	3	0	5	0	3	0	4	0	5	0	2	0	26 a
st			0	5	0	4	0	5	0	1	0	3	0	2	0	3	0	23

Table 2. *Cont.*

			wk	day	wk	Day	wk	day	wk	day	wk	day	wk	day	wk	day	Total	
			0	2	10	69	14	100	17	117	19	134	22	153	28	193		
Water	Leaf	Inoculum	*Pr*	*Pg*	*Pr*	*Pg*	*Pr*	*Pg*	*Pr*	*Pg*	*Pr*	*Pg*	*Pr*	*Pg*	*Pr*	*Pg*	*Pr*	*Pg*
nst	Brown		0	5	0	5	0	5	0	4	0	1	0	4	0	1	0	25
st			1	5	0	5	0	5	0	4	0	1	0	4	0	1	1	25

a: Out of 34 total attempts in each row per *Phytophthora* species.

Table 3. Count of *P. ramorum* (*Pr*) and *P. gonapodyides* (*Pg*) recovery from single leaf disc baits deployed for three to seven days in a total of four microcosms per treatment (i.e., out of four possible colonization events per sampling, 28 in total. Dash indicates not inoculated and not recovered) with sterile water only in the yellow leaf experiment. Grey shading indicates results when all leaves had been removed from the microcosms (116 days).

	wk	Day	wk	day	wk	day	wk	day	wk	day	wk	day	wk	day	Total	
	8	54	12	85	14	97	15	102	17	119	20	138	25	178		
Inoculum	*Pr*	*Pg*	*Pr*	*Pg*	*Pr*	*Pg*	*Pr*	*Pg*	*Pr*	*Pg*	*Pr*	*Pg*	*Pr*	*Pg*	*Pr*	*Pg*
Pr	3	-	4	-	4	-	4	-	4	-	1	-	0	-	20	-
Pr + *Pg*	0	4	0	4	0	4	0	4	0	4	0	4	0	4	0	28

3.3. Leaf Decomposition

In the experiment with green and brown leaves maintained in the same microcosm, only green leaves in microcosms with no *Phytophthora* inoculum decomposed at a slower rate than all other treatments (Figure 2). In fact, on average, they did not lose significant biomass throughout the 16 weeks. Leaves in all other treatments, including brown leaves in non-inoculated microcosms, decomposed at similar rates (Figure 2). The interaction of leaf type and *Phytophthora* inoculation was, therefore, a highly significant predictor in the model ($p < 0.0001$, Table S4). Estimated decay constants are listed in Table S5.

In the experiment where green and brown leaves were maintained in different microcosms, green leaves in microcosms with *Phytophthora* inoculum decomposed faster than all other treatments (Figure 2). In this experiment, all treatments with brown leaves and green leaf treatments with no *Phytophthora* decomposed at similar rates. Notably, in contrast to the other experiment, green leaves in non-inoculated treatments in this experiment did decompose over the 16 weeks, ultimately achieving a similar level of biomass loss as green leaf treatments with *Phytophthora* inoculum. Nevertheless, reflecting the difference in decomposition rate for green leaves in inoculated and non-inoculated treatments, the effect of the leaf type by *Phytophthora* inoculum interaction was significant in the model ($p = 0.0094$, Table S6). Estimated decay constants are listed in Table S7.

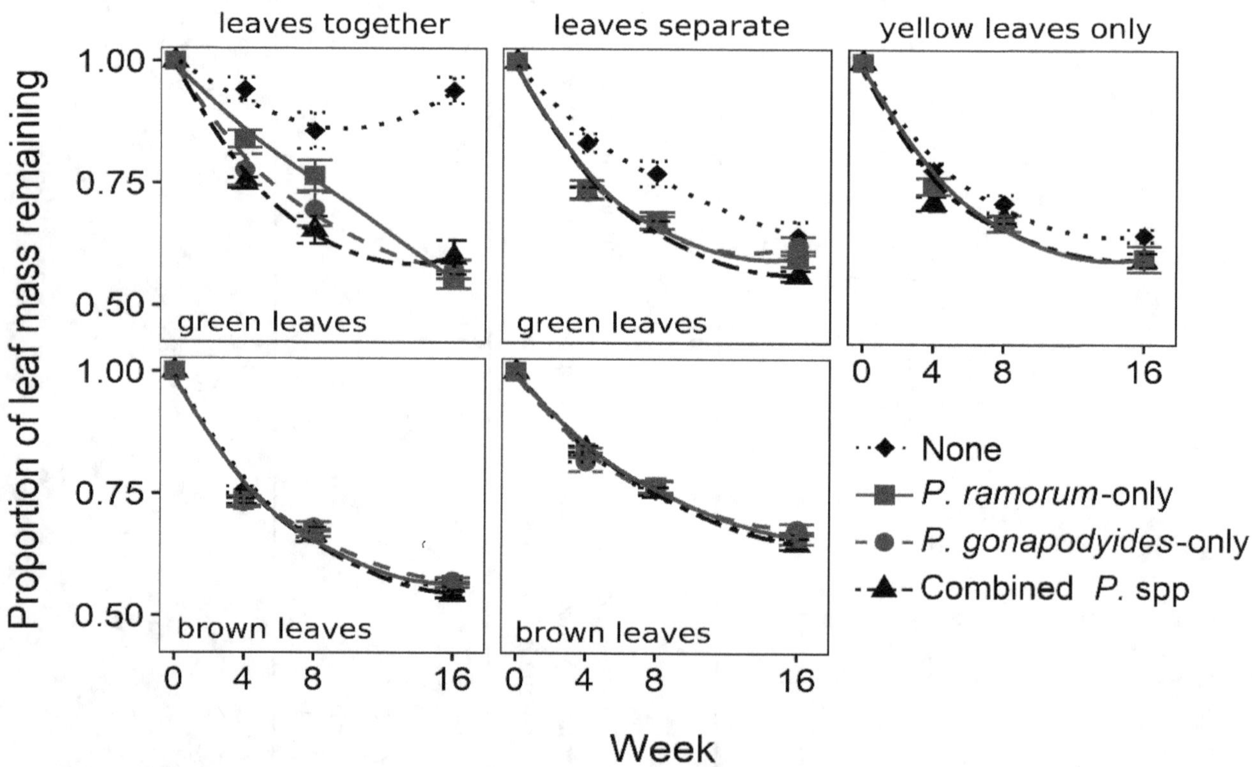

Figure 2. Leaf decomposition represented as a proportion of leaf mass remaining at sampling intervals over 16 weeks for green (top left two panels), brown (bottom panels) and yellow leaves exposed to three *Phytophthora* inoculum treatments and one non-inoculated control in three different experiments (vertical panels). Two experiments included green and brown leaves, the first with both leaf types together in the same microcosm and the second with each leaf type separate in different microcosms. One experiment included yellow leaves only with only *P. ramorum* and combined *Phytophthora* species inoculum treatments. Results are averaged over stream water treatments which did not have a significant effect, except for the experiment with yellow leaves which used only sterile nutrient solution. Bars represent ± standard error, $n = 10$.

In the experiment with yellow leaves only, leaves in the non-inoculated treatment decomposed at a slightly but significantly lower rate than *Phytophthora*-inoculated treatments ($p = 0.0292$, Figure 2, Table S8), of which the decomposition rates were not statistically different from one another (Table S9). The decomposition rate of yellow leaves in *Phytophthora*-inoculated treatments was similar to that of green leaves in *Phytophthora*-inoculated treatments of the second experiment with which it was essentially concurrent, though the results of the different experiments were not statistically compared.

4. Discussion

The goal of these experiments, broadly, was to better understand how the previously observed differences in trophic specialization between *P. ramorum* and *P. gonapodyides* [45], the latter as a representative of stream-resident clade 6 *Phytophthora* species, affected their ability to utilize different kinds of leaf litter available in streams. More specifically, we sought to determine if the previously observed decline of *P. ramorum* in green leaves decomposing in streams [45] was due to competitive displacement by saprotrophic organisms or due to an intrinsic inability of this pathogenic species to persist on colonized but decomposing leaf tissue, and to discover if the observed specialization of each species as pathogen or saprotroph would be consequential for the colonization of senescent or fully senesced leaves, a factor that has important implications regarding the prevalence of suitable leaf litter substrate for these organisms. Additionally, we wanted to test the contribution by each *Phytophthora* species to leaf decomposition and to determine if there was any difference depending on leaf senescence based on their differing trophic adaptations. While the inclusion of natural stream water in these experiments is an imperfect approximation of natural conditions, namely in excluding both shredder organisms and other microorganisms eliminated in the holding period, it had the potential to reflect the interaction of the inoculated *Phytophthora* species with bacteria, protozoa, fungal communities, and possibly micro-invertebrates that persisted in stream water. Though the effect of stream water treatment was not statistically significant in the models, a noticeably higher occurrence of *P. ramorum* on brown leaves in sterile stream water treatments compared with non-sterilized stream water additions (Supplemental Figure S2) and also higher detection of *P. ramorum* spores by baiting in sterile compared to non-sterile stream water treatments (Tables 1–3) both indicated that there was some difference between the two treatments. Though a much greater diversity of organisms likely influences this system under natural conditions, our previous research exposing leaves in natural streams demonstrated both *P. ramorum* and clade 6 *Phytophthora* species effectively colonize California bay leaves under natural conditions [45].

As expected, based on previous work [45], both *Phytophthora* species rapidly colonized more than 60% of the leaf area of green leaves in both experiments. That *P. ramorum* also persisted on green leaves at high levels for the entire 16 weeks despite the loss of approximately 40% of leaf biomass stands in contrast to our previous findings where its colonization of leaves peaked within a few weeks after exposure in natural streams, but then rapidly dropped to very low levels as colonization by clade 6 *Phytophthora* species rose and persisted at high levels [45]. This is evidence that the reduced recovery of *P. ramorum* from green leaves in natural streams as decomposition progressed was due to displacement from saprotrophic organisms like clade 6 *Phytophthora* species. Unfortunately, *P. ramorum* was completely suppressed from colonizing leaves in combined inoculations with *P. gonapodyides* and it could not be determined if the pattern observed in field experiments would occur under these simulations when both species were present. The suppression of *P. ramorum* colonization of green leaves in combined *Phytophthora* inoculations—consistent across all three experiments—was surprising because both species were effective at colonizing leaves when inoculated alone. One explanation could be that sporulation of *P. gonapodyides* from mycelial mats occurred earlier than that of *P. ramorum* and that the latter was therefore precluded from leaves because in all experiments, full colonization of green leaves by *P. gonapodyides* occurred very rapidly. Indeed, in the first experiment, colonization of *P. gonapodyides* occurred more rapidly on green leaves than that of *P. ramorum* (Figure 1). However, baiting two days after inoculation in the second experiment showed that *P. ramorum* spores were active in the microcosms where it was inoculated alone, but almost absent in the combined inoculation microcosms. This suggests that the presence of *P. gonapodyides* itself may have suppressed sporulation by *P. ramorum*. The rapid leaf colonization by *P. gonapodyides* in these microcosms also contrasts with the slower colonization that was observed in natural streams [45] and may be an artifact of high inoculum loads and the relative abundance of substrate. The aim of these experiments was to characterize the capacity of each organism for growing and reproducing from each type of leaf rather than estimating typical colonization and decomposition in streams. Though logistically more difficult to prepare and standardize for an experiment of this magnitude, using sporangia or zoospore inoculum rather than mycelial mats may overcome the problem of uneven inoculum activation, the success of which we have experienced in smaller scale experiments [45]. Alternatively, the use of colonized plant tissue (e.g., leaf discs) instead of mycelial mats as a source of inoculum may also produce a different outcome from the suppression of *P. ramorum* that we found with this approach in mixed inoculations. Interestingly, the kind of succession observed in field experiments did occur in a few control microcosms into which both *Phytophthora* species were accidentally contaminated (data not shown). However, the limited occurrence and unknown relative quantity of original inoculum precluded more substantial evaluation. In any case, the suppression of *P. ramorum* sporulation in treatments where *P. gonapodyides* was present raises the question of what mechanism was responsible for the effect. It also furthers the impression that *P. gonapodyides* and other clade 6 *Phytophthora* species may have a moderating effect on the presence of *P. ramorum* in streams.

The green leaves that we used were of mature cuticle and collected in midwinter and late summer for the first and second experiments, respectively. While some seasonal variation in susceptibility to *P. ramorum* infection has been reported in California bay leaves [60,61], the physical and chemical properties of mature leaves have also been reported to be relatively consistent throughout the year [62]. Our results were similar for both experiments, and therefore, any variation in the leaves was overcome by experiment factors.

The extensive colonization of brown leaves by *P. gonapodyides* and their limited colonization by *P. ramorum* is consistent with previous work where we showed that the former is a competent saprotroph while the latter is relatively ineffective at colonizing dead tissue [45]. A significant discovery in this work was that *P. ramorum* colonized yellow, senescent leaves that were still fresh and had an intact cuticle to nearly the same degree as it did green leaves. At this stage, though chloroplasts and most of the protein content are gone from leaves, the cells are expected to be still alive, while in brown leaves that have dried the cells are no longer biologically active [30,43,44]. In fact, colonization of the yellow leaves by *P. ramorum* was not quite as extensive as its colonization of green leaves in the second experiment, which ran more or less concurrently and in which green and brown leaves were maintained in separate microcosms (Figure 1), though the difference between the separate experiments was not analyzed statistically. Though green leaves are shed into streams as a relatively low proportion of total litter, yellow leaves, often shed directly into streams from trees, constitute a much greater proportion of leaf litter in streams (Aram, personal observation, see also [29]). This indicates that a great proportion of leaf litter in the streams is suitable for colonization by *P. ramorum*, and conforms to the regular recovery of this pathogen from natural leaf litter [45]. Furthermore, the degree of colonization of yellow leaves by both *Phytophthora* species remained persistent throughout the 16 weeks, as with green leaves in the other experiments, suggesting that the same kind of succession may be expected in these leaves as seen with green leaves in natural streams [45].

Also consistent with previous findings with leaves colonized in naturally infested streams [45], leaves colonized by both *Phytophthora* species were generally conducive to sporulation as detected by baiting from the microcosms. *Phytophthora gonapodyides* was consistently recovered from *P. gonapodyides*-only and combined *Phytophthora* species inoculation treatments where it had colonized all green and brown leaves at all sampling points. The results from baiting of *P. ramorum* spores from microcosms were less regular, but nonetheless, mostly successful from microcosms containing colonized green or yellow leaves and occurred minimally from microcosms containing brown leaves which were colonized at only very low levels. The relatively less frequent recovery of *P. ramorum* by baiting from microcosms with non-sterilized stream water, not observed for *P. gonapodyides*, may be the consequence of *P. ramorum* not being well adapted to sporulation in biologically active aquatic environments or relying on different environmental signals. Nevertheless, these results confirm that both of these *Phytophthora* species can sporulate from colonized, decomposing leaves, whether green, yellow or brown leaves. Furthermore, at least under these conditions, their spores persisted for weeks and even months after any visible substrate was available, though the effect occurred more definitively and for longer with *P. gonapodyides*. As *P. gonapodyides* is not known to produce long-term survival structures, the question arises of how *P. gonapodyides* persisted so long in the microcosms in the absence of leaves. This observation also stands in contrast to our successful elimination of *Phytophthora* spores from original stream water collections simply by holding the water at cool temperatures for approximately three weeks. The observed persistence of spores of both *Phytophthora* species may be the result of an abundance of zoospore cysts due to the compact nature of the microcosms, or perhaps because the spores originated from propagules that would not have been suspended in the water column of the flowing streams.

While oomycetes have been acknowledged as decomposers in aquatic environments until recently they have primarily been regarded as acting on non-cellulosic detritus such as insect and animal tissue [63]. As most *Phytophthora* species are known as plant pathogens, the recent evidence that they may also degrade plant tissue in detritus is not surprising [64–66]. Parasitism is considered an early characteristic in the evolution of oomycetes, [67], but the possible evolution of a saprotrophic lifestyle from parasitic precursors has been considered for fungi and oomycetes [67,68]. Clade 6 *Phytophthora* are known to be opportunistic pathogens [2,69–71]. Stradling saprotrophic and parasitic lifestyles, stream-resident *Phytophthora* may play an important role in the early breakdown of leaves and vegetative matter that still contain living cells. As facultative pathogens, [2,69–71] clade 6 *Phytophthora* species can enter living cells and open intact tissues to further colonization by other saprotrophic organisms with less ability to penetrate living tissue. This is analogous to the paradigm of 'conditioning' of vegetative litter by pioneer microbial species [29,30], though in this case with respect to secondary saprotrophic microorganisms that could not on their own overcome physical and chemical protections still present in senescent but still alive leaf tissue. Our results were consistent with this hypothesis, as green leaves decayed more slowly in the absence of *Phytophthora*. It is uncertain why in the first experiment green leaves in the treatments with no *Phytophthora* inoculation decomposed very little over the entire 16 weeks of the experiment. In this experiment, both green and brown leaves were maintained together in microcosms, and it is possible that leachate from the leaves, particularly the brown leaves, may have had an inhibitory effect on some microorganisms. In the second experiment, leaves were leached prior to being deployed in the experiment, and also green and brown leaves were kept in separate microcosms. Green leaves in non-inoculated controls in the second experiment lost biomass to a degree ultimately similar to that of inoculated treatments, albeit at a slower rate. This indicates that other organisms were present that could initiate the decomposition of green leaves through the presence of *Phytophthora* accelerated it. We attempted additional isolations from some samples of leaves on acidified potato dextrose agar medium and found that the leaves in both controls and inoculated treatments were generally well colonized by a multitude of fungi (data not shown). The fact that similar fungi occurred on leaves from microcosms prepared with both sterile and non-sterilized stream water suggests that many of these fungi were present on the leaves before entering streams as leaf litter (e.g., [72]). A diversity of fungi have been reported from bay leaves in coastal California forests [73]. Additionally, overall there were no differences in decomposition rates between treatments with sterile or non-sterilized stream water added. Decomposition was also similar for leaves colonized by either *Phytophthora* species, indicating that, though *P. gonapodyides* is a better adapted saprotroph, both species had a similar effect on the decomposition of live, green and yellow leaves. This would be consistent with *Phytophthora* having the effect of opening integral tissue to colonization by other saprotrophs that then push decomposition forward. Finally, it is interesting that the presence of fungi in these leaves did not affect the persistence of *P. ramorum* throughout the experiments, suggesting that they are using different resources and that the successive displacement of *P. ramorum* in previous work may be specific to competition with other *Phytophthora* species or similar organisms. Under natural conditions, leaves would be exposed to a greater diversity of organisms, including other oomycetes such as *Phytopythium* species [74].

As *P. gonapodyides* can colonize dead leaf tissue, it could be expected that it would contribute to leaf decay in brown leaves as well. This was not observed, as loss of biomass in brown leaves was the same in all treatments unaffected by *Phytophthora* colonization. The fact that *P. gonapodyides* substantially colonized brown, senesced leaves, but did not increase the rate of biomass loss raises the question as to what resources the organism uses in this substrate. Though biomass loss is a useful measure of decomposition [29,53], it does not offer a complete picture and other measures, such as changes in leaf toughness or chemical properties may offer a fuller picture of decomposition [29,75] that could account for the effects of *Phytophthora* colonization. Moreover, decomposition of brown leaves proceeded more slowly in the second experiment than the first. This may be due to lower nitrogen and other nutrient availability both because in the first experiment green and brown leaves were maintained together in microcosms and also that in the second experiment, the leaves were leached prior to being introduced into microcosms at the start of the experiment [28]. This may also be the reason that colonization of brown leaves by *P. gonapodyides* was significantly less than that of green leaves when leaves were kept in separate microcosms, while the levels were similar when leaves were maintained in the same microcosms. Another possibility is that sporulation from green leaves allowed greater colonization of brown leaves where the leaves were kept in the same microcosm.

Our results demonstrate that green and yellow California bay leaves are suitable substrates for the growth, colonization, and sporulation of *P. ramorum* in streams where they constitute a significant proportion of vegetative litter, they likely play an important part of supporting the inoculum load in streams. Yellow leaves resemble green ones in that, in contrast with brown leaves, they have an intact cuticle, and their cells are essentially still alive. California bay leaves infected by *P. ramorum* have been shown to senesce and abscise from trees more frequently than uninfected leaves [37]. In infested forests, a great portion of senescent leaves probably enters the stream already colonized by *P. ramorum* [37,45]. As leaves that fall into the water do not dry out, their cells likely remain alive for an extended period, allowing further colonization by *P. ramorum*. However, stream resident clade 6 *Phytophthora* species also compete for this substrate and may limit the extent to which *P. ramorum* can grow on, persist, and reproduce from them [45]. As dry, brown, senesced California bay leaves begin to make up a greater proportion of leaf litter in late summer and fall, the ability of clade 6 *Phytophthora* species to exploit these, while *P. ramorum* cannot, may be one explanation for why the latter is recovered less regularly and with lower frequency from these and other California streams in the fall and early winter [46]. Moreover, as the summer progresses, green and yellow leaves will be more decomposed and less suitable for *P. ramorum*. The warming of streams late in the summer may additionally favor clade 6 *Phytophthora* species that are known to have generally higher optimal growth temperatures than most other species [5]. We maintained temperatures constant for experimental purposes, but the persistence and sporulation of these *Phytophthora* species, and *P. ramorum* in particular, may be significantly affected by temperature fluctuations and extremes.

We have isolated *P. ramorum* and clade 6 *Phytophthora* from leaf litter of other tree species in naturally infested streams, including leaves of coast redwood, madrone, white alder, big leaf maple, and coast live oak (authors' unpublished data). Occasionally, we have found portions of other submerged riparian plants, such as chain fern (*Woodwardia fimbriata* Sm.) or elk clover (*Aralia californica* S. Watson), to be colonized (authors' unpublished data). It is well-established that California bay leaves are an optimal substrate for *P. ramorum*, and though clade 6 *Phytophthora* species are known from a great variety of vegetative litter, it is uncertain how conducive other vegetative litter would be to survival and sporulation of either species. Stamler et al. [20] recovered primarily clade 6 and 9 *Phytophthora* species from rivers in the southwestern USA using leaves of *Salix* and *Populus* species, common as riparian trees, as bait. It would be expected that natural leaf litter in such ecosystems would also harbor these organisms. Themann et al. [76] recovered primarily *P. gonapodyides* but also *P. cinnamomi* from vegetative litter in sediments in an irrigation reservoir. Therefore, leaf and other vegetative litter should be considered as potential sources of *Phytophthora*, including pathogenic species, whether they are found in natural streams or other surface waters. Alternatively, the suitability of local vegetation may be a determinant of what *Phytophthora* species become established or prominent in streams.

5. Conclusions

With these studies, we have demonstrated that the trophic specializations of *Phytophthora* species in coastal California streams determine what leaf litter is available to them, but that nevertheless, suitable leaf litter is available throughout much of the year for both *P. ramorum* and clade 6 *Phytophthora* species. The role of stream resident *Phytophthora* species in leaf decay is probably one analogous to "conditioning" of fresh leaf litter (i.e., opening biologically integral tissues), essentially accelerating the earliest stages of decomposition. Nevertheless, they continue to persist and sporulate even as leaves become substantially decomposed. Green and yellow California bay leaves were similarly conducive to colonization and sporulation by both *P. ramorum* and *P. gonapodyides*, and the effects of both *Phytophthora* species on the decomposition of these leaves were similar. *Phytophthora ramorum* could not, however, colonize brown, biologically dead leaves, and though *P. gonapodyides* colonized brown leaves, it did not contribute to leaf decomposition as measured by loss of biomass. These studies expand the current knowledge about the ecological role of *Phytophthora* in streams.

Author Contributions: Project conceptualization and methodology were developed by K.A. and D.M.R. Funding acquisition and procuring of resources by K.A. and D.M.R. Project administration, investigation, formal analysis, and writing of original manuscript draft, including figures, by K.A. The manuscript was reviewed, revised and edited by K.A. and D.M.R.

Acknowledgments: The authors are grateful to Heather Mehl, Kero Agaiby, Nicole Beadle, Rino Oguchi, Grace Scott and Valerie Wuerz for technical assistance; to Christina Freeman (California State Parks) and Suzanne Decoursey (Sonoma State University) for site access and permitting; to Neil Willits (UC Davis) for statistical consulting.

References

1. Erwin, D.C.; Ribeiro, O.K. *Phytophthora diseases worldwide.*; APS Press: St. Paul, Minnesota, USA, 1996; ISBN 978-0-89054-212-.
2. Hansen, E.M.; Reeser, P.W.; Sutton, W. *Phytophthora* beyond agriculture. *Annu. Rev. Phytopathol.* **2012**, *50*, 359–378. [CrossRef]
3. Hansen, E.M. Alien forest pathogens: *Phytophthora* species are changing world forests. *Boreal Environ. Res.* **2008**, *13*, 33–44.
4. Kroon, L.P.; Brouwer, H.; de Cock, A.W.; Govers, F. The genus *Phytophthora* anno 2012. *Phytopathology* **2012**, *102*, 348–364. [CrossRef] [PubMed]
5. Brasier, C.M.; Cooke, D.E.L.; Duncan, J.M.; Hansen, E.M. Multiple new phenotypic taxa from trees and riparian ecosystems in *Phytophthora gonapodyides-P. megasperma* ITS Clade 6, which tend to be high-temperature tolerant and either inbreeding or sterile. *Mycol. Res.* **2003**, *107*, 277–290. [CrossRef]
6. Brazee, N.J.; Wick, R.L.; Hulvey, J.P. *Phytophthora* species recovered from the Connecticut River Valley in Massachusetts, USA. *Mycologia* **2016**, *108*, 6–19. [CrossRef] [PubMed]
7. Chastagner, G.; Oak, S.; Omdal, D.; Ramsey-Kroll, A.; Coats, K.; Valachovic, Y.; Lee, C.; Hwang, J.; Jeffers, S.; Elliott, M.; et al. Spread of *Phytophthora ramorum* from nurseries into waterways—implications for pathogen establishment in new areas. In *Proceedings of the Sudden Oak Death Fourth Science Symposium*; Gen. Tech. Rep. PSW-GTR-229; U.S. Department of Agriculture, Forest Service, Pacific Southwest Research Station: Albany, CA, USA, 2010; pp. 22–28.

8. Dunstan, W.A.; Howard, K.; Hardy, G.E.S.; Burgess, T.I. An overview of Australia's *Phytophthora* species assemblage in natural ecosystems recovered from a survey in Victoria. *IMA Fungus* **2016**, *7*, 47–58. [CrossRef]
9. Greslebin, A.G.; Hansen, E.M.; Winton, L.M.; Rajchenberg, M. *Phytophthora* species from declining *Austrocedrus chilensis* forests in Patagonia, Argentina. *Mycologia* **2005**, *97*, 218–228. [CrossRef]
10. Hansen, E.M.; Reeser, P.W.; Sutton, W. *Phytophthora borealis* and *Phytophthora riparia,* new species in *Phytophthora* ITS Clade 6. *Mycologia* **2012**, 11–349. [CrossRef] [PubMed]
11. Hong, C.; Gallegly, M.E.; Richardson, P.A.; Kong, P.; Moorman, G.W. *Phytophthora irrigata*, a new species isolated from irrigation reservoirs and rivers in Eastern United States of America. *FEMS Microbiol. Lett.* **2008**, *285*, 203–211. [CrossRef]
12. Huai, W.-X.; Tian, G.; Hansen, E.M.; Zhao, W.-X.; Goheen, E.M.; Gruenwald, N.J.; Cheng, C. Identification of Phytophthora species baited and isolated from forest soil and streams in northwestern Yunnan province, China. *For. Pathol.* **2013**, *43*, 87–103.
13. Hwang, J.; Oak, S.W.; Jeffers, S.N. Recovery of *Phytophthora* species from drainage points and tributaries within two forest stream networks: a preliminary report. *N. Z. J. For. Sci.* **2011**, *41S*, S83–S87.
14. Jung, T.; Stukely, M.J.C.; Hardy, G.E.S.T.J.; White, D.; Paap, T.; Dunstan, W.A.; Burgess, T.I. Multiple new *Phytophthora* species from ITS Clade 6 associated with natural ecosystems in Australia: evolutionary and ecological implications. *Persoonia - Mol. Phylogeny Evol. Fungi* **2011**, *26*, 13–39. [CrossRef] [PubMed]
15. Nagel, J.H.; Slippers, B.; Wingfield, M.J.; Gryzenhout, M. Multiple *Phytophthora* species associated with a single riparian ecosystem in South Africa. *Mycologia* **2015**, *107*, 915–925. [CrossRef] [PubMed]
16. Oh, E.; Gryzenhout, M.; Wingfield, B.D.; Wingfield, M.J.; Burgess, T.I. Surveys of soil and water reveal a goldmine of *Phytophthora* diversity in South African natural ecosystems. *IMA Fungus* **2013**, *4*, 123–131. [CrossRef]
17. Reeser, P.W.; Sutton, W.; Hansen, E.M.; Remigi, P.; Adams, G.C. *Phytophthora* species in forest streams in Oregon and Alaska. *Mycologia* **2011**, *103*, 22–35. [CrossRef]
18. Shrestha, S.K.; Zhou, Y.; Lamour, K. Oomycetes baited from streams in Tennessee 2010–2012. *Mycologia* **2013**, *105*, 1516–1523. [CrossRef]
19. Sims, L.L.; Sutton, W.; Reeser, P.; Hansen, E.M. The *Phytophthora* species assemblage and diversity in riparian alder ecosystems of western Oregon, USA. *Mycologia* **2015**, *107*, 889–902. [CrossRef]
20. Stamler, R.A.; Sanogo, S.; Goldberg, N.P.; Randall, J.J. *Phytophthora* Species in Rivers and Streams of the Southwestern United States. *Appl. Environ. Microbiol.* **2016**, *82*, 4696–4704. [CrossRef]
21. Bily, D.S.; Diehl, S.V.; Cook, M.; Wallace, L.E.; Sims, L.L.; Watson, C.; Baird, R.E. Temporal and Locational Variations of a Phytophthora spp. Community in an Urban Forested Water Drainage and Stream-runoff System. *Southeast. Nat.* **2018**, *17*, 176–201. [CrossRef]
22. Jung, T.; Durán, A.; von Stowasser, E.S.; Schena, L.; Mosca, S.; Fajardo, S.; González, M.; Ortega, A.D.N.; Bakonyi, J.; Seress, D.; et al. Diversity of Phytophthora species in Valdivian rainforests and association with severe dieback symptoms. *For. Pathol.* **2018**, *48*, e12443. [CrossRef]
23. Hansen, E.M.; Wilcox, W.F.; Reeser, P.W.; Sutton, W. *Phytophthora rosacearum* and *P. sansomeana*, new species segregated from the *Phytophthora megasperma* "complex". *Mycologia* **2009**, *101*, 129–135. [CrossRef]
24. Hüberli, D.; Hardy, G.S.J.; White, D.; Williams, N.; Burgess, T.I. Fishing for Phytophthora from Western Australia's waterways: a distribution and diversity survey. *Australas. Plant Pathol.* **2013**, *42*, 251–260. [CrossRef]
25. Hulvey, J.; Gobena, D.; Finley, L.; Lamour, K. Co-occurrence and genotypic distribution of *Phytophthora* species recovered from watersheds and plant nurseries of eastern Tennessee. *Mycologia* **2010**, *102*, 1127–1133. [CrossRef]
26. Hwang, J.; Jeffers, S.N.; Oak, S.W. Aquatic habitats-a reservoir for population diversity in the genus *Phytophthora*. *Phytopathology* **2010**, *100*, S150–S151.
27. Marano, A.V.; Jesus, A.L.; de Souza, J.I.; Jerônimo, G.H.; Gonçalves, D.R.; Boro, M.C.; Rocha, S.C.O.; Pires-Zottarelli, C.L.A. Ecological roles of saprotrophic *Peronosporales* (Oomycetes, Straminipila) in natural environments. *Fungal Ecol.* **2016**, *19*, 77–88. [CrossRef]
28. Hardham, A.R.; Blackman, L.M. Molecular cytology of *Phytophthora*-plant interactions. *Australas. Plant Pathol.* **2010**, *39*, 29–35. [CrossRef]
29. Gessner, M.O.; Chauvet, E.; Dobson, M. A Perspective on Leaf Litter Breakdown in Streams. *Oikos* **1999**, *85*, 377–384. [CrossRef]

30. Gessner, M.O.; Schwoerbel, J. Leaching kinetics of fresh leaf-litter with implications for the current concept of leaf-processing in streams. *Arch. Für Hydrobiol.* **1989**, *115*, 81–90.
31. Bärlocher, F.; Boddy, L. Aquatic fungal ecology–How does it differ from terrestrial? *Fungal Ecol.* **2016**, *19*, 5–13. [CrossRef]
32. Wallace, J.B.; Eggert, S.L.; Meyer, J.L.; Webster, J.R. Effects of Resource Limitation on a Detrital-Based Ecosystem. *Ecol. Monogr.* **1999**, *69*, 409–442. [CrossRef]
33. Cooper, W.S. The Broad-Sclerophyll Vegetation of California. In *World Vegetation Types*; Eyre, S.R., Ed.; The Geographical Readings series; Palgrave Macmillan: Basingstoke, UK, 1971; pp. 149–156. ISBN 978-0-333-11031-7.
34. Cobb, R.C.; Eviner, V.T.; Rizzo, D.M. Mortality and community changes drive sudden oak death impacts on litterfall and soil nitrogen cycling. *New Phytol.* **2013**, *200*, 422–431. [CrossRef]
35. Shreve, F. The Vegetation of a Coastal Mountain Range. *Ecology* **1927**, *8*, 27–44. [CrossRef]
36. Rundel, P.W.; Sturmer, S.B. Native plant diversity in riparian communities of the Santa Monica mountains, California. *Madroño* **1998**, *45*, 93–100.
37. Davidson, J.M.; Patterson, H.A.; Wickland, A.C.; Fichtner, E.J.; Rizzo, D.M. Forest type influences transmission of *Phytophthora ramorum* in California oak woodlands. *Phytopathology* **2011**, *101*, 492–501. [CrossRef] [PubMed]
38. Rizzo, D.M.; Garbelotto, M.; Hansen, E.M. *Phytophthora ramorum*: Integrative Research and Management of an Emerging Pathogen in California and Oregon Forests. *Annu. Rev. Phytopathol.* **2005**, *43*, 309–335. [CrossRef]
39. Davidson, J.M.; Wickland, A.C.; Patterson, H.A.; Falk, K.R.; Rizzo, D.M. Transmission of *Phytophthora ramorum* in Mixed-Evergreen Forest in California. *Phytopathology* **2005**, *95*, 587–596. [CrossRef] [PubMed]
40. Davidson, J.M.; Patterson, H.A.; Rizzo, D.M. Sources of Inoculum for *Phytophthora ramorum* in a Redwood Forest. *Phytopathology* **2008**, *98*, 860–866. [CrossRef]
41. DiLeo, M.V.; Bostock, R.M.; Rizzo, D.M. *Phytophthora ramorum* does not cause physiologically significant systemic injury to California bay laurel, its primary reservoir host. *Phytopathology* **2009**, *99*, 1307–1311. [CrossRef] [PubMed]
42. Elosegi, A.; Pozo, J. Litter Input. In *Methods to Study Litter Decomposition*; Graça, M.A.S., Bärlocher, F., Gessner, M.O., Eds.; Springer: Dordrecht, Netherlands, 2005; pp. 3–11. ISBN 978-1-4020-3348-3.
43. Avila-Ospina, L.; Moison, M.; Yoshimoto, K.; Masclaux-Daubresse, C. Autophagy, plant senescence, and nutrient recycling. *J. Exp. Bot.* **2014**, *65*, 3799–3811. [CrossRef] [PubMed]
44. Lim, P.O.; Kim, H.J.; Gil Nam, H. Leaf Senescence. *Annu. Rev. Plant Biol.* **2007**, *58*, 115–136. [CrossRef]
45. Aram, K.; Rizzo, D.M. Distinct Trophic Specializations Affect How *Phytophthora ramorum* and Clade 6 *Phytophthora* spp. Colonize and Persist on *Umbellularia californica* Leaves in Streams. *Phytopathology* **2018**, *108*, 858–869. [CrossRef]
46. Murphy, S.K.; Lee, C.; Valachovic, Y.; Bienapfl, J.; Mark, W.; Jirka, A.; Owen, D.R.; Smith, T.F.; Rizzo, D.M. Monitoring *Phytophthora ramorum* distribution in streams within California watersheds. In *Proceedings of the Sudden Oak Death Third Symposium*; US Forest Service, Pacific Southwest Research Station: Albany, CA, USA, 2005; Vol. PSW-GTR-214, pp. 409–411.
47. Werres, S.; Marwitz, R.; De Cock, A.W.; Bonants, P.J.; De Weerdt, M.; Themann, K.; Ilieva, E.; Baayen, R.P. *Phytophthora ramorum* sp. nov., a new pathogen on *Rhododendron* and *Viburnum*. *Mycol. Res.* **2001**, *105*, 1155–1165. [CrossRef]
48. Hansen, E.M.; Parke, J.L.; Sutton, W. Susceptibility of Oregon Forest Trees and Shrubs to *Phytophthora ramorum*: A Comparison of Artificial Inoculation and Natural Infection. *Plant Dis.* **2005**, *89*, 63–70. [CrossRef]
49. DiLeo, M.V.; Bostock, R.M.; Rizzo, D.M. Microclimate Impacts Survival and Prevalence of *Phytophthora ramorum* in *Umbellularia californica*, a Key Reservoir Host of Sudden Oak Death in Northern California Forests. *PLoS ONE* **2014**, *9*, e98195. [CrossRef]
50. Maloney, P.E.; Lynch, S.C.; Kane, S.F.; Jensen, C.E.; Rizzo, D.M. Establishment of an emerging generalist pathogen in redwood forest communities. *J. Ecol.* **2005**, *93*, 899–905. [CrossRef]
51. Wood, S.E.; Gaskin, J.F.; Langenheim, J.H. Loss of monoterpenes from *Umbellularia californica* leaf litter. *Biochem. Syst. Ecol.* **1995**, *23*, 581–591. [CrossRef]
52. Medeiros, A.O.; Pascoal, C.; Graça, M.A.S. Diversity and activity of aquatic fungi under low oxygen conditions. *Freshw. Biol.* **2009**, *54*, 142–149. [CrossRef]

53. Bärlocher, F. Leaf Mass Loss Estimated by Litter Bag Technique. In *Methods to Study Litter Decomposition*; Graça, M.A.S., Bärlocher, F., Gessner, M.O., Eds.; Springer: Dordrecht, Netherlands, 2005; pp. 37–42. ISBN 978-1-4020-3348-3.
54. Adair, E.C.; Hobbie, S.E.; Hobbie, R.K. Single-pool exponential decomposition models: potential pitfalls in their use in ecological studies. *Ecology* **2010**, *91*, 1225–1236. [CrossRef]
55. Newell, S.Y.; Fell, J.W. Do halophytophthoras (marine Pythiaceae) rapidly occupy fallen leaves by intraleaf mycelial growth. *Can. J. Bot.* **1995**, *73*, 761–765. [CrossRef]
56. Warton, D.I.; Hui, F.K. The arcsine is asinine: the analysis of proportions in ecology. *Ecology* **2011**, *92*, 3–10. [CrossRef]
57. Pinheiro, J.; Bates, D.; Sarkar, D.; R Core Team. nlme: Linear and Nonlinear Mixed Effects Models. R package version 3.1-128. 2016. Available online: http://CRAN.R-project.org/package=nlme (accessed on 28 March 2019).
58. R Core Team R: A language and environment for statistical computing. R Foundation for Statistical Computing: Vienna, Austria, 2015. Available online: https://www.R-project.org/ (accessed on 28 March 2019).
59. Lenth, R.V. Least-Squares Means: The R Package lsmeans. *J. Stat. Softw.* **2016**, *69*(1), 1–33. [CrossRef]
60. Johnston, S.F.; Cohen, M.F.; Torok, T.; Meentemeyer, R.K.; Rank, N.E. Host Phenology and Leaf Effects on Susceptibility of California Bay Laurel to *Phytophthora ramorum*. *Phytopathology* **2015**, *106*, 47–55. [CrossRef] [PubMed]
61. Meshriy, M.; Hüberli, D.; Harnik, T.Y.; Miles, L.; Reuther, K.D.; Garbelotto, M. Variation in susceptibility of *Umbellularia californica* (Bay Laurel) to *Phytophthora ramorum*. In *Proceedings of the Sudden oak death second science symposium*; Gen. Tech. Rep. PSW-GTR-196; Pacific Southwest Research Station, Forest Service, U.S. Department of Agriculture: Pacific Southwest Research Station: Albany, CA, USA, 2006; pp. 87–89.
62. Goralka, R.J.L.; Schumaker, M.A.; Langenheim, J.H. Variation in chemical and physical properties during leaf development in California bay tree (*Umbellularia californica*): Predictions regarding palatability for deer. *Biochem. Syst. Ecol.* **1996**, *24*, 93–103. [CrossRef]
63. Wong, M.K.; Goh, T.-K.; Hodgkiss, I.J.; Hyde, K.D.; Ranghoo, V.M.; Tsui, C.K.; Ho, W.-H.; Wong, W.S.; Yuen, T.-K. Role of fungi in freshwater ecosystems. *Biodivers. Conserv.* **1998**, *7*, 1187–1206. [CrossRef]
64. Richards, T.A.; Soanes, D.M.; Jones, M.D.M.; Vasieva, O.; Leonard, G.; Paszkiewicz, K.; Foster, P.G.; Hall, N.; Talbot, N.J. Horizontal gene transfer facilitated the evolution of plant parasitic mechanisms in the oomycetes. *Proc. Natl. Acad. Sci.* **2011**, *108*, 15258–15263. [CrossRef] [PubMed]
65. Wu, C.-H.; Yan, H.-Z.; Liu, L.-F.; Liou, R.-F. Functional characterization of a gene family encoding polygalacturonases in *Phytophthora parasitica*. *Mol. Plant. Microbe Interact.* **2008**, *21*, 480–489. [CrossRef] [PubMed]
66. Blackman, L.M.; Cullerne, D.P.; Torreña, P.; Taylor, J.; Hardham, A.R. RNA-Seq Analysis of the Expression of Genes Encoding Cell Wall Degrading Enzymes during Infection of Lupin (*Lupinus angustifolius*) by *Phytophthora parasitica*. *PloS ONE* **2015**, *10*, e0136899. [CrossRef]
67. Beakes, G.W.; Glockling, S.L.; Sekimoto, S. The evolutionary phylogeny of the oomycete "fungi". *Protoplasma* **2011**, *249*, 3–19. [CrossRef] [PubMed]
68. Cooke, R.C.; Whipps, J.M. The Evolution of Modes of Nutrition in Fungi Parasitic on Terrestrial Plants. *Biol. Rev.* **1980**, *55*, 341–362. [CrossRef]
69. Navarro, S.; Sims, L.; Hansen, E. Pathogenicity to alder of *Phytophthora* species from riparian ecosystems in western Oregon. *For. Pathol.* **2015**, *45*, 358–366. [CrossRef]
70. Parke, J.L.; Knaus, B.J.; Fieland, V.J.; Lewis, C.; Grünwald, N.J. *Phytophthora* Community Structure Analyses in Oregon Nurseries Inform Systems Approaches to Disease Management. *Phytopathology* **2014**, *104*, 1052–1062. [CrossRef]
71. Yakabe, L.E.; Blomquist, C.L.; Thomas, S.L.; MacDonald, J.D. Identification and frequency of *Phytophthora* species associated with foliar diseases in California ornamental nurseries. *Plant Dis.* **2009**, *93*, 883–890. [CrossRef] [PubMed]
72. Voříšková, J.; Baldrian, P. Fungal community on decomposing leaf litter undergoes rapid successional changes. *ISME J.* **2013**, *7*, 477–486. [CrossRef]
73. Andrews, J.; Douhan, L.; Douhan, G.; Rizzo, D.M. Fungi associated with leaves of California bay laurel. *Phytopathology* **2005**, *95*, S4.

74. Aram, K. The Ecology of *Phytophthora ramorum* and Resident *Phytophthora* in California Streams. Ph.D. Dissertation, University of California, Davis, Davis, California, 2017.
75. Graça, M.A.S.; Bärlocher, F.; Gessner, M.O. *Methods to study litter decomposition a practical guide*; Springer: Dordrecht, The Netherlands; New York, NY, USA, 2005; ISBN 978-1-4020-3466-4.
76. Themann, K.; Werres, S.; Lüttmann, R.; Diener, H.A. Observations of *Phytophthora* spp. in Water Recirculation Systems in Commercial Hardy Ornamental Nursery Stock. *Eur. J. Plant Pathol.* **2002**, *108*, 337–343. [CrossRef]

8

Phytophthora Species from Xinjiang Wild Apple Forests in China

Xiaoxue Xu [1], Wenxia Huai [1], Hamiti [2], Xuechao Zhang [3] and Wenxia Zhao [1,*]

1 The Key Laboratory of National Forestry and Grassland Administration on Forest Protection, Research Institute of Forest Ecology Environment and Protection, Chinese Academy of Forestry, Beijing 100091, China; xuxiaoxue@caf.ac.cn (X.X.); huaiwx@caf.ac.cn (W.H.)

2 Forest Bureau of Emin County, Xinjiang Uygur Autonomous Region, Emin 834600, China; hamit@126.com

3 Institute of Agricultural Sciences of Ili Kazakh Autonomous Prefecture, Yining 835000, China; zhangxuechao678@163.com

* Correspondence: zhaowenxia@caf.ac.cn.

Abstract: *Phytophthora* species are well-known destructive forest pathogens, especially in natural ecosystems. The wild apple (*Malus sieversii* (Ledeb.) Roem.) is the primary ancestor of *M. domestica* (Borkh.) and important germplasm resource for apple breeding and improvement. During the period from 2016 to 2018, a survey of *Phytophthora* diversity was performed at four wild apple forest plots (Xin Yuan (XY), Ba Lian (BL), Ku Erdening (KE), and Jin Qikesai (JQ)) on the northern slopes of Tianshan Mountain in Xinjiang, China. *Phytophthora* species were isolated from baiting leaves from stream, canopy drip, and soil samples and were identified based on morphological observations and the rDNA internal transcribed spacer (ITS) sequence analysis. This is the first comprehensive study from Xinjiang to examine the *Phytophthora* communities in wild apple forests The 621 resulting *Phytophthora* isolates were found to reside in 10 different *Phytophthora* species: eight known species (*P. lacustris* being the most frequent, followed by *P. gonapodyides*, *P. plurivora*, *P. gregata*, *P. chlamydospora*, *P. inundata*, *P. virginiana*, and *P. cactorum*) and two previously unrecognized species (*P.* sp. CYP74 and *P.* sp. forestsoil-like). The highest species richness of *Phytophthora* occurred at BL, followed by XY. *P. lacustris* was the dominant species at BL, XY, and JQ, while *P. gonapodyides* was the most common at KE. In the present paper, the possible reasons for their distribution, associated implications, and associated diseases are discussed.

Keywords: *Phytophthora*; diversity; wild apple forest; decline

---

1. Introduction

Xinjiang wild apple (*Malus sieversii* (Ledeb.) Roem.), the wild ancestor of the domesticated apple, is mainly distributed in the Tian Shan mountains in Central Asia, including the Ili River Valley in northwest China's Xinjiang Uygur Autonomous Region and southeast and east Kyrgyzstan [1–3]. It is the dominant species in the relict wild fruit wood forests of inner Eurasia and is protected as a vulnerable species among the endangered rare germplasm resources of China [4–6]. However, the wild populations of *M. sieversii* have experienced a dramatic decrease in recent years, and symptoms of the decline can be observed in many wild apple forests. Affected trees show higher canopy loss, branch dieback, bark and cambium necrosis, and growth reduction. Similar to other countries in Central Asia, several abiotic and biotic factors negatively affect the health status of wild apple forests in China [7], including environmental and climate impacts, insect pests, cambium feeders such as *Agrilus mali* Matsumura, and infection by pathogenic fungi [8–13] (Figure 1).

Figure 1. Declining symptoms of wild apple forests in Xinjiang.

The presence of *Phytophthora* species in forests and natural ecosystems is considered to be an important biotic factor responsible for the decline, dieback, and mortality of trees [14]. Belonging to the class Oomycetes, or "water molds", in the kingdom Chromista, these fungus-like organisms can cause root rot, bark cankers, and diseases leading to the decline and dieback of a wide range of plant species worldwide [15].

Over the past two decades, numerous surveys have shown that many known and previously unknown *Phytophthora* species are present in a variety of natural and semi-natural forests and river systems in Europe, America, Australia, South Africa, and more recently, Asia [16]. Some of these *Phytophthora* species, including *P. cactorum*, *P. cinnamomi*, and *P. plurivora*, have shown strong involvement in the decline and dieback of forests, while the exact role in forest ecosystems of many other species, such as *P. cryptogea*, *P. chlamydospora*, and *P. gonapodyides*, is unclear [16]. In a recent study, *P. plurivora* was found to cause damage to the fine roots and stems of *M. sieversii* in the declining wild fruit forests of Xinjiang Province, China [17].

However, the findings described in that report were based on a limited number of samples. Furthermore, the distribution and ecological roles of *P. plurivora* and other *Phytophthora* species in wild apple forests are still unknown.

In June to October 2016–2018, a survey of *Phytophthora* diversity was performed at four plots in wild apple forests on the northern slope of Tianshan Mountain in Xinjiang using baiting assays from rhizosphere soil, streams, and canopy drip. This study presents the results of this survey related to the *Phytophthora* species associated with the decline or dieback of *Malus sieversii*.

2. Material and Methods

2.1. Study Area

The study was carried out in wild apple forests on the northern slopes of Tianshan Mountain in Xinjiang, China. We chose Xin Yuan (XY) (83°33′ E, 43.25′ N), Ba Lian (BL) (82°50′ E, 43°15′ N), Ku Erdening (KE) (82°51′ E, 43°13′ N), and Jin Qikesai (JQ) (83°25′ E, 43°18′ N) as the studied trap plots, as these locations are where the wild apple trees mostly live [18] (Figure 2). In these four plots, the decline of wild apple trees in XY is the most serious. A total of 10 stream sites that flow though the declining wild apple trees were set at each of the 4 areas; 10 canopy drip sites under the declining wild apple trees were set at each of the 4 areas. Fifteen soil sites under the declining trees were set at XY and at BL. In total, 40 stream sites, 40 forest sites, and 30 soil sampling sites were set in these 4 plots to investigate the presence of *Phytophthora* in the wild apple forests.

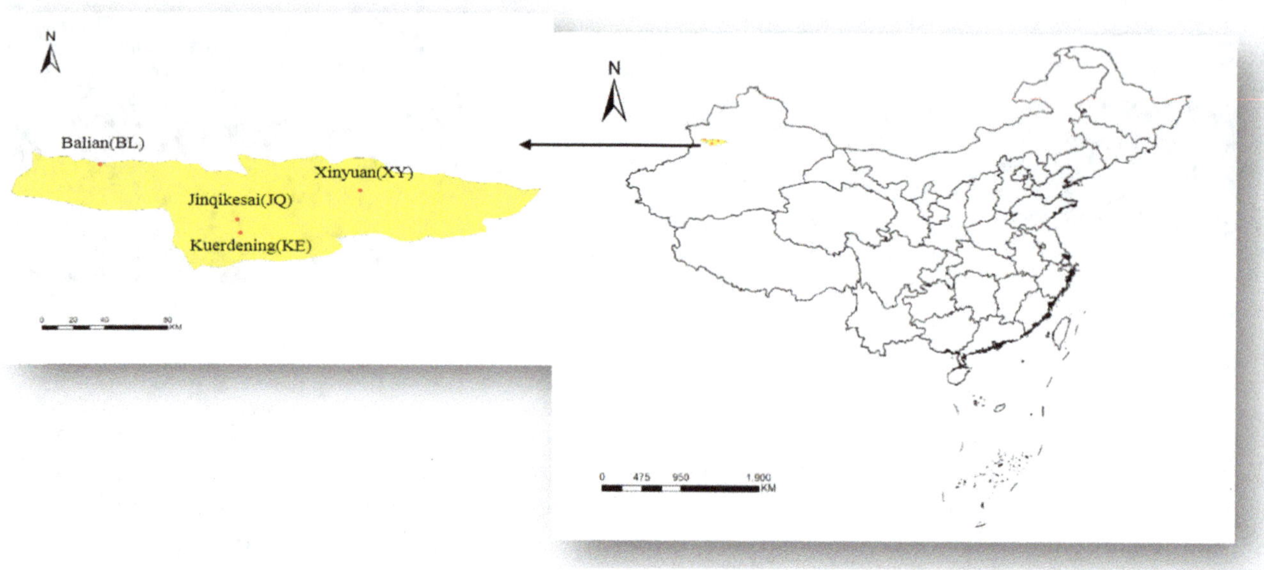

Figure 2. The four plots located in Xinyuang, Gongliu county, China.

2.2. *Sampling and Phytophthora Isolation*

This research used stream, canopy drip, and soil sampling baiting [19–22] at the 4 plots from 2016 to 2018. All isolates of *Phytophthora* spp. were recovered from sites by baiting with leaves of wax (*Fraxinus chinensis* Roxb). The baiting leaves were placed in the surveyed sites for 1 week, retrieved, and brought to the laboratory from June to October each year. Pieces of approximately 2 mm^2 were cut from the margins of the brown spots and plated in CARP+. Colonies of suspected *Phytophthora* species growing from plated baits were transferred to CARP (CARP+ without Benlate and hymexazol) to confirm purity and then to CMA for characterization and storage [23].

2.3. *Classical Identification of Isolates*

Morphospecies were first identified by colony patterns in V8 agar (V8A). Colony growth patterns of 7 day old cultures grown at 20 °C in the dark in V8A and morphological features of sporangia, oogonia, antheridia, chlamydospores, hyphal swellings, and aggregations were studied, described, and photographed. The formation of sporangia was stimulated by flooding small V8A disks (0.5 mm diameter) with sterile water at room temperature, and this was observed by microscope after 12, 24, 48, and 72 hours. The results were compared with those of known isolates and species descriptions present in the literature [24].

2.4. *Molecular Identification of Isolates*

DNA isolation and amplification was carried out. For all *Phytophthora* isolates obtained in this study, mycelial DNA was extracted from pure cultures grown in corn meal agar (CMA). DNA extraction and amplification were performed in accordance with Huai et al. [25], using the primer pairs ITS6 (5′-GAA GGT GAA GTC GTA ACA AGG-3′) [26] and ITS4 (5′-TCC TCC GCT TAT TGA TAT GC-3′) [27] to amplify both ITS1 and ITS2 regions, including the 5.8S rDNA. The PCR reaction was conducted in a 25 μL reaction mixture consisting of 12.5 μL of 2× Taq Master Mix, 9.5 μL of double-distilled H_2O (both supplied by TIANGEN Biotech, Beijing, China), 0.75 μL of rDNA internal transcribed spacer (ITS) primers or 0.25 μL of Btub or EF1a primer (all primers were sourced from Sangon Biotech, Shanghai, China), and 1.5 μL of DNA template. The PCR thermal cycling program was as follows: initial denaturation at 95 °C for 3 min, followed by 35 cycles consisting of denaturation at 95 °C for 1 min, annealing at 58 °C for 30 s, extension at 72 °C for 1 min, and a final extension at 72 °C for 10 min.

Sequence data were initially assembled using STADEN Package 2.0.0. and compared with other closely related species, including some undescribed taxa obtained from GenBank [28].

2.5. Phylogenetic Analysis

Sequences were aligned using MEGA version 6.0. No further editing was done on the alignment to ensure reproducibility and to prevent the introduction of bias. Sequences were concatenated for phylogenetic analyses. Isolates represented each of the known species; the undescribed taxa obtained in this study were compiled into a single data set, and Maximum Likelihood (ML) analysis was performed using MEGA version 6.0. The bootstrap support values were obtained with 1000 bootstrap replicates for each. To ensure general reproducibility, the analysis was repeated using MrBayes to build a Bayesian tree. DANMAN 8 was also used to align multiple sequences to compare the sequences of the same species.

3. Results

3.1. Phytophthora Species Identification

In total, 621 *Phytophthora* isolates were recovered from this study: 261 from BL, 199 from XY, 121 from KE, and 40 from JQ. Most isolates were recovered from streams and rivers. A total of 597 isolates were from stream baiting, 15 were from soil sampling, and nine were from canopy drip.

Ten morphospecies were identified on the basis of colony patterns and micromorphological features. The features of the sporangium were recorded and photographed. There were two unknown *Phythphthora* taxa. One was *P.* sp. CYP74, which is heterothallic, and no oogonia were observed. Another one was *P.* sp. forestsoil-like which is self-sterile. Species identification was confirmed by comparing ITS rDNA sequences. ITS sequence data were obtained for all isolates, and their identities were confirmed by conducting a BLAST search in GenBank. The results of morphological observations were consistent with ITS analysis; 578 isolates were of eight known species, which, in order of frequency, were *P. lacustris*, *P. gonapodyides*, *P. plurivora*, *P. gregata*, *P. chlamydospora*, *P. inundata*, *P. virginiana*, and *P. cactorum*. In addition, 43 isolates of two unknown species were identified; in order of frequency, these were *P.* sp. CYP74 and *P.* sp. forestsoil-like (Table 1).

Table 1. The information of 621 *Phytophthora* strains relating to clades, species, methods, numbers, and plots including Xin Yuan (XY), Ba Lian (BL), Ku Erdening (KE), and Jin Qikesai (JQ).

Clade	Species	Plot	Method	Number
clade1a	*P. cactorum*	BL	canopy drip	1
clade2c	*P. plurivora*	BL	soil	1
		XY	soil	2
		XY	stream	3
		KE	stream	5
clade6b	*P. lacustris*	BL	stream	175
		BL	canopy drip	2
		BL	soil	4
		XY	stream	84
		XY	canopy drip	1
		XY	soil	1
		KE	stream	48
		JQ	stream	28
	P. gonapodyides	BL	stream	56
		XY	stream	74
		XY	canopy drip	5
		XY	soil	3
		KE	stream	63
		JQ	stream	7
	P. chlamydospora	BL	stream	3
		XY	stream	3

Table 1. *Cont.*

Clade	Species	Plot	Method	Number
	P. gregata	BL	stream	2
		XY	stream	4
		XY	soil	1
	P. inundata	XY	stream	1
	P. sp. CYP74	BL	stream	13
		XY	stream	15
		XY	canopy drip	2
		KE	stream	3
		JQ	stream	2
	P. sp. forestsoil-like	BL	stream	3
		KE	stream	2
		JQ	stream	3
clade9a	*P. virginiana*	BL	stream	1

3.2. Phylogenetic Analysis

The aligned ITS dataset consisted of approximately 913 characters. Phylogenetic analysis of Bayes and ML trees revealed sequences that were grouped within four of the 10 main clades of the genus *Phytophthora* [29,30] ascribed to 10 different species (Table 2). Several isolates corresponded to known species as follows: *P. cactorum*—clade 1; *P. plurivora*—clade 2; *P. lacustris*, *P. gonapodyides*, *P. chlamydospora*, *P. gregata* and *P. inundata*—clade 6; *P. virginiana*—clade 9.

Table 2. GenBank number and rDNA internal transcribed spacer (ITS) clade of 10 *Phytophthora* species in Xinjiang wild apple forests.

Species	Isolate No.	ITS Clade	Genbank Number
Phytophthora cactorum	8BLL3	1	MN175469
Phytophthora plurivora	1KEX3(6)	2	MN175458
Phytophthora lacustris	4GLX9(3)	6	MN175455
	8KEX3(2)	6	MN175463
Phytophthora gregata	9XYX6(5)	6	MN175456
	7XYT6(2)	6	MN175459
Phytophthora gonapodyides	9XYX7(2)	6	MN175457
	2KEX5(5)	6	MN175462
Phytophthora sp. CYP74	1KEX9(1)	6	MN175460
	15XYX2(1)	6	MN175465
Phytophthora chlamydospora	9XYX5(4)	6	MN175461
	5BLX9(2)	6	MN175466
Phytophthora sp. forestsoil-like	1BLX1(7)	6	MN175464
	8JQX4(1)	6	MN175468
Phytophthora inundata	15XYX3(1)	6	MN209784
Phytophthora virginiana	1BLX1(3)	9	MN175467

The phylogenetic analysis revealed two undescribed *Phytophthora* taxa, *P.* sp. CYP74 and *P.* sp. forestsoil-like, both from clade 6 (Figure 3). *P.* sp. CYP74 was found to be similar to *P. mississippiae* and *P. ornamentata*. Compared with the isolate *P.* sp. CYP74, the ITS sequence identity of *P. mississippiae* and *P. omamentata* was shown to be 99.66%. *P.* sp. forestsoil-like was found to be similar to *P.* sp. forestsoil-like (TW55), which was reported in Taiwan [24], with an ITS sequence identity of 96.38%.

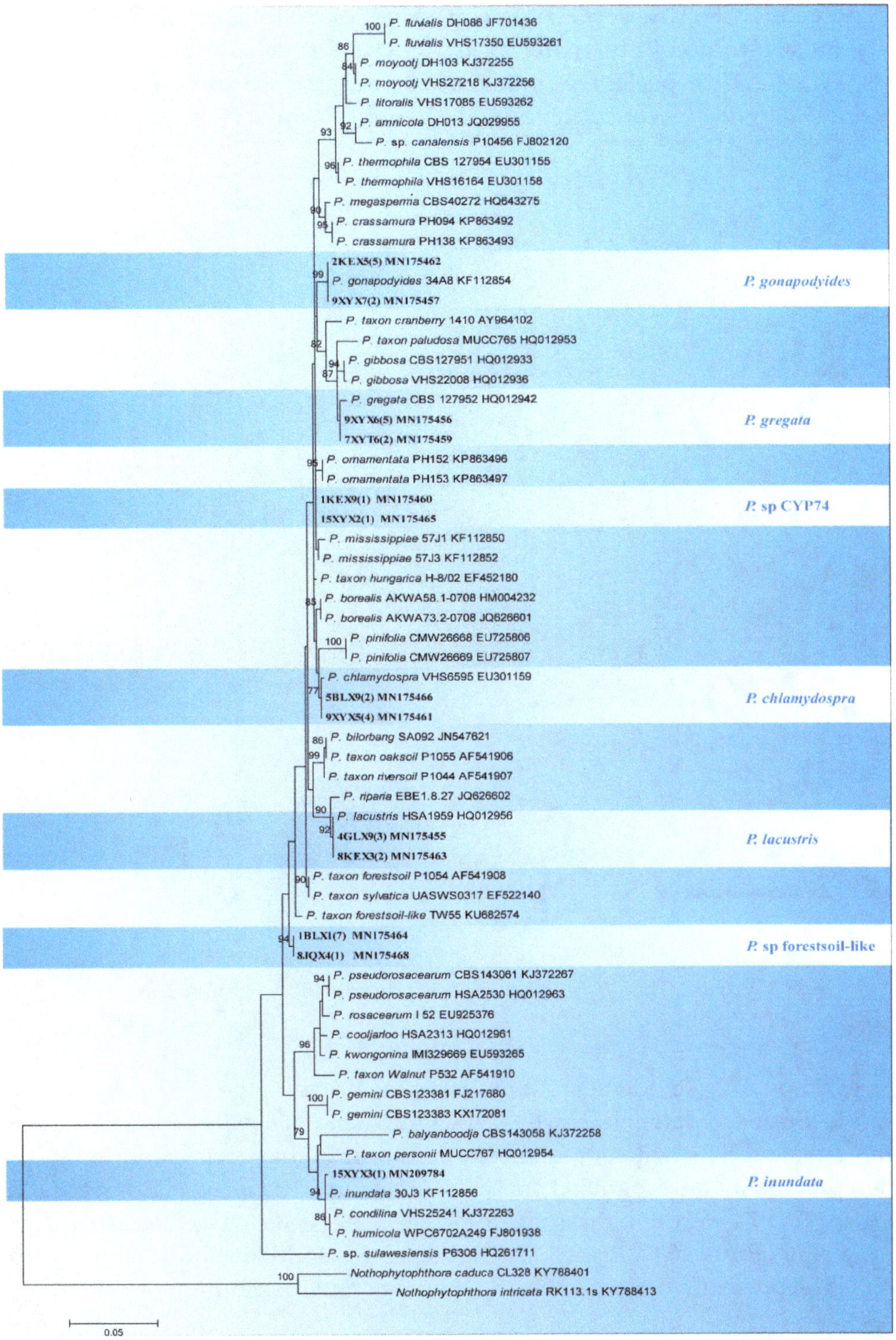

Figure 3. A phylogram based on ITS sequence data indicating the placement of clade 6 and the undescribed *Phytophthora* taxa recovered in this study. The topological structures of Bayes and Maximum likelihood (ML) trees are the same. Bootstrap support is given above the line. Numbers above the branches represent the bootstrap support based on the maximum likelihood analysis.

3.3. The Distribution of Phytophthora Species

In this research, four plots with varying *Phytophthora* diversity were investigated (Figure 4). Ba Lian (BL), with nine species, had the greatest diversity, and Jin Qikesai (JQ) had the least diversity with five species. *P. lacustris* was the most widespread species and the main species at JQ and BL (70% and

69%). It also accounted for the largest proportion at XY, followed by *P. gonapodyides*. Overall, these two species were the most common in the wild apple forests. The third most common species, with 35 strains, was *P.* sp. CYP74; this species was baited at all plots and was predominantly found at XY and BL. *P. cactorum* and *P. virginiana* were only found at BL, and *P. inundata* was only found at XY.

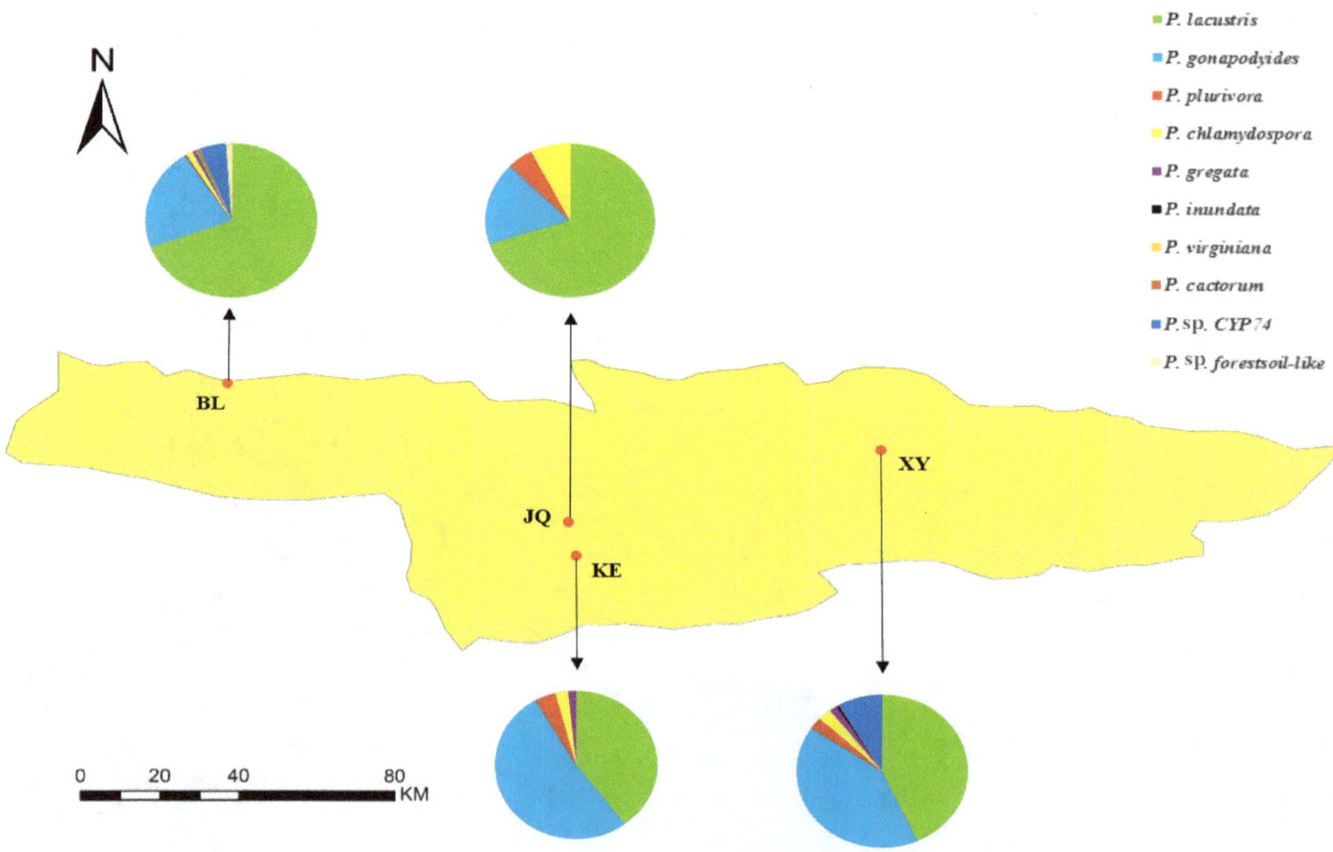

Figure 4. The *Phytophthora* species distribution at four plots.

4. Discussion

The results of this study provide the first record of the broad range of *Phytophthora* spp. associated with the wild apple forest ecosystem in Northwest China. Ten *Phytophthora* species belonging to clades 1, 2, 6, and 9 were isolated in the wild apple forests, including eight species reported for the first time in Xinjiang and two previously unrecognized species. *P. cactorum*, *P. plurivora*, *P. lacustris*, *P. gonapodyides*, *P. gregata*, and *P.* sp. CYP74 were caught in the forest stands. *P. plurivora*, *P. lacustris*, *P. gonapodyides*, *P. chlamydospora*, *P. gregata*, *P. inundata*, *P.* sp. CYP74, *P.* sp. forestsoil-like, and *P. virginiana* were caught in the natural rivers. The present work indicates the diversity and distribution of *Phytophthora* in Xinjiang wild apple forests.

From clade 6, *P. lacustris* and *P. gonapodyides* were caught in 4 plots from June to October. The number of these two species took up 88.6% of all *Phytophthora* species in this survey. In particular, *P. lacustris* represented more than half of the total number of strains. *P. lacustris* and *P. gonapodyides* were obtained from canopy drip samples, soil samples, and mostly stream samples. These species often co-exist in river systems in the temperate regions of North America, Europe, and Asia [20,31,32]. *P. lacustris*, which like *P. gonapodyides* belongs to clade 6, is widely distributed globally. Initially identified as a saprotroph that infects plant detritus, it has now been shown to cause significant damage to fine roots and weak-to-moderate bark lesions in *Alnus glutinosa* and *Prunus persica* in Portugal, Italy, and Turkey, among other places [33–37]. Samples in previous research on this species were from soil, trees, and roots, while in the present study, they were mostly taken from streams, with a few collected by canopy drip and from the soil by baiting. In the present study, *P. lacustris* was acquired from all four site plots, demonstrating that a large number of *P. lacustris* live in the wild apple forests of Xinjiang, especially in the riparian habitats of streams. *P. gonapodyides* was described in 1927 as a global species, appearing in almost every *Phytophthora* survey and demonstrating weak pathogenicity [32,38–43].

P. gregata was obtained from stream samples at BL in July and stream, soil samples at XY in September. It was reported in China in 2013 by stream baiting [20] and was shown to cause significant reduction of shoot and root growth but was not found to kill plants [44].

The present survey is the first report of *P. inundata* in China but was previously reported as a pathogen of shrubs and trees in Europe and South America [45] and the cause of Viburnum latent infection in Australia and Virginia [46]. All *Phytophthora* species have the potential to disturb natural ecosystems, particularly those of exotic origin, provided that environmental conditions are conducive to disease development [14,32,47].

P. chlamydospora was caught at BL in August by stream baiting and stream samples at XY in September. It has been recovered in Europe, North America, Argentina, and Taiwan from cankers on trees, roots, and foliage of horticultural nursery stock [16,24,48,49].

In this research, we baited two undescribed species: *P.* sp. CYP74, which is heterothallic like *P. mississippiae*, and *P.* sp. forestsoil-like, which is self-sterile like *P.* sp. forestsoil-like (TW55), reported in Taiwan in 2017 [24]. Detailed information for these two species will be shown in future studies.

P. plurivora from clade 2 is known to be a serious pathogen of many forest trees, including oak, beech, and *Alnus glutinosa* seedlings. This species can cause dieback and root loss and is most frequently associated with cankers in Europe, North America, and Asia [31–33,50–55]. Via the examination of plant tissues and soil samples, it has been reported to cause cankers in wild apple forests in Xinjiang [17], corroborating its discovery in stream water and soil samples in the present study.

From clade 1, we collected *P. cactorum* at BL by canopy drip in September. First described in 1886, this clade1a species is similar to the notorious pathogenic species *P. infestans*, which can cause damping-off of seedlings, fruits, leaf stems, and roots, as well as collar and crown rot and stem canker on an extremely wide host range of more than 200 species from 160 generas of plants, including many fruits, ornamental plants, and forest trees. It has been reported to be the cause of aerial cankers on European beech trees and is a major problem in apple orchards, causing the death of apple trees [56]. A previous study of staple crops in Xinjiang showed that *P. cactorum* was isolated from the fruits, root crowns, and diseased soils of strawberry, safflower, apple, and pear plants [47]. The present study represents the first time that *P. cactorum* has been found in the forest system in Xinjiang. It may have spread to the forests from nearby farms and is a probable reason for the decline in apple trees [15,57–59].

In clade 9, *P. virginiana* was obtained at BL by stream baiting in July. It was first isolated from irrigation water at several ornamental nurseries in 2013 in Virginia [60]. No pathogenicity of this species has yet been detected.

In this study, all plots were situated in areas of declining wild apple trees. Although the diversity of *Phytophthora* was different at different plots, it may be related to the population density of wild apple forests and human activities. We will set some survey plots in healthy stands and investigate the pathogenicity of these *Phytophthora* species in future studies to confirm their effects on this wild apple forest ecological system and better understand the reasons for its decline.

5. Conclusions

1) This first extensive survey demonstrated 10 *Phytophthora* species in a Xinjing wild apple forest ecosystem. Discussing the potential pathogenicity of these *Phytophthora*, this is a basic study to find out the reasons why the wild apple trees have declined.

2) *P. lacustris* and *P. gonapodyides* are the most widespread species, and BL has the highest number of *Phytophthora* species. Two undescribed species were also detected in this research. These results form a foundation for the study of the genetic diversity of *Phytophthora*.

Author Contributions: Conceptualization, W.Z. and W.H.; methodology, X.X.; formal analysis, X.X.; investigation, X.Z.; resources, H.; writing—original draft preparation, X.X.; writing—review and editing, W.H.; supervision, W.Z.; project administration, W.Z.; funding acquisition, W.Z. and W.H.

Acknowledgments: The authors are grateful to the Zhongsheng Li and Jinlong Zhang for material support, Jiaqing Jiang for strains saved during this work. In particular, thank Krishan Andradige for his kind encouragements and supports.

References

1. Liu, X.S.; Lin, P.J.; Zhong, J.P. Analysis of habitat for wild fruit forests in Ili and discussion on its occurrence. *Arid Zone Res.* **1993**, *3*, 28–30.
2. Yan, G.; Long, H.; Song, W.; Chen, R. Genetic polymorphism of *Malus sieversii* populations in Xinjiang, China. *Genet. Resour. Crop Evol.* **2008**, *55*, 171–181. [CrossRef]
3. Zhang, H.X.; Zhang, M.L.; Wang, L.N. Genetic structure and historical demography of *Malus sieversii* in the Yili Valley and the western mountains of the Junggar Basin, Xinjiang, China. *J. Arid Land* **2015**, *7*, 264–271. [CrossRef]
4. Zhang, X.S. On the eco-geographical characters and the problems of classification of the wild fruit-tree forest in the Ili Valley of Sinkiang. *Acta Bot. Sin.* **1973**, *15*, 239–253.
5. Chen, L.Z. *Present Situations of and Conservation Strategies for Biodiversity in China*; Science China Press: Beijing, China, 1993.
6. Wang, N.; Jiang, S.; Zhang, Z.; Fang, H.; Xu, H.; Wang, Y.; Chen, X. *Malus sieversii*: The origin, flavonoid synthesis mechanism, and breeding of red-skinned and red-fleshed apples. *Hortic. Res.* **2018**, *5*, 70. [CrossRef] [PubMed]
7. Panyushkina, I.; Mukhamadiev, N.; Lynch, A.; Ashikbaev, N.; Arizpe, A.; O'Connor, C.; Sagitov, A. Wild Apple Growth and Climate Change in Southeast Kazakhstan. *Forests* **2017**, *8*, 406. [CrossRef]
8. Zhang, P.; Lü, Z.Z.; Zhang, X.; Zhao, X.P.; Zhang, Y.G.; Gulzhanat, T.; Maisupova, B.; Adilbayeva, Z.; Cui, Z.J. Age Structure of *Malus sieversii* Population in Ili of Xinjiang and Kazakhstan. *Arid Zone Res.* **2019**, *36*, 844–853.
9. Wang, Z.Y.; Yang, Z.Q.; Zhang, Y.L.; Wang, X.Y.; Tang, Y.L. Biological control of agrilus mali (coleoptera: Buprestidae) by applying four species of bethylid wasp (hymenoptera: Bethylidae) on *malus sieversii* in Xinjiang. *Sci. Silvae Sin.* **2014**, *50*, 97–101.
10. Yan, G.; Zheng, X.U. Study on the wild fruit tree diseases of Tianshan mountains and their distribution in Xinjiang. *Arid Zone Res.* **2001**, *18*, 47–49.
11. Liu, A.H.; Zhang, X.P.; Wen, J.B.; Yue, C.Y.; Alimu, J.; Zhang, S.P.; Kereman, J.W. Preliminary research on the composite damage of *Agrilus mali* matsumura and Valsa mali miyabe et yamada in wild apple trees in tianshan mountain. *Xinjiang Agric. Sci.* **2014**, *51*, 2240–2244.
12. Niu, C.; Wang, J.; Zhu, X.; Chen, X.; Guo, L. Brown rot pathogens on stone and pome fruit trees in Xinjiang wild forest. *Mycosystema* **2016**, *35*, 1514–1525.
13. Cheng, Y.; Zhao, W.; Lin, R.; Yao, Y.; Yu, S.; Zhou, Z.; Huai, W. *Fusarium* species in declining wild apple forests on the northern slope of the Tianshan Mountains in north-western China. *For. Pathol.* **2019**. [CrossRef]
14. Jung, T.; Pérez-Sierra, A.; Durán, A.; Horta, M.J.; Balci, Y.; Scanu, B. Canker and decline diseases caused by soil- and airborne *Phytophthora* species in forests and woodlands. *Persoonia Mol. Phylogeny Evol. Fungi* **2018**, *40*, 182–220. [CrossRef] [PubMed]
15. Erwin, D.C.; Ribeiro, O.K. *Phytophthora Diseases-Worldwide*; APS Press: St. Paul, MN, USA, 1996.
16. Jung, T.; Durßn, A.; Sanfuentes von Stowasser, E.; Schena, L.; Mosca, S.; Fajardo, S.; Tomšovskř, M. Diversity of *Phytophthora* species in Valdivian rainforests and association with severe dieback symptoms. *For. Pathol.* **2018**, *48*, e12443. [CrossRef]
17. Liu, A.H.; Shang, J.; Zhang, J.W.; Kong, T.T.; Yue, Z.Y.; Wen, J.B. Canker and fine-root loss of *Malus sieversii* (Ldb.) Roem. caused by *Phytophthora plurivora* in Xinjiang Province in China. *For. Pathol.* **2018**, *48*, e12462. [CrossRef]
18. Lin, P.Y.; Cui, N.R. *Wild Fruit Forest Resources in Tianshan Mountains—Comprehensive Research on Wild Fruit Forests in Ili. Xinjiang, China*; China Forestry Publishing House: Beijing, China, 2000.
19. Hüberli, D.; Hardy, G.E.S.J.; White, D.; Williams, N.; Burgess, T.I. Fishing for *Phytophthora* from Western

Australia's waterways: A distribution and diversity survey. *Australas. Plant Pathol.* **2013**, *42*, 251–260. [CrossRef]

20. Huai, W.X.; Tian, G.; Hansen, E.M.; Zhao, W.X.; Goheen, E.M.; Grünwald, N.J.; Cheng, C. Identification of *Phytophthora* species baited and isolated from forest soil and streams in northwestern Yunnan province, China. *For. Pathol.* **2013**, *43*, 87–103. [CrossRef]
21. Reeser, P.; Sutton, W.; Hansen, E. *Phytophthora* species in tanoak trees, canopy-drip, soil, and streams in the sudden oak death epidemic plot of south-western Oregon, USA. *N. Z. J. For. Sci.* **2011**, *41*, S65–S73.
22. Wingfield, B.D.; Oh, E.; Gryzenhout, M.; Burgess, T.I.; Wingfield, M.J. Surveys of soil and water reveal a goldmine of *Phytophthora* diversity in South African natural ecosystems. *IMA Fungus* **2013**, *4*, 123–131.
23. Li, W.W.; Zhao, W.X.; Huai, W.X. *Phytophthora pseudopolonica* sp. Nov., a new species recovered from stream water in subtropical forests of China. *Int. J. Syst. Evol. Microbiol.* **2017**, *67*, 3666–3675. [CrossRef]
24. Jung, T.; Chang, T.T.; Bakonyi, J.; Seress, D.; Pérez-Sierra, A.; Yang, X.; Maia, C. Diversity of *Phytophthora* species in natural ecosystems of Taiwan and association with disease symptoms. *Plant Pathol.* **2017**, *66*, 194–211. [CrossRef]
25. Huai, W.X.; Guo, L.D.; He, W. Genetic diversity of an ectomycorrhizal fungus Tricholoma terreum in a Larix principis-rupprechtii stand assessed using random amplified polymorphic DNA. *Mycorrhiza* **2003**, *13*, 265–270. [CrossRef] [PubMed]
26. Cooke, D.E.L.; Drenth, A.; Duncan, J.M.; Wagels, G.; Brasier, C.M. A molecular phylogeny of *Phytophthora* and related oomycetes. *Fungal Genet. Biol.* **2000**, *30*, 17–32. [CrossRef] [PubMed]
27. White, T.J.; Bruns, T.; Lee, S.; Taylor, J. *Amplification and Direct Sequencing of Fungal Ribosomal Rna Genes for Phylogenetics*; Academic Press Inc.: Cambridge, MA, USA, 1990.
28. Altschul, S.F.; Madden, T.L.; Schäffer, A.A.; Zhang, J.; Zhang, Z.; Miller, W.; Lipman, D.J. Gapped-BLAST and PSI-BLAST: A new generation of protein database search programs. *Nucleic Acids Res.* **1997**, *25*, 3389–3402. [CrossRef]
29. Yang, X.; Tyler, B.M.; Hong, C. An expanded phylogeny for the genus *Phytophthora*. *IMA Fungus* **2017**, *8*, 355–384. [CrossRef]
30. Jung, T.; Horta Jung, M.; Cacciola, S.O.; Cech, T.; Bakonyi, J.; Seress, D.; Mosca, S.; Schena, L.; Seddaiu, S.; Pane, A.; et al. Multiple new cryptic pathogenic Phytophthora species from Fagaceae forests in Austria, Italy and Portugal. *IMA Fungus* **2017**, *8*, 219–244. [CrossRef]
31. Zeng, H.C.; Ho, H.H.; Zheng, F.C. A survey of *Phytophthora* species on Hainan Island of South China. *J. Phytopathol.* **2009**, *157*, 33–39. [CrossRef]
32. Sutton, W.; Adams, G.C.; Remigi, P.; Reeser, P.W.; Hansen, E.M. *Phytophthora* species in forest streams in Oregon and Alaska. *Mycologia* **2011**, *103*, 22–35.
33. Akilli, S.; Ulubaş Serçe, Ç.; Katircioğlu, Y.Z.; Maden, S. *Phytophthora* dieback on narrow leaved ash in the black sea region of turkey. *For. Pathol.* **2013**, *43*, 252–256. [CrossRef]
34. Aday Kaya, A.G.; Lehtijärvi, A.; Şaşmaz, Y.; Nowakowska, J.A.; Oszako, T.; Doğmuş Lehtijärvi, H.T.; Woodward, S. *Phytophthora* species detected in the rhizosphere of *Alnus glutinosa* stands in the Floodplain Forests of Western Turkey. *For. Pathol.* **2018**, *48*, 11–14. [CrossRef]
35. Tkaczyk, M.; Nowakowska, J.A.; Oszako, T. *Phytophthora* species isolated from ash stands in Białowieża Forest nature reserve. *For. Pathol.* **2016**, *46*, 660–662. [CrossRef]
36. Kanoun-Boulé, M.; Vasconcelos, T.; Gaspar, J.; Vieira, S.; Dias-Ferreira, C.; Husson, C. *Phytophthora ×alni* and *Phytophthora lacustris* associated with common alder decline in Central Portugal. *For. Pathol.* **2016**, *46*, 174–176. [CrossRef]
37. Nechwatal, J.; Bakonyi, J.; Cacciola, S.O.; Cooke, D.E.L.; Jung, T.; Nagy, Z.A.; Brasier, C.M. The morphology, behaviour and molecular phylogeny of *Phytophthora* taxon *Salixsoil* and its redesignation as *Phytophthora lacustris* sp. nov. *Plant Pathol.* **2012**, *62*, 355–369. [CrossRef]
38. Jung, T.; La Spada, F.; Pane, A.; Aloi, F.; Evoli, M.; Horta Jung, M.; Magnano di San Lio, G. Diversity and Distribution of *Phytophthora* Species in Protected Natural Plots in Sicily. *Forests* **2019**, *10*, 259. [CrossRef]
39. Stamler, R.A.; Sanogo, S.; Goldberg, N.P.; Randall, J.J. *Phytophthora* Species in Rivers and Streams of the Southwestern United States. *Appl. Environ. Microbiol.* **2016**, *82*, 4696–4704. [CrossRef]
40. Ghimire, S.R.; Richardson, P.A.; Kong, P.; Hu, J.; Lea-Cox, J.D.; Ross, D.S.; Hong, C. Distribution and Diversity of *Phytophthora* species in Nursery Irrigation Reservoir Adopting Water Recycling System During Winter Months. *J. Phytopathol.* **2011**, *159*, 713–719. [CrossRef]

41. Jung, T.; Blaschke, H.; Neumann, P. Isolation, identification and pathogenicity of *Phytophthora* species from declining oak stands. *For. Pathol.* **1996**, *26*, 253–272. [CrossRef]
42. Balcì, Y.; Halmschlager, E. *Phytophthora* species in oak ecosystems in Turkey and their association with declining oak trees. *Plant Pathol.* **2003**, *52*, 694–702. [CrossRef]
43. Orlikowski, L.B.; Ptaszek, M.; Rodziewicz, A.; Nechwatal, J.; Thinggaard, K.; Jung, T. *Phytophthora* root and collar rot of mature Fraxinus excelsior in forest stands in Poland and Denmark. *For. Pathol.* **2011**, *41*, 510–519. [CrossRef]
44. Belhaj, R.; McComb, J.; Burgess, T.I.; Hardy, G.E.S.J. Pathogenicity of 21 newly described *Phytophthora* species against seven Western Australian native plant species. *Plant Pathol.* **2017**, *67*, 1140–1149. [CrossRef]
45. Brasier, C.M.; Sanchez-Hernandez, E.; Kirk, S.A. *Phytophthora inundata* sp. nov., a part heterothallic pathogen of trees and shrubs in wet or flooded soils. *Mycol. Res.* **2003**, *107*, 477–484. [CrossRef] [PubMed]
46. Parkunan, V.; Johnson, C.S.; Bowman, B.C.; Hong, C.X. First report of *Phytophthora* inundata associated with a latent infection of tobacco (Nicotiana tabacum) in Virginia. *Plant Pathol.* **2010**, *59*, 1164. [CrossRef]
47. Jung, T.; Orlikowski, L.; Henricot, B.; Abad-Campos, P.; Aday, A.G.; Aguín Casal, O.; Bakonyi, J.; Cacciola, S.O.; Cech, T.; Chavarriaga, D.; et al. Widespread *Phytophthora* infestations in European nurseries put forest, semi-natural and horticultural ecosystems at high risk of *Phytophthora* diseases. *For. Pathol.* **2016**, *46*, 134–163. [CrossRef]
48. O'Hanlon, R.; Choiseul, J.; Corrigan, M.; Catarame, T.; Destefanis, M. Diversity and detections of *Phytophthora* species from trade and non-trade environments in Ireland. *EPPO Bull.* **2016**, *46*, 594–602. [CrossRef]
49. Hansen, E.M.; Reeser, P.; Sutton, W.; Brasier, C.M. Redesignation of *Phytophthora* taxon *Pgchlamydo* as *Phytophthora chlamydospora* sp. nov. *N. Am. Fungi* **2015**, *10*, 1–14.
50. Brasier, C.M.; Cooke, D.E.L.; Duncan, J.M.; Hansen, E.M. Multiple new phenotypic taxa from trees and riparian ecosystems in *Phytophthora gonapodyides-P. megasperma* ITS Clade 6, which tend to be high-temperature tolerant and either inbreeding or sterile. *Mycol. Res.* **2003**, *107*, 277–290. [CrossRef]
51. Li, H. Identification of *Phytophthora* species infecting staple crops in Xinjiang. *Acta Phytopathol. Sin.* **1999**, *29*, 364–371.
52. Jung, T.; Blaschke, M. *Phytophthora* root and collar rot of alders in Bavaria: Distribution, modes of spread and possible management strategies. *Plant Pathol.* **2004**, *53*, 197–208. [CrossRef]
53. Kovács, J.; Lakatos, F.; Szabó, I. Occurrence and diversity of soil borne *Phytophthoras* in a declining black walnut stand in Hungary. *Acta Silv. Lignaria Hung.* **2015**, *9*, 57–69. [CrossRef]
54. Ankowiak, R.; St, H.; Bila, P.; Kola, M. Occurrence of *Phytophthora plurivora* and other *Phytophthora* species in oak forests of southern Poland and their association with site conditions and the health status of trees. *Folia Microbiol.* **2014**, *59*, 531–542. [CrossRef]
55. Zamora-Ballesteros, C.; Haque, M.M.U.; Diez, J.J.; Martín-García, J. Pathogenicity of *Phytophthora alni* complex and *P. plurivora* in *Alnus glutinosa* seedlings. *For. Pathol.* **2017**, *47*, e12299. [CrossRef]
56. Mircetich, S.M. *Phytophthora* Root and Crown Rot of Apricot Trees. *Acta Hortic.* **2015**, *121*, 385–396. [CrossRef]
57. Day, W.R. Root- rot of sweet chestnut and beech caused by species of *Phytophthora*. I. Cause and symptoms of disease: Its relation to soil conditions. *Forestry* **1938**, *12*, 101–116. [CrossRef]
58. Mircetich, S.M.; Matheron, M.E. *Phytophthora* root and crown rot of walnut trees. *Phytopathology* **1983**, *73*, 1481–1488. [CrossRef]
59. Wilcox, W.F.; Ellis, M.A. *Phytophthora* root and crown rots of peach trees in the eastern Great Lakes region. *Plant Dis.* **1989**, *73*, 794–798. [CrossRef]
60. Yang, X.; Hong, C. *Phytophthora virginiana* sp. nov., a high-temperature tolerant species from irrigation water in Virginia. *Mycotaxon* **2014**, *126*, 167–176. [CrossRef]

9

An Overview of *Phytophthora* Species Inhabiting Declining *Quercus suber* Stands in Sardinia (Italy)

Salvatore Seddaiu [1], Andrea Brandano [2], Pino Angelo Ruiu [3], Clizia Sechi [1] and Bruno Scanu [2,4,*]

1 Settore Difesa Delle Piante Forestali, Agris Sardegna, Via Limbara 9, 07029 Tempio Pausania (SS), Italy; saseddaiu@agrisricerca.it (S.S.); csechi@agrisricerca.it (C.S.)
2 Dipartimento di Agraria, Sezione di Patologia Vegetale ed Entomologia, Università degli Studi di Sassari, Viale Italia 39, 07100 Sassari, Italy; abrandano@uniss.it
3 Settore Sughericoltura e Selvicoltura, Agris Sardegna, Via Limbara 9, 07029 Tempio Pausania (SS), Italy; paruiu@agrisricerca.it
4 Nucleo Ricerca Desertificazione, Università degli Studi di Sassari, Viale Italia 39, 07100 Sassari, Italy
* Correspondence: bscanu@uniss.it

Abstract: Cork oak forests are of immense importance in terms of economic, cultural, and ecological value in the Mediterranean regions. Since the beginning of the 20th century, these forests ecosystems have been threatened by several factors, including human intervention, climate change, wildfires, pathogens, and pests. Several studies have demonstrated the primary role of the oomycete *Phytophthora cinnamomi* Ronds in the widespread decline of cork oaks in Portugal, Spain, southern France, and Italy, although other congeneric species have also been occasionally associated. Between 2015 and 2019, independent surveys were undertaken to determine the diversity of *Phytophthora* species in declining cork oak stands in Sardinia (Italy). Rhizosphere soil samples were collected from 39 declining cork oak stands and baited in the laboratory with oak leaflets. In addition, the occurrence of *Phytophthora* was assayed using an in-situ baiting technique in rivers and streams located throughout ten of the surveyed oak stands. Isolates were identified by means of both morphological characters and sequence analysis of internal transcribed spacer (ITS) regions of ribosomal DNA. In total, 14 different *Phytophthora* species were detected. *Phytophthora cinnamomi* was the most frequently isolated species from rhizosphere soil, followed by *P. quercina*, *P. pseudocryptogea*, and *P. tyrrhenica*. In contrast, *P. gonapodyides* turned out to be the most dominant species in stream water, followed by *P. bilorbang*, *P. pseudocryptogea*, *P. lacustris*, and *P. plurivora*. Pathogenicity of the most common *Phytophthora* species detected was tested using both soil infestation and log inoculation methods. This study showed the high diversity of *Phytophthora* species inhabiting soil and watercourses, including several previously unrecorded species potentially involved in the decline of cork oak forests.

Keywords: cork oak; oak decline; oomycetes; *Phytophthora cinnamomi*

---

1. Introduction

Cork oak (*Quercus suber* L.) represents an important component of the Mediterranean forests landscape, covering more than 2 million ha across southern European and northern African countries [1]. This type of forest ecosystem has great socio-economic value, providing a range of non-timber forest products, such as cork, firewood, grazing, honey, and mushrooms, playing a key role in the rural economy in less favorable regions [2,3]. In particular, cork production represents a highly sustainable non-wood product derived from forests in the western Mediterranean countries and an additional source of income for farmers. In Sardinia (Italy), which hosts more than 80% of the Italian distribution area, cork oak is the second most important production chain of the island [4]. Moreover, these forest systems provide a wide range of several un-costed ecosystem services, including biodiversity conservation and desertification control [1,3,5]. For all of these reasons, many cork oak forests are recognized as protected ecosystems under the Pan-European network of protected areas (www.natura.org), Sites of Community Importance and Special Protection Areas for biodiversity conservation (Council Directive 92/43/EEC).

Despite performing these important functions, Mediterranean cork oak forests are currently under large scale reduction due to a wide range of drivers, such as pathogens and pests, climate change, wildfires, overgrazing, degradation, and fragmentation [6–8]. Over the last three decades, the role of pathogens in such forest ecosystems has gained increased attention due to the exponential emergence of forest diseases worldwide, particularly in Mediterranean ecosystems [9–13]. Several studies have demonstrated the involvement of the oomycete *Phytophthora cinnamomi* in the widespread decline of Mediterranean oaks, including cork oak, in Portugal, Spain, southern France, and Italy [14–19]. Although *P. cinnamomi* appears to be the most dominant species, other congeneric species can also be associated with Mediterranean oak decline [12,20,21]. The diversity of *Phytophthora* species in Mediterranean oak ecosystems has been further explored in recent years using metagenomic approaches based on high-throughput sequencing (HTS), which, through the use of species-specific primers, allow the amplification of a high number of target organisms from environmental DNA [22–24]. However, most of these studies are related to holm oak (*Quercus ilex* L.), while cork oak forests still remain poorly investigated, and to the best of our knowledge, only three *Phytophthora* species have been formally reported [12]. In the extensive surveys on reforestations and afforestations across Europe between 1998 and 2009 by Jung and collaborators [25], *P. cryptogea* (now known as *P. pseudocryptogea*) and *P. quercina* were reported from only two cork oak plantations in Spain. More recently, three previously unrecorded *Phytophthora* species have been associated with episodic events of cork oak decline in Italy, including the newly described *P. tyrrhenica* and the exotic pathogenic *P. megasperma* and *P. multivora* [21,26].

The main objective of the present work was to investigate the diversity of *Phytophthora* species in declining cork oak forest ecosystems in Sardinia. As many *Phytophthora* species have a specific aquatic lifestyle, their occurrence was also explored along rivers and streams within the surveyed sites. Moreover, the pathogenicity of the most frequently isolated *Phytophthora* species detected was tested using both soil infestation and log inoculation methods.

2. Materials and Methods

2.1. Soil Sampling and Phytophthora Isolation

Between 2015 and 2019, independent surveys were undertaken to investigate the diversity of *Phytophthora* species from 39 declining cork oak forests in Sardinia: 25 natural forests and 14 afforestation sites (Figure 1). These sites were selected over the years on the basis of different reports, submitted by private and public entities, of problems affecting the health of cork oak trees. Tree health monitoring and surveillance work were made by an initial pre-screening in the field based on visible symptoms of decline, such as yellowing leaves, crown transparency, epicormic shoots, branch dieback, bleeding cankers, as well as necrotic lesions at the collar and root levels. Samplings were conducted in the autumn and spring seasons, and in some areas, these were repeated twice. A total of 295 symptomatic cork oak trees were sampled. Rhizosphere soil samples consisted of a mixture of four subsamples taken from around the stem base of selected trees, scraping away the litter and taking about 200–300 g of roots and soil. *Phytophthora* isolations were made using an adaptation of the baiting methods described by Jung et al. [27]. In the laboratory, roots and soil were flooded in 12 × 10 × 22 cm glass trays with 500 mL of distilled water, then young leaflets taken from 1–2-month-old cork oak seedlings were used as baits floated over the water. After 3–5 days, leaves with black spots were checked under the microscope for the presence of sporangia, dried on filter paper, and plated onto synthetic mucor agar (SMA), a selective medium for *Phytophthora* [28]. All Petri dishes were incubated in the dark at 20 °C and checked daily for *Phytophthora*-like hyphae development, which was subsequently transferred to

Petri dishes containing carrot-agar (CA; 16 g agar technical no.3, Oxoid Ltd., Basingstoke, UK, 200 g carrots and 1000 mL distilled water) [29].

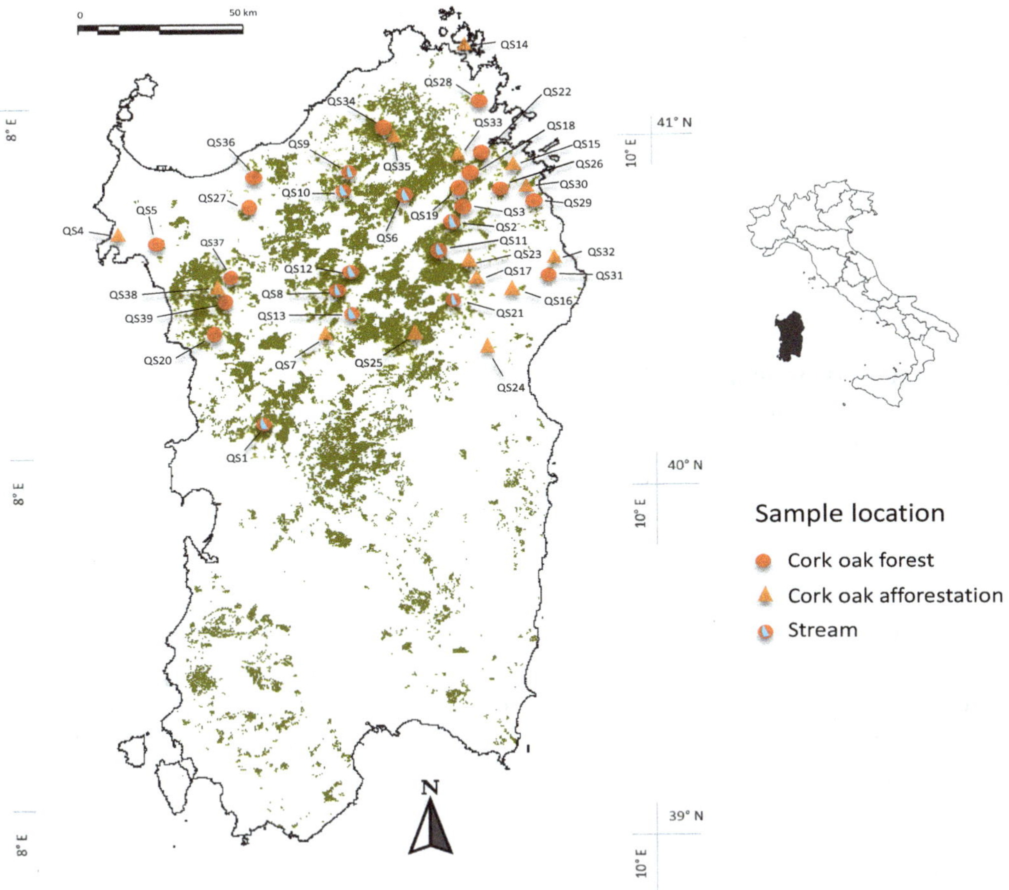

Figure 1. The geographic location of the 39 *Quercus suber* stands investigated in this study (QS1–QS39). The green area represents the geographic distribution of cork oak in Sardinia. A map of Italy is inserted into the top right corner showing the location of Sardinia.

2.2. Stream Baiting and Phytophthora Isolation

In addition to soil samples, river and stream water within ten surveyed cork oak forests (Figure 1) were assayed using the in-situ baiting technique in spring 2018. Watercourses were chosen depending on their water flow and on the capacity to collect water from the bordering forests. They were subdivided into two different groups, mainly based on the altimetric gradient, including valley floor rivers with permanent water flow, and mountain or hill streams or water catchments into the forests, with water flow strictly correlated with the seasonal rainfall and very often drying up in summer. River baiting was made using an adaptation of the method described by Reeser et al. [30] and Hüberli et al. [31], which consisted of two squared layers of fly mesh or metallic net, sealed together, with young leaves of different plant species, such as *Alnus glutinosa* Mill., *Arbutus unedo* L., *Buxus sempervirens* L., *Laurus nobilis* L., *Fraxinus ornus* L., *Hedera helix* L., *Parthenocissus quinquefolia* L., *Q. ilex* L., *Q. suber* L., and *Taxus baccata* L. placed between the two mesh layers. Rafts were placed in-situ at the same time as their set up, and each raft contained about 20 leaves. Cork stoppers were used to float the raft over the water surface. Three rafts were placed every 200 m from each other along the watercourse. The rafts were fastened to natural restraints, such as branches and rocks, and left floating over the water for 3–4 days and then brought to the laboratory where the leaves were washed with sterile water and blotted dry on filter paper. *Phytophthora* isolations were made by placing small fragments cut from necrotic lesions detected on the leaf baits on SMA. Any developing colonies were sub-cultured on CA for further analyses.

2.3. Morphological and Molecular Identification

The isolates obtained from soil samples and stream water were first grouped based on their colony growth patterns after 5–7 days at 20 °C in the dark on CA. In addition, morphological features of sporangia, oogonia, antheridia, chlamydospores, hyphal swellings, and aggregations were examined under the Leitz Diaplan compound microscope (Leitz, Wetzlar, Germany) and compared with species descriptions in the literature [21,32–38]. A subset of representative morphotypes (88 isolates) was selected for molecular analyses, which consisted of DNA extraction, amplification, sequencing, and analysis of sequences of the entire region of the internal transcribed spacers (ITS1 and ITS2) and the 5.8 S gene of the rDNA.

DNA was extracted from mycelium fragments, using the extraction kit InstaGene™ Matrix (BioRad Laboratories, Hercules, CA, USA). The amplification of the ITS region was carried out with a thermocycler (Hybaid PCR Express), using the forward primers ITS1 or ITS6 and the reverse primer ITS4 [39,40]. A total volume of 50 μL, consisting of 18.2 μL of water, 5 μL of BSA, 5 μL o f dNTPs, 5 μL of both ITS6 and ITS4 primers, 10 μL of the buffer, 0.3 μL of Go Taq polymerase, and 1.5 μL of DNA from each morphotype, was used for standard PCR (polymerase chain reaction). The cycle used for the amplification of the ITS genes regions was as follows: initial denaturation of 1 min at 94 °C, followed by 35 cycles of 1 min at 94 °C, 1 min at 56 °C, 1 min at 72 °C, followed by a final elongation phase of 7 min at 72 °C. The PCR products were purified using the EUROGOLD gel extraction kit (EuroClone S.p.A., Pero, Italy). After quantification, purified PCR amplicons and the sequencing primers were sent to BMR Genomics sequencing service (https://www.bmr-genomics.it).

DNA sequence chromatograms were viewed and edited using BioEdit v. 5.0.6 software [41]. Heterozygous sites observed were labeled according to the IUPAC coding system. Isolates were assigned to a species when sequence identities were above a 99% cut-off with respect to those of ex-type isolates or key isolates. All sequences were deposited at GenBank (http://www.ncbi.nlm.nih.gov/), and the accession numbers are given in Table 1.

2.4. Pathogenicity Tests

Pathogenicity of the most frequent *Phytophthora* species isolated was assayed using the soil infestation method described by Jung et al. [27], with some modification as reported by Scanu et al. [29]. In particular, two isolates of *P. cinnamomi*, *P. gonapodyides*, *P. pseudocryptogea*, *P. psychrophila*, *P. quercina*, and *P. tyrrhenica* were grown in individual 500 mL Erlenmeyer flasks containing an autoclaved mixture of 250 mL of vermiculite and 150 mL of *Lolium italicum* seeds thoroughly moistened with 100 mL of carrot juice (200 mL/L carrot juice, 3 g/L $CaCO_3$, and 800 mL/L distilled water). Flasks were incubated at 20 °C for 1 month, then 20 mL of inoculum was collected and inserted inside the soil of 2-year-old cork oak seedlings (provided by the Regional Agency Fo.Re.S.T.A.S.).

The substrate in the controls received a sterile mixture of vermiculite/seeds-carrot juice at the same ratio. To stimulate the production of sporangia and pathogen spread and infection via zoospores, pots were flooded immediately after inoculation for 48 h, and flooding was repeated at three-week intervals by immerging pots in 10 L buckets just to 1 cm above the soil surface.

There were eight replicates per isolate and controls. After 5 months of incubation at 20 °C(±2 °C), 70% relative humidity with a 12/12 h photoperiod, seedlings were visually assessed for symptoms, and the mortality rate was recorded; then each plant was removed from the pot, and the root system gently washed under tap water. Single roots were cut off at the collar, and after scanning, the total root length of all the plant root systems was measured using the APS Assess 2.0 software (The American Phytopathological Society, St. Paul, MN, USA). The remaining soil was baited following the method described above to determine whether the pathogen was still viable. Re-isolations were also made directly from necrotic roots using SMA.

Pathogenicity of the above isolates (except for those of *P. tyrrhenica*) was further tested using freshly cut logs of cork oak following the method described by Brasier and Kirk [42]. Four logs (1.4 m long and 20 cm in diameter) were cut from stems of living cork oak trees 24 h before the experiment, and the cut ends were sealed with a liquid waterproofing membrane. In each log, three bands were marked around the log circumference, 30 cm apart from each other, with 5 inoculation points per band, about 15 cm apart. After sterilizing the bark with 70% ethanol, a 7 mm diameter hole was punched through the bark to the wood surface with a steel cork borer.

The same-sized plug was taken from the edge of an actively growing colony on CA and used as inoculum by inserting into the hole replacing the bark plug. Three control inoculation points per log were inoculated with a sterile CA plug and covered with the removed piece of bark. Moist cotton wool was placed over the wounds, covered with a 5 × 5 cm piece of aluminum foil, and sealed with an adhesive PVC tape. There were four replicates per isolate.

Inoculated logs were covered individually in loose polythene sleeves (sealed at both ends) and incubated at 20 °C (±2 °C) in an air-conditioned laboratory and checked weekly for the appearance of symptoms. After 45 days, the experiment was finished, and logs were destructively sampled by removing the periderm with a drawknife to expose the phloem. Each lesion's outline was then recorded on tracing paper and scanned on an Epson Perfection V30 photo scanner, and the lesion area calculated using APS Assess software, as described by Scanu and Webber [43]. Re-isolation of all the inoculated *Phytophthora* species onto SMA was attempted from the lesion margins.

Statistical analyses for both pathogenicity tests were performed using XLSTAT software (Addinsoft). Data were first checked for normality and then subjected to analysis of variance (ANOVA). Statistical differences among mean values of root lengths and lesion areas were determined using Fisher's protected least significant difference (LSD) test. Differences with $p < 0.05$ were considered significant.

3. Results

3.1. Symptomatology

A wide range of symptoms of decline was observed on cork oak trees across all the investigated sites. These included rapid dieback of the crown in both mature (Figure 2a) and young oak trees (Figure 2b), which was frequently observed in early autumn, especially after a long summer and drought conditions. In the case of afforestation sites (10 to 20-year-old), the infections could reach epidemic levels and cause extensive mortality of oak trees.

Other symptoms included shoot dieback and increased transparency of the whole crown, leaf chlorosis, and abundant proliferation of epicormic shoots on stems and branches (Figure 2c,d). At the collar level, trees showed necrotic bark lesions frequently associated with black exudation and very often girdling the stem (Figure 2e). In the root system of declining oak trees, an extensive loss of both lateral small woody roots and fine roots and callusing or open cankers on suberized roots were observed.

Figure 2. Symptoms caused by *Phytophthora* species on *Quercus suber.* (**a**) The sudden death of mature trees; (**b**) severe dieback and mortality of trees in 10-year-old afforestation; (**c**) trees showing a chronic decline with increasing transparency and wilting on the crown; (**d**) widespread dead and declining trees in a natural stand; (**e**) bleeding cankers at the stem base of a young tree.

3.2. Soilborne Phytophthora Species

Phytophthora species were recovered from 68.5% of the 295 soil samples tested. The highest level of soil infestation was detected in the afforestation sites (80.4%), while in natural forests, the percentage of positive trees was 61.7%. In total, 224 isolates were obtained from rhizosphere soil samples collected from around symptomatic trees in declining cork oak stands (Table 1). All isolates conformed morphologically to previously known *Phytophthora* species. ITS sequence analysis of the isolates confirmed the morphological identification of all *Phytophthora* species. BLAST searches in GenBank showed 99–100% similarity with reference sequences, including those of ex-type cultures or representative isolates (Table 1). In total, eight *Phytophthora* species belonging to five (clade 3, 6, 7, 8, and 12) out of the twelve known phylogenetic clades were isolated, including *P. cinnamomi*, *P. gonapodyides*, *P. pseudocryptogea*, *P. psychrophila*, *P. quercina*, *P. syringae*, *P. tyrrhenica*, and *P.* ×*cambivora* (Table 1 and Figure 3a,c). *Phytophthora cinnamomi* from clade 7c was the most frequent species isolated from both natural forests and afforestation stands. It was detected from almost all investigated afforestation sites (from 30 out of 39 sites) with an infection rate of 80.2%, while its incidence was markedly lower in natural cork oak forests (45.7% of 122 investigated trees). At one afforestation site (QS35), *P. cinnamomi* was recovered from 27 out of 30 cork oak trees sampled (Table 1). It was the only species recovered in 10 investigated stands. Similarly, the second most common species, *P. quercina*, from clade 12 occurred in both afforestation and natural stands, with an infestation rate ranging from 8.1% to 30.2%, respectively. It was the only species isolated from declining trees in sites QS6, QS13, and QS25. All *P. quercina* isolates had identical ITS sequences; however, a certain phenotypic variation among the isolates was observed. Both *P. pseudocryptogea* (clade 8a) and *P. tyrrhenica* (clade 7a) were recovered from eight declining cork oak stands, with an infestation rate of around 10%. The ITS sequences of all *P. pseudocryptogea* isolates from rhizosphere soil matched the ex-holotype isolate (GenBank no. KP288376). However, they had a unique polymorphism at position 56 (C instead of Y) and were heterozygous at position 650 (Y instead of T). *Phytophthora gonapodyides* from clade 6b occurred only in natural contexts, isolated from 12 declining trees in seven sites, and always in association with other *Phytophthora* species. Among the less frequently isolated species, *P. psychrophila* (clade 3) and *P.* ×*cambivora* (clade 7a) were detected only from natural forests at a very low infestation rate (2.5% and 1%, respectively), while *P. syringae* was exclusively found at one afforestation site from two symptomatic trees. In 18 soil samples from declining trees and 12 sites, multiple *Phytophthora* species were detected. *Phytophthora cinnamomi* was isolated along with *P. quercina* in four samples and along with both *P. quercina* and *P. tyrrhenica* (two samples) or *P. pseudocryptogea* (one sample). In four cases, *P. tyrrhenica* was isolated together with *P. cinnamomi* and, in one case, with *P. quercina*.

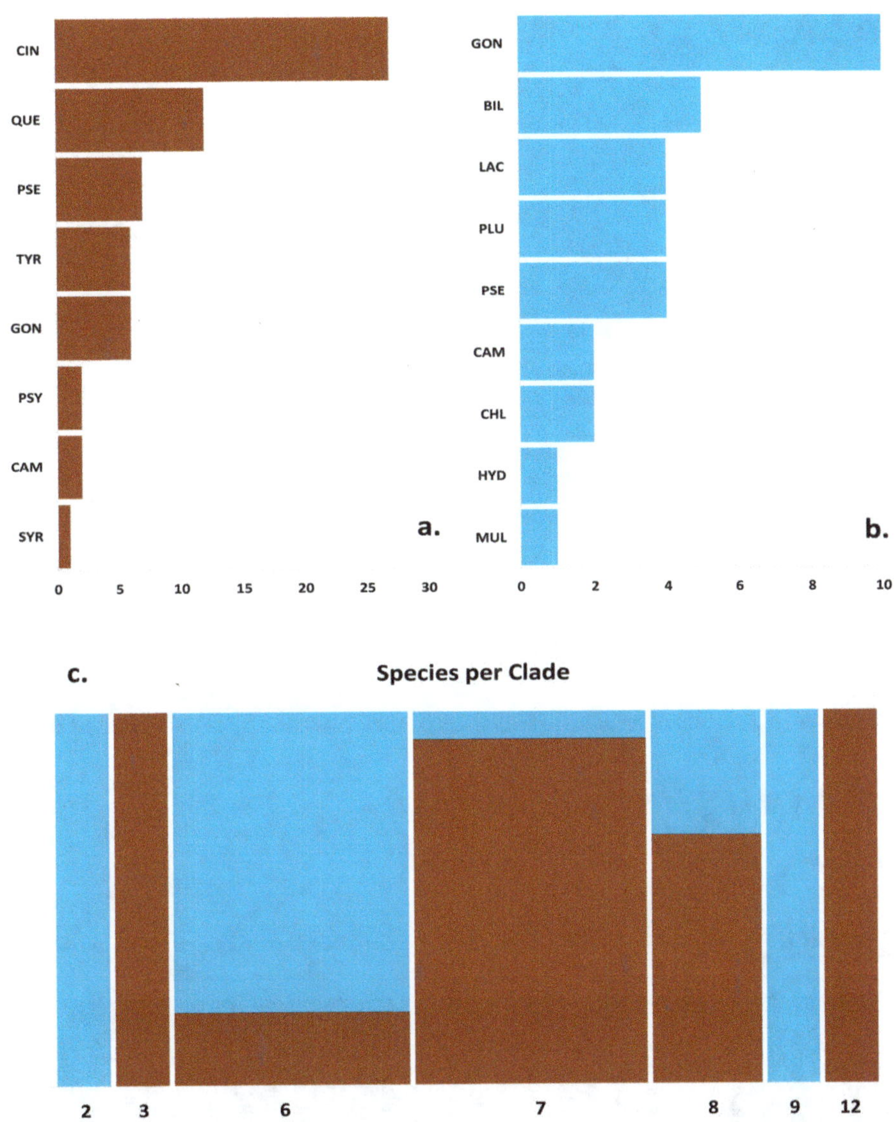

Figure 3. Diversity and frequency of the 8 soilborne (**a**) and 9 waterborne (**b**) *Phytophthora* species detected in this study; the horizontal axis is the count of sites. (**c**) Mosaic plot showing the distribution of soilborne (brown bars) and waterborne (blue bars) *Phytophthora* species grouped for phylogenetic clades. The bar width is proportional to the number of species, while bar heights show the relative proportion of soilborne and waterborne *Phytophthora* species per clade. CIN = *P. cinnamomi*, QUE = *P. quercina*, PSE = *P. pseudocryptogea*, TYR = *P. tyrrhenica*, GON = *P. gonapodyides*, PSY = *P. psychrophila*, CAM = *P.* ×*cambivora*, SYR = *P. syringae*, BIL = *P. bilorbang*, LAC = *P. lacustris*, PLU = *P. plurivora*, CHL = *P. chlamydospora*, HYD = *P. hydropathica*, MUL = *P.* ×*multiformis*.

Table 1. Location, forest type, and altitude of the 39 declining cork oak stands sampled in Sardinia and *Phytophthora* taxa isolated from the rhizosphere soil samples collected in this study.

Site	Location (Municipality)	Forest Type [a]	Altitude (m a.s.l)	Trees Sampled (No.)	Positive Trees (No.)	*Phytophthora* spp. [b]							
						CAM	CIN	GON	PSE	PSY	QUE	SYR	TYR
QS1	Abbasanta	For	352	4	2		1	1					
QS2	Alà dei Sardi	For	620	4	2		1						
QS3	Alà dei Sardi	For	580	5	3		1						2
QS4	Alghero	Aff	80	6	5		4		1				1
QS5	Alghero	For	122	4	2				2				
QS6	Berchidda	For	201	8	2						2		
QS7	Bolotana	Aff	340	4	4		4						
QS8	Bono	For	841	6	4		2	1			1		
QS9	Bortigiadas	For	158	14	10		5	2	4				
QS10	Bortigiadas	For	157	12	11	1	4	4	2	2	5		3
QS11	Buddusò	For	762	11	3						3		
QS12	Bultei	For	503	6	4	1	2			1			
QS13	Illorai	For	214	4	2						2		
QS14	La Maddalena	Aff	118	6	4		3				2		
QS15	Loiri P.S.P.	Aff	240	16	12		8						
QS16	Lula	Aff	560	4	3		2						3
QS17	Lula	Aff	380	4	4		2						4
QS18	Monti	For	268	8	7		4	1			2		
QS19	Monti	For	260	4	2		1				2		
QS20	Montresta	For	480	10	8		3	2	2		4		
QS21	Nuoro	For	490	10	3						3		
QS22	Olbia	For	60	4	4		4						
QS23	Onanì	Aff	455	4	1		1						
QS24	Orgosolo	Aff	568	6	6		4				2		
QS25	Orotelli	Aff	410	4	4		4				3		
QS26	Padru	For	363	8	6		4				2		
QS27	Ploaghe	For	320	11	5						3		3
QS28	San Pantaleo	For	180	4	4		4						
QS29	San Teodoro	For	13	8	5		4		2				
QS30	San Teodoro	Aff	118	6	4		4						
QS31	Siniscola	For	125	12	6		6						
QS32	Siniscola	Aff	248	8	8		6		2				
QS33	Telti	Aff	210	6	2							2	
QS34	Tempio P.	For	438	4	3		2				1		

Table 1. *Cont.*

Site	Location (Municipality)	Forest Type [a]	Altitude (m a.s.l)	Trees Sampled (No.)	Positive Trees (No.)	*Phytophthora* spp. [b]							
						CAM	CIN	GON	PSE	PSY	QUE	SYR	TYR
QS35	Tempio P.	Aff	465	30	27		27						
QS36	Tergu	For	346	12	12		1	1		2	5		4
QS37	Villanova M.	For	538	10	4		4						
QS38	Villanova M.	Aff	440	3	2				2				
QS39	Villanova M.	For	428	5	2								2
GenBank accession numbers						MT823269	MT328694 MT328695 MT328696	MT823270	MT328706	MT328708	MT328709 MT328710	MT328711	MT328712

[a] For = forest, Aff = afforestation. [b] CAM = *P.* ×*cambivora*, CIN = *P. cinnamomi*, GON = *P. gonapodyides*, PSE = *P. pseudocryptogea*, PSY = *P. psychrophila*, QUE = *P. quercina*, SYR = *P. syringae*, TYR = *P. tyrrhenica*.

3.3. Waterborne *Phytophthora* Species

In total, 115 *Phytophthora* isolates were detected in all watercourses monitored through ten selected declining cork oak stands. Based on morphological analyses and molecular identification, these isolates belonged to five phylogenetic clades (clade 2, 6, 7, 8, and 9) corresponding to nine formally known *Phytophthora* species, including *Phytophthora bilorbang*, *Phytophthora chlamydospora*, *P. gonapodyides*, *P. hydropathica*, *P. lacustris*, *P. plurivora*, *P. pseudocryptogea*, *P.* ×*cambivora*, and *P.* ×*multiformis* (Table 2 and Figure 3b,c). Overall, more than 50% of the isolates obtained were identified as *P. gonapodyides*, which was recovered from all watercourses surveyed.

Interestingly, eight isolates detected from stands QS1, QS8, and QS9 were heterozygous at position 106 (R instead of G). Another isolate from stand QS1 (GON3) shared the heterozygous position 106 and had unique polymorphisms at positions 106 (A instead of R or G) and 517 (T instead of G), respectively. Two isolates, PH255 (QS10) and PH267 (QS9), also differed from the other *P. gonapodyides* isolates by having unique polymorphisms at positions 145 (T instead of C) and 517 (T instead of G), respectively.

Phytophthora bilorbang was the second most widespread species isolated from five investigated streams, four of which only flow seasonally. All isolates from stand QS21 were heterozygous at position 106 (Y instead of T). *Phytophthora pseudocryptogea* and *P. plurivora* were isolated only from permanent water bodies, while *P. lacustris* occurred from both permanent and intermittent watercourses (Table 2). The ITS sequences of all *P. lacustris* isolates differed from the ex-type culture (GenBank no. AF266793) having a heterozygous site at position 783 (S instead of C). In addition, one isolate was heterozygous at position 458 (Y instead of C).

Almost all isolates identified as *P. pseudocryptogea* from river water had identical ITS sequences than those isolates obtained from rhizosphere soil, differing from the ex-holotype by 2 bp. Moreover, one isolate (PH269) from stand Q12 differed from the ex-holotype isolate of *P. pseudocryptogea* and the other isolates obtained in this study by 4–5 bp at positions 56, 601, 650, 728, and 733. *Phytophthora chlamydospora* and *P.* ×*cambivora* were recovered from two streams, while *P.* ×*multiformis* and *P. hydropathica* were exclusively isolated from QS10 and QS13, respectively.

The two *P. chlamydospora* isolates from stand QS9 were heterozygous at position 666 (Y instead of T). All isolates identified as *P. hydropathica* differed from the ex-type culture (GenBank no. EU583793) by having two heterozygous sites at positions 413 (S instead of C) and 665 (K instead of T) and by a unique polymorphism at position 628 (G instead of C). The ITS sequences of *P.* ×*cambivora* often generated overlapping ITS sequences starting at position 396 in both directions. This was a consistent pattern observed in all isolates of *P.* ×*cambivora* obtained from river water. The non-overlapping sequences up the indel positions were identical to that of the neotype culture of *P*· ×*cambivora* (GenBank no. KU899179).

Looking at the diversity of *Phytophthora* species across rivers, the two geographically close rivers at sites QS9 and QS10 hosted the highest number of *Phytophthora* species, followed by a stream in stand QS12 with five species detected. Only two species were isolated from water bodies at stands QS1–3, QS13, and QS21, and they were all from clade 6, except for a clade 9 species at site QS13 detected together with *P. gonapodyides*.

Table 2. Location, name, and typology of the 10 watercourses sampled across declining cork oak stands in Sardinia and *Phytophthora* taxa identified.

Site No.	River/Stream	Description	*Phytophthora* spp. [a]								
			BIL	CAM	CHL	GON	HYD	LAC	MUL	PLU	PSE
QS1	Pizziu	Permanent river	+			+					
QS2	Sa Labia	Permanent river				+		+			
QS6	Berchidda	Permanent river				+		+			
QS8	Monte Pisanu	Intermittent stream		+		+				+	+
QS9	Santu Brancazzu	Intermittent stream	+	+	+	+				+	+
QS11	Sos Canales	Intermittent stream	+			+					
QS12	Olletto	Intermittent stream	+			+		+		+	+
QS13	Tirso	Permanent river				+	+				
QS21	Errede	Intermittent stream	+			+					
QS10	Puddina	Intermittent stream			+	+		+	+	+	+
GenBank accessions			MT328690 MT328691 MT328692	MT328713	MT328693	MT328697 MT328698 MT328699	MT822885	MT328700 MT328701	MT822886	MT328704 MT328705	MT328707

[a] BIL = *P. bilorbang*, CAM = *P.* ×*cambivora*, CHL = *P. chlamydospora*, GON = *P. gonapodyides*, HYD = *P. hydropathica*, LAC = *P. lacustris*, MUL = *P.* ×*multiformis*, PLU = *P. plurivora*, PSE = *P. pseudocryptogea*.

3.4. Pathogenicity Test

The soil infestation experiment showed that all *Phytophthora* species tested were able to cause a significant reduction of the root system in 2-year-old cork oak seedlings (Figure 4a). The mean root length was significantly higher in control seedlings ($p < 0.05$) than in seedlings infected with *Phytophthora* isolates. *Phytophthora cinnamomi* was the most aggressive species causing a root length reduction near to 65% compared to the control seedlings, followed by *P. pseudocryptogea* (56.2%), *P. tyrrhenica* (44.6%), and *P. quercina* (40.2%). *Phytophthora gonapodyides* and *P. psychrophila* caused a root length reduction below 35%. *Phytophthora cinnamomi* was the only species associated with extensive lesions on the mother root, with lesions in some cases reaching 15 mm in length. Apart from *P. psychrophila*, all the other *Phytophthora* species were re-isolated from limited necrotic lesions on taproot. No symptoms of pathogen infection could be seen on the roots of control seedlings.

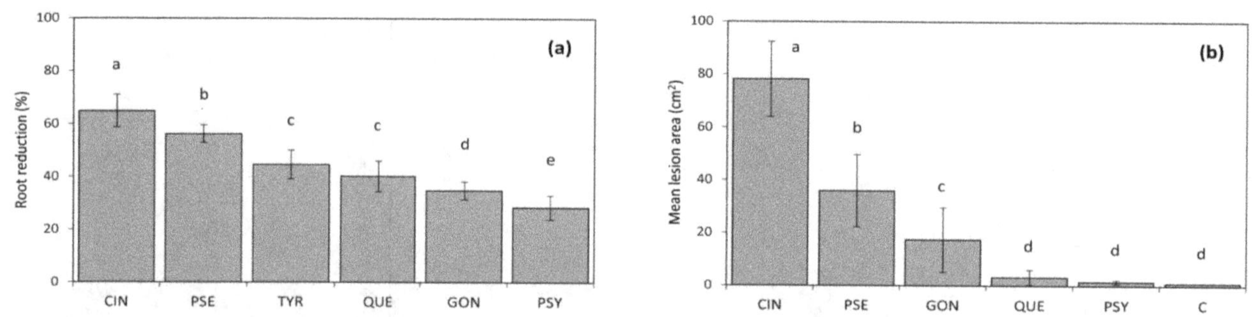

Figure 4. Root length reduction (%) compared to control seedling roots of 2-year-old seedlings of *Quercus suber* after 5 months of growth in soil infested with *Phytophthora* spp. obtained in this study (**a**). Mean lesion sizes, caused by isolates of *Phytophthora* following inoculation and incubation of logs for 4 weeks (**b**). Different letters above bars indicate significant differences according to Fisher's protected least significant difference (LSD) test ($p = 0.05$). Bars represent standard errors. CIN = *P. cinnamomi*, PSE = *P. pseudocryptogea*, TYR = *P. tyrrhenica*, QUE = *P. quercina*, GON = *P. gonapodyides*, PSY = *P. psychrophila*, C = control.

In the log inoculation tests, lesions in the phloem tissue of cork oak caused by *P. cinnamomi* were significantly larger ($p < 0.0001$) compared with the negative controls, with a mean necrosis area of 80 cm^2 (Figure 4b). *Phytophthora pseudocryptogea* and *P. gonapodyides* also showed considerable aggressiveness on inoculated logs. The mean lesion area formed by *P. cinnamomi* was approximately two to three times larger ($p < 0.0001$) than that developed following inoculation with *P. pseudocryptogea* and *P. gonapodyides*, respectively (Figure 4b). *Phytophthora psychrophila* and *P. quercina* were not able to colonize phloem tissues producing lesions that did not differ significantly from the negative controls ($p > 0.05$). Apart from *P. psychrophila* and *P. quercina*, all *Phytophthora* species were readily re-isolated from the necrotic lesions. In contrast, the controls developed only limited discoloration around the inoculation point and never yielded any *Phytophthora*.

4. Discussion

The extensive surveys made over four years across declining cork oak stands in Sardinia, together with morphological and ITS sequences analyses, have revealed the occurrence of 14 *Phytophthora* taxa from seven of the 12 known phylogenetic clades [21]. These include species common in Mediterranean oak soil, such as *P. cinnamomi*, *P. gonapodyides*, and *P. quercina*, and the less widespread species *P. pseudocryptogea*, *P. psychrophila*, *P. syringae*, *P. tyrrhenica*, and *P. ×cambivora* [12,17,20,44,45]. The detection of nine *Phytophthora* species from stream and river water represents the first attempt to look at the diversity of aquatic species in such forest ecosystems. Apart from *P. gonapodyides* and *P. pseudocryptogea*, all the other species identified from watercourses, *P. bilorbang*, *P. chlamydospora*, *P. hydropathica*, *P. lacustris*, *P. plurivora*, *P. ×cambivora*, and *P. ×multiformis*, were never reported in cork oak ecosystems.

Most of the previous surveys on Mediterranean oak decline have focused on the association of *P. cinnamomi* and *Q. ilex* (ssp. *ballota* and *rotundifolia*) [16,17,46–49], while cork oak has been less studied as it appears to be more tolerant to the disease due to its defense response mechanisms to *Phytophthora* infection [14,15,50]. Our study represents the first extensive survey on the distribution of *P. cinnamomi* and other congeneric species in declining cork oak stands.

Among the *Phytophthora* species detected from soil samples, *P. cinnamomi* was the most common species encountered. Listed as one of the 100 worst invasive alien species, *P. cinnamomi* is considered one of the most devastating plant pathogens worldwide [51,52]. It was first associated with the severe dieback and mortality of Mediterranean oaks, including both cork and holm oak, in the Iberian peninsula by Brasier in 1992 [53]; and since then, *P. cinnamomi* has been reported across the Mediterranean basin, and this is well documented for European countries [14–17,44,54]. In this study, the pathogen was detected only from rhizosphere soil; however, it was occasionally isolated from bleeding lesions on the stem (data not shown), as reported by Robin et al. [15]. Although cork oak has been shown to be less susceptible than holm oak [46,55], the high ability of *P. cinnamomi* to colonize phloem tissues, as exhibited in the log inoculation trial in this study and previously [56], together with its widespread occurrence across Sardinian stands, could suggest less tolerance of the Sardinian cork oak population [57]. This hypothesis, however, needs further investigation.

Phytophthora quercina is the second most prevalent species from soil samples. This oak-specific pathogen has been previously reported in central and southern Europe, causing a chronic decline in *Quercus cerris*, *Quercus faginea*, *Q. ilex*, *Quercus petrea*, *Quercus pubescens*, and *Quercus robur* [20,44,58,59]. Although *P. quercina* was recorded from two cork oak plantations in Spain [25], this appears to be the first widespread occurrence in natural cork oak forests. Interestingly, two distinct phenotypes were observed amongst the isolates detected from cork oak trees, supporting the hypothesis that *P. quercina* originated from Europe [21]. Multigene sequencing and phylogenetic analyses are currently underway to investigate the genetic population structure of a large number of *P. quercina* isolates from different oaks and various geographic provenances (Scanu and Jung, unpublished).

Two other slow-growing and homothallic species are detected at low frequency, and these are identified as *P. psychorphila* and *P. syringae*. Together with *P. quercina*, both species were previously reported from Mediterranean oaks in Spain, and their pathogenicity was demonstrated on both *Q. ilex* and *Q. faginea* [20]. Pathogenicity tests in the present study showed both species were not able to invade the inner bark of cut logs of cork oak; however, this did not correlate with root susceptibility. Previous results obtained by Perez-Sierra et al. [20] suggest these species are well adapted to the Mediterranean climate and may act as fine root nibblers, the incidence of which varies depending on the occurrence of extreme climatic events, such as recurrent drought and wet seasons [60]. Both *P. psychrophila* and *P. quercina* are already reported in Sardinia from declining *Q. ilex* trees, while *P. syringae* is associated with dieback and mortality of *Juniperus phoenicea* on Caprera Island [28]. Due to their low maximum temperature for growth (around 25 °C), typical of *Phytophthora* species from cool temperate regions, a potential seasonal activity, as suggested for *P. cinnamomi* and other cool-temperature pathogens, may occur [61–64]. Of note, *P. syringae* was detected only from a new plantation, suggesting its possible new introduction through infected plant material [25,65]. This could also happen for the other *Phytophthora* species considering the massive afforestation effort in Sardinia between 1990 and 2010 through EU programs, like EEC Regulation 2080/92 [4,25].

The finding of the recently described *P. tyrrhenica* confirmed its original description from declining cork oak trees [21]. Apart from one site (QS39), it was detected at nine stands and always coexisted with other congeneric species. So far, this cryptic species has been recovered from cork and holm oak trees in Sardinia and Sicily (Italy), respectively, and it is considered endemic to the Mediterranean basin [21,66]. Similarly, *P. pseudocryptogea* from clade 8 was the only species obtained from QS5 and QS38; otherwise, it co-occurred with *P. cinnamomi* or other species. It is noteworthy that all the previous isolates identified as *Phytophthora cryptogea* in Sardinia from forest trees, including oaks and *Pinus radiate*, are indeed *P. pseudocryptogea* [28,37,48,67]. *Phytophthora pseudocryptogea* was recently reported as one of the most widespread species in riparian thermo-Mediterranean forest stands and from five rivers in Sicily [66]. The clade 6 species *P. gonapodyides* had a more scattered distribution (seven stands and 12 trees), and its finding on cork oak confirmed results obtained by Jung et al. [66]. Finally, for the first time, we recorded the hybrid species *P.* ×*cambivora* on *Q. suber*. In Mediterranean regions, this species is frequently associated with "chestnut ink disease" [68], as well as from other Fagaceae [66,69] and Pinaceae trees [70]. In a recent survey in Sicily, *P.* ×*cambivora* was isolated from three different Mediterranean oaks, namely *Q. cerris*, *Q. ilex*, and *Q. pubescens* [66].

The detection of nine *Phytophthora* species from five phylogenetic clades in 10 rivers within declining cork oak stands was unexpected since similar diversity rates are often reported in more diverse forest ecosystems and with higher sampling rates from surveys in Australia, Europe, the USA, and South Africa [31,32,66]. The *Phytophthora* assemblage from watercourses was different from that detected from soil samples at the same sites, which was consistent with previous surveys [38,66,71,72]. Only three species were shared between terrestrial and aquatic environments, and these were *P. pseudocryptogea*, *P. gonapodyides*, and *P.* ×*cambivora*. Interestingly, the most common *Phytophthora* species isolated from rhizosphere soil, *P. cinnamomi*, was never detected from streams running through declining cork oak stands. As reported by previous similar studies, clade 6 species were the most common inhabitants of streams and rivers, highlighting their specific lifestyle to aquatic environments [73,74]. *Phytophthora gonapodyides* occurred in all investigated sites, followed by *P. bilorbang*, *P. lacustris*, and *P. chlamydospora* isolated from five, four, and two watercourses, respectively. Both *P. bilorbang* and *P. gonapodyides* have been previously reported in Sardinia on Caprera Island, detected from both rhizosphere soil beneath declining Mediterranean maquis vegetation and holm oak trees [28]. In Italy, *P. bilorbang* has also been reported from riparian ecosystems of *Alnus glutinosa*, in Sardinia, and on 15-year-old olive trees in Calabria [75,76], while *P. gonapodyides* is generally encountered in Mediterranean forest ecosystems, including holm oak [20,44,45]. In the pathogenicity tests, *P. gonapodyides* caused significant lesions in cork oak logs, confirming its ability to colonize bark and xylem tissues in both artificial and natural infections [77], behaving as a weak pathogen able to survive as a saprophyte on twigs and leaves playing a role in the breakdown of trees debris [78,79]. The other two clade 6 species, *P. lacustris* and *P. chlamydospora*, are common species previously reported in Italy from water bodies in forest ecosystems [36,66,80]. *Phytophthora plurivora* and *P. pseudocryptogea* are cosmopolitan pathogens with a broad host range [34,37,81]. The DNA of both species was recently detected from holm oak stands across different regions in Spain [23,24], and the pathogenicity tests showed they are amongst the most aggressive species on inoculated holm oak seedlings [82]. The detection of the clade 9 *P. hydropathica* is not surprising since it was recently detected from river water in Sicily and again using metabarcoding in Spain [24,66]. The origin of *P. hydropathica* is unknown but considering its low frequency (only one river) and occurrence from ornamental plants in a nursery in Italy [83], a recent introduction into wild environments is most likely.

Of note is the detection of *P.* ×*cambivora* and *P.* ×*multiformis*, two stable hybrid species from clade 7a that have evolved elsewhere. While *P.* ×*cambivora* has been previously reported from Mediterranean oaks in Italy [44], and *Q. suber* soil in the present study, the isolation of *P.* ×*multiformis* most likely occurred due to the presence of *Alnus glutinosa* trees along the river where the hybrid was isolated. The occurrence of multiple heterozygous sites in the ITS sequences of some isolates, including *P. chlamydospora*, *P. gonapodyides*, *P. hydropathica*, and *P. lacustris*, together with mixed unreadable ITS sequences generated for some of these isolates could indicate their possible hybrid nature. However, since the ITS region is not a particularly useful locus for studying interspecific hybrids due to the presence of the highly variable non-coding regions ITS1 and ITS2 [83], further molecular analyses, such as cloning, sequencing of other nuclear and mitochondrial genes, estimation of nuclear DNA content by flow cytometry, as well as morphological characterization of the isolates are required [38,84,85].

5. Conclusions

This unexpected high diversity of *Phytophthora* species in cork oak stands in such a small geographic area underlines how limited our current knowledge of oomycete diversity in Mediterranean oak forests is [86]. Recent molecular studies in Spanish oak forests have revealed the presence of several *Phytophthora* species besides *P. cinnamomi* [12,22,24,87], most of which were detected also in this survey. Although the high-throughput amplicon pyrosequencing of environmental DNA represents a very useful tool for assessing *Phytophthora* diversity in environmental samples, very little is known on the biological status of the detected microorganisms; therefore, specific baiting technique and metagenomic approaches should be carried out in parallel [72,88].

Pathogenicity tests and results obtained on the susceptibility of Mediterranean oaks to *Phytophthora* taxa from this and other prior studies [15,46,55,56,62] suggest these previously unrecorded species may play a relevant role in the aetiology of cork oak decline either acting as fine root "nibblers" [20,33,58] or shaping recruitment patterns due to their negative effects on seedling establishment [89,90]. Future studies will be based around understanding the ecological role of all *Phytophthora* species recovered in such ecosystems as well as their possible interactions with hosts and a changing environment that could promote the establishment of invasive *Phytophthora* in cork oak forests.

Author Contributions: Conceptualization: B.S., S.S.; data curation: B.S., S.S., A.B.; formal analysis: B.S., S.S.; investigation: B.S., S.S., A.B.; methodology: B.S., S.S., A.B.; project administration: B.S., S.S., C.S., P.A.R.; resources: B.S., C.S., P.A.R.; Supervision: B.S.; writing—original draft: S.S., B.S. writing—review and editing: B.S., S.S., A.B., C.S., P.A.R. All authors have read and agreed to the published version of the manuscript.

Acknowledgments: The Regional Agency Fo.Re.S.T.A.S. is gratefully acknowledged for providing cork oak seedlings used in the pathogenicity tests. The authors would like to thank Carolyn Riddel for the English language editing.

References

1. Aronson, J.; Pereira, J.S.; Pausas, J.G. *Cork Oak Woodlands on the Edge: Ecology, Adaptive Management, and Restoration*; Island Press: Washington, DC, USA, 2009.
2. Merlo, M.; Croitoru, L. *Valuing Mediterranean Forests: Towards Total Economic Value*; CABI Publishing: Cambridge, MA, USA, 2005; p. 406.
3. Bugalho, N.M.; Caldeira, C.M.; Pereira, J.S.; Aronson, J.; Pausas, J.G. Mediterranean cork oak savannas require human use to sustain biodiversity and ecosystem services. *Front. Ecol. Environ.* **2011**, *5*, 278–286. [CrossRef]
4. Sedda, L.; Delogu, G.; Dettori, S. Forty-four years of land use changes in a Sardinian cork oak agro-silvopastoral system: A qualitative analysis. *Open For. Sci. J.* **2011**, *4*, 57–66.
5. Myers, N.; Mittermeier, R.A.; Mittermeier, C.G.; da Fonseca, G.A.B.; Kent, J. Biodiversity hotspots for conservation priorities. *Nature* **2000**, *403*, 853–858. [CrossRef] [PubMed]
6. Gomez-Aparicio, L.; Ibanez, B.; Serrano, M.S.; De Vita, P.; Avila, J.M.; Perez-Ramos, I.M.; Garcia, L.V.; Sanchez, M.E.; Maranon, T. Spatial patterns of soil pathogens in declining Mediterranean forests: Implications for tree species regeneration. *New Phytol.* **2012**, *194*, 1014–1024. [CrossRef] [PubMed]
7. Tomaz, C.; Alegria, C.; Abrantes Massano Monteiro, J.A.; Canavarro Teixeira, M.C. Land cover change and afforestation of marginal and abandoned agricultural land: A 10-year analysis in a Mediterranean region. *For. Ecol. Manag.* **2013**, *308*, 40–49. [CrossRef]
8. Kim, M.; Lee, W.K.; Choi, G.M.; Song, C.; Lim, C.H.; Moon, J.; Piao, D.; Kraxner, F.; Shividenko, A.; Forsell, N. Modeling stand-level mortality based on maximum stem number and seasonal temperature. *For. Ecol. Manag.* **2017**, *386*, 37–50. [CrossRef]

9. Garbelotto, M.; Pautasso, M. Impacts of exotic forest pathogens on Mediterranean ecosystems: Four case studies. *Eur. J. Plant Pathol.* **2012**, *133*, 101–116. [CrossRef]
10. Cacciola, S.O.; Gullino, M.L. Emerging and re-emerging fungus and oomycete soil-borne plant diseases in Italy. *Phytopathol. Mediterr.* **2019**, *58*, 451–472.
11. Moricca, S.; Linaldeddu, B.T.; Ginetti, B.; Scanu, B.; Franceschini, A.; Ragazzi, A. Endemic and emerging pathogens threatening cork oak trees: Management options for conserving a unique forest ecosystem. *Plant Dis.* **2016**, *100*, 2184–2193. [CrossRef]
12. Jung, T.; Pérez-Sierra, A.; Durán, A.; Horta Jung, M.; Balci, Y.; Scanu, B. Canker and decline diseases caused by soil- and airborne *Phytophthora* species in forests and woodlands. *Persoonia* **2018**, *40*, 182–220. [CrossRef]
13. Sena, K.; Crocker, E.; Vincelli, P.; Barton, C. *Phytophthora cinnamomi* as a driver of forest change: Implications for conservation and management. *For. Ecol. Manag.* **2018**, *409*, 799–807. [CrossRef]
14. Brasier, C.M.; Robredo, F.; Ferraz, J.F.P. Evidence for *Phytophthora cinnamomi* involvement in Iberian oak decline. *Plant Pathol.* **1993**, *42*, 140–145. [CrossRef]
15. Robin, C.; Desprez-Loustau, M.L.; Capron, G.; Delatour, C. First record of *Phytophthora cinnamomi* on cork and holm oaks in France and evidence of pathogenicity. *Ann. Sci. For. INRA/EDP Sci.* **1998**, *55*, 869–883. [CrossRef]
16. Gallego, F.J.; de Algaba, A.P.; Fernandez-Escobar, R. Etiology of oak decline in Spain. *Eur. J. For. Pathol.* **1999**, *29*, 17–27. [CrossRef]
17. Sánchez, M.E.; Caetano, P.; Ferraz, J.; Trapero, A. Phytophtora disease of *Quercus ilex* in south-western Spain. *For. Pathol.* **2002**, *32*, 5–18. [CrossRef]
18. Camilo-Alves, C.S.P.; da Clara, M.I.E.; Ribeiro, N.A. Decline of Mediterranean oak trees and its association with *Phytophthora cinnamomi*: A review. *Eur. J. For. Res.* **2013**, *132*, 411–432. [CrossRef]
19. Scanu, B.; Linaldeddu, B.T.; Franceschini, A.; Anselmi, N.; Vannini, A.; Vettraino, A.M. Occurrence of *Phytophthora cinnamomi* in cork oak forests in Italy. *For. Pathol.* **2013**, *43*, 340–343. [CrossRef]
20. Pérez-Sierra, A.; López-García, C.; León, M.; García-Jiménez, J.; Abad-Campos, P.; Jung, T. Previously unrecorded low temperature *Phytophthora* species associated with *Quercus* decline in a Mediterranean forest in Eastern Spain. *For. Pathol.* **2013**, *43*, 331–339. [CrossRef]
21. Jung, T.; Horta Jung, M.; Cacciola, S.O.; Cech, T.; Bakonyi, J.; Seress, D.; Mosca, S.; Schena, L.; Seddaiu, S.; Pane, A.; et al. Multiple new cryptic pathogenic *Phytophthora* species from Fagaceae forests in Austria, Italy and Portugal. *IMA Fungus* **2017**, *8*, 219–244. [CrossRef]
22. Ruiz-Gómez, F.J.; Navarro-Cerrillo, R.M.; de Luque, A.P.; Owald, W.; Vannini, A.; Morales-Rodríguez, C. Assessment of functional and structural changes of soil fungal and oomycete communities in holm oak declined dehesas through metabarcoding analysis. *Sci. Rep.* **2019**, *9*, 5315. [CrossRef]
23. Català, S.; Pérez-Sierra, A.; Abad-Campos, P. The use of genus-specific amplicon pyrosequencing to assess *Phytophthora* species diversity using eDNA from soil and water in northern Spain. *PLoS ONE* **2015**, *10*, e0119311. [CrossRef] [PubMed]
24. Mora-Sala, B.; Gramaje, D.; Abad-Campos, P.; Berbegal, M. Diversity of *Phytophthora* species associated with *Quercus ilex* L. in Three Spanish Regions Evaluated by NGS. *Forests* **2019**, *10*, 979. [CrossRef]
25. Jung, T.; Orlikowski, L.; Henricot, B.; Abad-Campos, P.; Aday, A.G.; Aguín Casal, O.; Bakonyi, J.; Cacciola, S.O.; Cech, T.; Chavarriaga, D.; et al. Widespread Phytophthora infestations in European nurseries put forest, semi-natural and horticultural ecosystems at high risk of Phytophthora diseases. *For. Pathol.* **2016**, *46*, 134–163. [CrossRef]
26. Vannini, A.; Osimani, L.; Morales-Rodríguez, C. Decline and mortality of evergreen oaks in a protected area in Central Italy driven by the pathogenic activity of the invasive *Phytophthora cinnamomi* and Phytophthora multivora. In Proceedings of the Book of Abstracts Proceedings of the 9th Meeting of the International Union of Forest Research Organizations (IUFRO) Working Party S07.02.09, Phytophthora Diseases in Forests and Natural Ecosystems, La Maddalena, Italy, 17–25 October 2019; p. 88.
27. Jung, T.; Blaschke, H.; Neumann, P. Isolation, identification and pathogenicity of *Phytophthora* species from declining oak stands. *Eur. J. For. Pathol.* **1996**, *26*, 253–272. [CrossRef]
28. Scanu, B.; Hunter, G.C.; Linaldeddu, B.T.; Franceschini, A.; Maddau, L.; Jung, T.; Denman, S. A taxonomic re-evaluation reveals that *Phytophthora cinnamomi* and *P. cinnamomi* var. *parvispora* are separate species. *For. Pathol.* **2014**, *44*, 1–20. [CrossRef]

29. Brasier, C.M. Physiology of Reproduction in *Phytophthora*. Ph.D. Thesis, University of Hull, Kingston upon Hull, UK, 1967.
30. Reeser, P.W.; Sutton, W.; Hansen, E.M.; Remigi, P.; Adams, G.C. *Phytophthora* species in forest streams in Oregon and Alaska. *Mycologia* **2011**, *103*, 22–35. [CrossRef]
31. Hüberli, D.; Hardy, G.E.S.J.; White, D.; Williams, N.; Burgess, T.I. Fishing for *Phytophthora* from Western Australia's waterways: A distribution and diversity survey. *Australas. Plant Path.* **2013**, *42*, 251–260. [CrossRef]
32. Scanu, B.; Linaldeddu, B.T.; Deidda, A.; Jung, T. Diversity of *Phytophthora* species from declining Mediterranean maquis vegetation, including two new species, *Phytophthora crassamura* and *P. ornamentata* sp. nov. *PLoS ONE* **2015**, *10*, e0143234. [CrossRef]
33. Jung, T.; Cooke, D.E.L.; Blaschke, H.; Duncan, J.M.; Oßwald, W. *Phytophthora quercina* sp. nov., causing root rot of European oaks. *Mycol. Res.* **1999**, *103*, 785–798. [CrossRef]
34. Jung, T.; Burgess, T.I. Re-evaluation of *Phytophthora citricola* isolates from multiple woody hosts in Europe and North America reveals a new species, *Phytophthora plurivora* sp. nov. *Persoonia* **2009**, *22*, 95–110. [CrossRef]
35. Aghighi, S.; Hardy, G.E.S.J.; Scott, J.K.; Burgess, T.I. *Phytophthora bilorbang* sp. nov., a new species associated with the decline of Rubus anglocandicans (European blackberry) in Western Australia. *Eur. J. Plant Pathol.* **2012**, *133*, 841–855. [CrossRef]
36. Nechwatal, J.; Bakonyi, J.; Cacciola, S.O.; Cooke, D.E.L.; Jung, T.; Nagy, Z.A.; Vannini, A.; Vettraino, A.M.; Brasier, C.M. The morphology, behaviour and molecular phylogeny of *Phytophthora* taxon Salixsoil and its redesignation as *Phytophthora lacustris* sp. nov. *Plant Pathol.* **2013**, *62*, 355–369. [CrossRef]
37. Safaiefarahani, B.; Mostowfizadeh-Ghalamfarsa, R.; Hardy, G.E.S.J.; Burgess, T.I. Re-evaluation of the *Phytophthora cryptogea* species complex and the description of a new species, *Phytophthora pseudocryptogea* sp. nov. *Mycol. Prog.* **2015**, *14*, 108. [CrossRef]
38. Jung, T.; Horta Jung, M.; Scanu, B.; Seress, D.; Kovács, G.M.; Maia, C.; Pérez-Sierra, A.; Chang, T.T.; Chandelier, A.; Heungens, K.; et al. Six new *Phytophthora* species from ITS Clade 7a including two sexually functional heterothallic hybrid species detected in natural ecosystems in Taiwan. *Persoonia* **2017**, *38*, 100–135. [CrossRef] [PubMed]
39. White, T.J.; Bruns, T.; Lee, S.; Taylor, J. Amplification and direct sequencing of fungal ribosomal DNA for phylogenetics. In *PCR Protocols: A Guide to Methods and Applications*; Innis, M.A., Gelfand, D.H., Sninsky, J.J., White, T.J., Eds.; Academic Press: San Diego, CA, USA, 1990; pp. 315–322.
40. Cooke, D.E.L.; Drenth, A.; Duncan, J.M.; Wagels, G.; Brasier, C.M. A molecular phylogeny of *Phytophthora* and related oomycetes. *Fungal Genet. Biol.* **2000**, *30*, 17–32. [CrossRef]
41. Hall, T. *Bio Edit Version 5.0.6*; Department of Microbiology, North Carolina State University: Raleigh, NC, USA, 2001.
42. Brasier, C.M.; Kirk, S.A. Designation of the EAN and NAN races of *Ophiostoma novo-ulmi* as subspecies: Their perithecial size differences and geographical distributions. *Mycol. Res.* **2001**, *105*, 547–554. [CrossRef]
43. Scanu, B.; Webber, J.F. Dieback and mortality of *Nothofagus* in Britain: Ecology, pathogenicity and sporulation potential of the causal agent *Phytophthora pseudosyringae*. *Plant Pathol.* **2016**, *65*, 26–36. [CrossRef]
44. Vettraino, A.M.; Barzanti, G.P.; Bianco, M.C.; Ragazzi, A.; Capretti, P.; Paoletti, E.; Luisi, N.; Anselmi, N.; Vannini, A. Occurrence of *Phytophthora* species in oak stands in Italy and their association with declining oak trees. *For. Pathol.* **2002**, *32*, 19–28. [CrossRef]
45. Corcobado, T.; Cubera, E.; Pérez-Sierra, A.; Jung, T.; Solla, A. First report of *Phytophthora gonapodyides* involved in the decline of *Quercus ilex* in xeric conditions in Spain. *New Dis. Rep.* **2010**, *22*, 33. [CrossRef]
46. Sánchez, M.E.; Andicoberry, S.; Trapero, A. Pathogenicity of three *Phytophthora* spp. causing late seedling rot of *Quercus ilex* ssp. *ballota*. *For. Pathol.* **2005**, *35*, 115–125.
47. Corcobado, T.; Cubera, E.; Juárez, E.; Moreno, G.; Solla, A. Drought events determine performance of *Quercus ilex* seedlings and increase their susceptibility to *Phytophthora cinnamomi*. *Agric. For. Meteorol.* **2014**, *192–193*, 1–8. [CrossRef]
48. Linaldeddu, B.T.; Scanu, B.; Maddau, L.; Franceschini, A. *Diplodia corticola* and *Phytophthora cinnamomi*: The main pathogens involved in holm oak decline on Caprera Island (Italy). *For. Pathol.* **2013**, *44*, 191–200. [CrossRef]
49. Frisullo, S.; Lima, G.; Magnano di San Lio, G.; Camele, I.; Melissano, L.; Puglisi, I.; Pane, A.; Agosteo, G.E.; Prudente, L.; Cacciola, S.O. *Phytophthora cinnamomi* involved in the decline of holm oak (*Quercus ilex*) stands in southern Italy. *For. Sci.* **2018**, *64*, 290–298. [CrossRef]

50. González, M.; Romero, M.; García, L.; Gomez-Aparicio, L.; Serrano, M.S. Unravelling the role of drought as predisposing factor for *Quercus suber* decline caused by *Phytophthora cinnamomi*. *Eur. J. Plant. Pathol.* **2020**, *156*, 1015–1021. [CrossRef]
51. Lowe, S.; Browne, M.; Boudjelas, S.; De Poortner, M. *One-Hundred of the World's Worst Invasive Alien Species. A Selection from the Global Invasive Species Database*; The Invasive Species Specialist Group, International Union for Conservation of Nature (IUCN): Gland, Switzerland, 2000; Available online: http://www.issg.org (accessed on 28 July 2020).
52. Burgess, T.I.; Scott, J.K.; Mcdougall, K.L.; Stukely, M.J.; Crane, C.; Dunstan, W.A.; Brigg, F.; Andjic, V.; White, D.; Rudman, T.; et al. Current and projected global distribution of *Phytophthora cinnamomi*, one of the world's worst plant pathogens. *Glob. Chang. Biol.* **2017**, *23*, 1661–1674. [CrossRef]
53. Brasier, C.M. Oak tree mortality in Iberia. *Nature* **1992**, *360*, 539. [CrossRef]
54. Moreira, A.C.; Martins, J.M.S. Influence of site factor on the impact of *Phytophthora cinnamomi* in cork oak stands in Portugal. *For. Pathol.* **2005**, *35*, 145–162. [CrossRef]
55. Robin, C.; Capron, G.; Desprez-Loustau, M.L. Root infection by *Phytophthora cinnamomi* in seedlings of three oak species. *Plant Pathol.* **2001**, *50*, 708–716. [CrossRef]
56. Moralejo, E.; García-Muñoz, J.A.; Descals, E. Susceptibility of Iberian trees to *Phytophthora ramorum* and P. cinnamomi. *Plant Pathol.* **2009**, *58*, 271–283. [CrossRef]
57. Magri, D.; Fineschi, S.; Bellarosa, R.; Buonamici, A.; Sebastiani, F.; Schirone, B.; Simeone, M.C.; Vendramin, G.G. The distribution of *Quercus suber* chloroplast haplotypes matches the palaeogeographical history of the western Mediterranean. *Mol. Ecol.* **2007**, *16*, 5259–5266. [CrossRef]
58. Jung, T.; Blaschke, H.; Oßwald, W. Involvement of soilborne *Phytophthora* species in Central European oak decline and the effect of site factors on the disease. *Plant Pathol.* **2000**, *49*, 706–718. [CrossRef]
59. Balci, Y.; Halmschlager, E. First report of *Phytophthora quercina* from oak forests in Austria. *Plant Pathol.* **2003**, *52*, 403. [CrossRef]
60. Ruiz Gómez, F.J.; Pérez-de-Luque, A.; Sánchez-Cuesta, R.; Quero, J.L.; Navarro Cerrillo, R.M. Differences in the response to acute drought and *Phytophthora cinnamomi* Rands Infection in *Quercus ilex* L. Seedlings. *Forests* **2018**, *9*, 634. [CrossRef]
61. Robin, C.; Dupuis, F.; Desprez-Loustau, M.L. Seasonal changes in Northern red oak susceptibility to *Phytophthora cinnamomi*. *Plant Dis.* **1994**, *78*, 369–374. [CrossRef]
62. Luque, J.; Parladé, J.; Pera, J. Seasonal changes in susceptibility of *Quercus suber* to *Botryosphaeria stevensii* and *Phytophthora cinnamomi*. *Plant Pathol.* **2002**, *51*, 338–345. [CrossRef]
63. Balci, Y.; Halmschlager, E. *Phytophthora* species in oak ecosystems in Turkey and their association with declining oak trees. *Plant Pathol.* **2003**, *52*, 694–702. [CrossRef]
64. Riddell, C.E.; Frederickson-Matika, D.; Armstrong, A.C.; Elliot, M.; Forster, J.; Hedley, P.E.; Morris, J.; Thorpe, P.; Cooke, D.E.L.; Pritchard, L.; et al. Metabarcoding reveals a high diversity of woody host-associated *Phytophthora* spp. in soils at public gardens and amenity woodlands in Britain. *PeerJ* **2019**, *7*, e6931. [CrossRef]
65. Garbelotto, M.; Frankel, S.; Scanu, B. Soil-and waterborne *Phytophthora* species linked to recent outbreaks in Northern California restoration sites. *Calif. Agric.* **2018**, *72*, 208–216. [CrossRef]
66. Jung, T.; La Spada, F.; Pane, A.; Aloi, F.; Evoli, M.; Horta Jung, M.; Scanu, B.; Roberto Faedda, R.; Rizza, C.; Puglisi, I.; et al. Diversity and distribution of *Phytophthora* species in protected natural plots in Sicily. *Forests* **2019**, *10*, 259. [CrossRef]
67. Sechi, C.; Seddaiu, S.; Linaldeddu, B.T.; Franceschini, A.; Scanu, B. Dieback and mortality of *Pinus radiata* trees in Italy associated with *Phytophthora cryptogea*. *Plant Dis.* **2014**, *98*, 159. [CrossRef]
68. Vettraino, A.M.; Natili, G.; Anselmi, N.; Vannini, A. Recovery and pathogenicity of *Phytophthora* species associated with resurgence of ink disease on *Castanea sativa* in Italy. *Plant Pathol.* **2001**, *50*, 90–96. [CrossRef]
69. Belisario, A.; Maccaroni, M.; Vettorazzo, M. First report of *Phytophthora cambivora* causing bleeding cankers and dieback on beech (Fagus sylvatica) in Italy. *Plant Dis.* **2006**, *90*, 1362. [CrossRef] [PubMed]
70. Talgø, V.; Herrero, M.L.; Toppe, B.; Klemsdal, S.S.; Stensvand, A. First report of root rot and stem canker caused by *Phytophthora cambivora* on noble fir (*Abies procera*) for bough production in Norway. *Plant Dis.* **2006**, *90*, 682. [CrossRef] [PubMed]
71. Oh, E.; Gryzenhout, M.; Wingfield, B.D.; Wingfield, M.J.; Burgess, T.I. Surveys of soil and water reveal a goldmine of *Phytophthora* diversity in South African natural ecosystems. *IMA Fungus* **2013**, *4*, 123–131. [CrossRef] [PubMed]

72. Jung, T.; Scanu, B.; Brasier, C.M.; Webber, J.; Milenkovic, I.; Corcobado, T.; Tomsovsky, M.; Panek, M.; Bakonyi, J.; Maia, C.; et al. A survey in natural forest ecosystems of Vietnam reveals high diversity of both new and described *Phytophthora* taxa including *P. ramorum*. *Forests* **2020**, *11*, 93. [CrossRef]
73. Brasier, C.M.; Cooke, D.E.L.; Duncan, J.M.; Hansen, E.M. Multiple new phenotypic taxa from trees and riparian ecosystems in *Phytophthora gonapodyides–P. megasperma* ITS Clade 6, which tend to be high-temperature tolerant and either inbreeding or sterile. *Mycol. Res.* **2003**, *107*, 277–290. [CrossRef]
74. Jung, T.; Stukely, M.J.C.; Hardy, G.E.S.J.; White, D.; Paap, T.; Dunstan, W.A.; Burgess, T.I. Multiple new *Phytophthora* species from ITS Clade 6 associated with natural ecosystems in Australia: Evolutionary and ecological implications. *Persoonia* **2011**, *26*, 13–39. [CrossRef]
75. Scanu, B.; Linaldeddu, B.T.; Peréz-Sierra, A.; Deidda, A.; Franceschini, A. *Phytophthora ilicis* as a leaf and stem pathogen of *Ilex aquifolium* in Mediterranean islands. *Phytopathol. Mediterr.* **2014**, *53*, 480–490.
76. Santilli, E.; Riolo, M.; La Spada, F.; Pane, A.; Cacciola, S.O. First Report of Root Rot Caused by *Phytophthora bilorbang* on *Olea europaea* in Italy. *Plants* **2020**, *9*, 826. [CrossRef]
77. Brown, A.V.; Brasier, C.M. Colonization of tree xylem by *Phytophthora ramorum*, *P. kernoviae* and other *Phytophthora* species. *Plant Pathol.* **2007**, *56*, 227–241. [CrossRef]
78. Brasier, C.M.; Sanchez-Hernandez, E.; Kirk, S.A. *Phytophthora inundata* sp. *nov.*, a part heterothallic pathogen of trees and shrubs in wet or flooded soils. *Mycol. Res.* **2003**, *107*, 477–484. [CrossRef] [PubMed]
79. Cline, E.T.; Farr, D.F.; Rossman, A.Y. Synopsis of *Phytophthora* with accurate scientific names, host range, and geographic distribution. *Plant Health Prog.* **2008**, *9*. [CrossRef]
80. Ginetti, B.; Moricca, S.; Squires, J.N.; Cooke, D.E.L.; Ragazzi, A.; Jung, T. Phytophthora acerina sp. nov., a new species causing bleeding cankers and dieback of Acer pseudoplatanus trees in planted forests in northern Italy. *Plant Pathol.* **2014**, *63*, 858–876. [CrossRef]
81. Seddaiu, S.; Linaldeddu, B.T. First Report of *Phytophthora acerina*, *P. plurivora*, and *P. pseudocryptogea* associated with declining common alder trees in Italy. *Plant Dis.* **2020**, *104*, 1874. [CrossRef]
82. Mora-Sala, B.; Abad-Campos, P.; Berbegal, M. Response of *Quercus ilex* seedlings to *Phytophthora* spp. root infection in a soil infestation test. *Eur. J. Plant Pathol.* **2018**, *154*, 2115–2225. [CrossRef]
83. Vitale, S.; Luongo, L.; Galli, M.; Belisario, A. First report of *Phytophthora hydropathica* causing wilting and shoot dieback on viburnum in Italy. *Plant Dis.* **2014**, *98*, 1582. [CrossRef]
84. Burgess, T.I. Molecular characterization of natural hybrids formed between five related indigenous clade 6 *Phytophthora* species. *PLoS ONE* **2015**, *10*, e0134225. [CrossRef]
85. Nagel, J.H.; Gryzenhout, M.; Slippers, B.; Wingfield, M.J.; Hardy, G.E.S.J.; Stukely, M.J.C.; Burgess, T.I. Characterization of Phytophthora hybrids from ITS clade 6 associated with riparian ecosystems in South Africa and Australia. *Fungal Biol.* **2013**, *117*, 329–347. [CrossRef]
86. Dickie, I.A.; Wakelin, A.M.; Martínez-García, L.B.; Richardson, S.J.; Makiola, A.; Tylianakis, J.M. Oomycetes along a 120,000-year temperate rainforest ecosystem development chrono sequence. *Fungal Ecol.* **2019**, *39*, 192–200. [CrossRef]
87. Català, S.; Berbegal, M.; Pérez-Sierra, A.; Abad-Campos, P. Metabarcoding and development of new real-time specific assays reveal *Phytophthora* species diversity in holm oak forests in eastern Spain. *Plant Pathol.* **2017**, *66*, 115–123. [CrossRef]
88. Bose, T.; Wingfield, M.J.; Roux, J.; Vivas, M.; Burgess, T.I. *Phytophthora* species associated with roots of native and non-native trees in natural and managed forests. *Microb. Ecol.* **2020**. [CrossRef] [PubMed]
89. Corcobado, T.; Miranda-Torres, J.J.; Martín-García, J.; Jung, T.; Solla, A. Early survival of *Quercus ilex* subspecies from different populations after infections and coinfections by multiple *Phytophthora* species. *Plant Pathol.* **2017**, *66*, 792–804. [CrossRef]
90. Domìnguez-Begines, J.; Avila, J.M.; Garcìa, L.V.; Gòmez-Aparicio, L. Soil-borne pathogens as determinants of regeneration patterns at community level in Mediterranean forests. *N. Phytol.* **2020**, *227*, 588–600. [CrossRef] [PubMed]

10

Small-Scale Abiotic Factors Influencing the Spatial Distribution of *Phytophthora cinnamomi* under Declining *Quercus ilex* Trees

Rafael Sánchez-Cuesta *,†, Rafael M. Navarro-Cerrillo †, José L. Quero and Francisco J. Ruiz-Gómez

Department of Forest Engineering, Laboratory of Dendrochronology, Silviculture and Global Change-DendrodatLab-ERSAF, University of Córdoba, Campus de Rabanales, Ctra. N-IV, km. 396, E-14071 Córdoba, Spain; rmnavarro@uco.es (R.M.N.-C.); jose.quero@uco.es (J.L.Q.); g72rugof@uco.es (F.J.R.-G.)

* Correspondence: rscuesta@uco.es

† Both authors contributed equally.

Abstract: *Phytophthora* root rot is considered one of the main factors associated with holm oak (*Quercus ilex* L.) mortality. The effectiveness and accuracy of soilborne pathogen and management could be influenced by soil spatial heterogeneity. This factor is of special relevance in many afforestation of southwestern Spain, which were carried out without phytosanitary control of the nursery seedlings. We selected a study area located in a 15 year-old afforestation of *Q. ilex*, known to be infested by *Phytophthora cinnamomi* Rands. Soil samples ($n_{\text{total}} = 132$) were taken systematically from a grid under 4 trees, and analysed to quantify 12 variables, the colony forming units (cfu) of *P. cinnamomi* plus 11 physical and chemical soil properties. The combined analysis of all variables was performed with linear mixed models (GLMM), and the spatial patterns of cfu were characterised using an aggregation index (I_a) and a clustering index (ν) by SADIE. Cfu values ranged from 0 to 211 cfu g^{-1}, and the GLMM built with the variables silt, P, K and soil moisture explained the cfu distribution to the greatest extent. The spatial analysis showed that 9 of the 12 variables presented spatial aggregation ($I_a > 1$), and the clustering of local patches ($\nu_i \geq 1.5$) for organic matter, silt, and Ca. The spatial patterns of the *P. cinnamomi* cfu under planted holm oak trees are related to edaphic variables and canopy cover. Small-scale spatial analysis of microsite variability can predict which areas surrounding trees can influence lower oomycetes cfu availability.

Keywords: GLMM; holm oak decline; tree mortality; root rot.; plantation; dehesas; montados; open forests

---

1. Introduction

Mediterranean-like savannas of the southern Iberian Peninsula (hereafter dehesas) are important ecosystems threatened by socioeconomic, climatic and phytosanitary factors. Currently, dehesas are affected by tree mortality caused by root rot [1,2]. After three decades of research and development, holm oak (*Quercus ilex* subsp. *ballota* (Desf.) Samp.) decline remains the most-important cause of tree loss in southern Spain and Portugal [3]. Oak decline has been related to management factors, as well as to the influence of climatic factors, such as severe droughts [4]. However, there is a broad consensus that biotic agents (pests and diseases) act as triggers for mortality episodes, in the context of a continuously-deteriorating ecosystem with limited regeneration capacity [5,6].

The root rot caused by soilborne pathogens from the genera *Phytophthora* spp. and *Pythium* spp. is considered one of the most-important causes triggering mortality of holm oak [2,7,8], causing rot of fine-roots, leading to water and nutritive stress [9], and therefore changes in tree physiology [10] which are visible in terms of defoliation and chlorosis, and in many cases tree death. These oomycetes remain as resistance structures either in soil, infected roots or debris under unfavourable conditions [11], waiting for suitable biotic and abiotic conditions for germination, leading to sporangia production and the subsequent release of zoospores which then infect new host roots [12].

Due to the heterogeneity of the holm oak fine root distribution, and thus the heterogeneity of the soil rhizosphere in dehesas, it is often necessary to carry out large sampling efforts to avoid false negative outcomes in diagnosis of root rot caused by *Phytophthora* spp. [13,14]. This soil heterogeneity has been proved to be related with the horizontal canopy distribution in several *Quercus* spp. [15,16]. Biotic and abiotic factors such as soil microbiota community [17–19], and soil moisture and mineral nutrients [20–23] are also related to the spatial distribution of the tree canopy, mainly due to the differences in the soil exposure to solar radiation and its contribution to processes involved in organic matter formation and mineral deposition [24].

The spatial distribution of the pathogen inoculum in the soil appears to be influenced by rhizosphere heterogeneity, but there is a lack of studies of soil microbiota spatial distribution in forest soils at a small-scale [25,26]. Colony forming units (hereafter, cfu) is the number of propagules which produce countable colonies after sowing in selective medium plates [27]. It is often used as an indicator of the abundance of the inoculum of *Phytophthora* spp. and other oomycetes in the soil. The density of cfu of *Phytophthora cinnamomi* Rands has been related to calcium content or plant diversity in Mediterranean oak forests [25,28]. Thus, it has been hypothesized that the spatial distribution of soil physicochemical properties associated with the tree crown may influence the density of the oomycete cfu concentration, and therefore the small-scale spatial distribution of the soil pathogen [20–23].

Furthermore, the afforestation programmes implemented in Spain at the end of the 20th century and the beginning of the 21st, in the context of the European Economic Community's aid scheme for forestry measures in agriculture (directives EEC-2080/92, and 1698/2005) have led to 232 000 ha of tree plantations in Andalusia, of which 82 775 ha are *Q. ilex* [29]. The afforestation programme funded plantation costs, conservation and maintenance, without paying attention to the phytosanitary control of the nursery seedlings at the time of planting. This lack of phytosanitary control is considered a threat to seedling survival and a reason for the spread of potential soil pathogens [11,30], especially when dealing with invasive alien pathogens such as *Phytophthora* spp., since biological invasion is one of the main drivers of global change in Mediterranean climates [31].

The main objective of this study was to evaluate the effect of several soil and plant parameters on the spatial distribution and aggregation of *P. cinnamomi* cfu in holm oak. For this purpose, we selected as a case-study an afforestation on former agricultural land, to take advantage of the homogeneity of plant genotype and age, climatic variables and soil conditions among the selected individuals. The specific objectives were: (i) to evaluate the effect of the tree crown cover and the soil physicochemical characteristics on the spatial distribution of the cfu; and (ii) to model the relationships between spatial patterns of cfu, tree crown cover and soil variables to evaluate the predictive capacity of these variables. Once these objectives were met, we intended to ascertain which microsite variables are potentially more important regarding the presence and spatial distribution of *P. cinnamomi*. This could help to guide the sampling effort for the diagnosis of root rot, and to determine the predisposition of soils to host the pathogen according to their physical and chemical characteristics.

2. Material and Methods

2.1. Study Zone

The study was carried out in a *Q. ilex* afforestation site located in Puebla de Guzmán (Andalusia - southern Spain, coordinates ETRS89, UTM30N: 120 500 mW, 4 167 500 mN, 185–188 masl). The area is characterised by a mean annual temperature of 16.8 °C and a mean annual rainfall of 570 mm (Meteorological Station of IFAPA, Puebla de Guzman, coordinates ETRS89, UTM30N: 124 659 mW, 4 164 620 mN; 195 masl. Institute of Agronomic and Fishery Research and Training-IFAPA, Junta de Andalusia), with a dry thermo-Mediterranean climate (120–150 biologically-dry days) with hot and dry summers and mild winters. The area has an undulating relief (slope between 5 and 10%, Supplementary Material, Table S1). Soils are shallow and acidic in nature, with rocky outcrops of slate and schists, an almost total absence of free calcium carbonate and sometimes a slight surface layer of organic matter. The site was cropped periodically before 1990 and has been left fallow since this date, supporting a mixture of native herbaceous species associated with former agricultural land and shrubs (*Cistus ladanifer* L.). A holm oak plantation was established in 1995. The planting area was sub-soiled, using a ripper with a single tine, to a depth of 60 cm and soil clods were broken up using a spring harrow and culta-mulcher, to provide a more-level surface for the plantation. The planting was done by hand following a systematic spatial pattern of distribution, with a density of 312 plants ha^{-1} (4×8 m spacing), using tree shelter. No additional soil treatments were carried out after the afforestation.

2.2. Soil Sampling Design

In spring (April) 2010, the plot (100 m^2) was established to include four *Q. ilex* trees with symptoms of oak decline (mean defoliation 30%) and previously shown to be infested by *P. cinnamomi* (Newbiotechnic S.A., NBT-No. 41/04/PR/PSX). The effects of practices carried out during cultivation and plantation management were homogeneous for the selected trees. The four trees were regular in size (DBH = 9.5 ± 0.5 cm; H = 2.5 ± 0.1 m). Visual phytosanitary assessment was carried out following the European Network of Damage in Forest Masses manual [32]. The trees ranged in crown damage (defoliation) from Class 1 ($10 <$ slight defoliation $\leq 25\%$) to Class 3 ($60 <$ severe defoliation $\leq 95\%$) (Supplementary Material, Table S1). Apart from *P. cinnamomi*, no other biotic agents contributing to the aboveground symptoms were identified during the visual inspection. The presence of root rot was confirmed in all cases through roots diagnosis (loss of secondary fine roots, discoloration and softness).

Under each tree crown, a sampling grid was established north facing, following the methodology proposed by Gallardo [20], and included two grid sizes (Figure 1): a general grid (G) 1×1 m ($n = 16$) in a 4×4 m quadrat centred in the trunk, and 0.5×0.33 m ($n = 24$) grid, which were concentrated within the general sampling quadrat according to the crown cover (crown position: O = Outside, T = Transition and I = Inside). Some of the points of the general and concentrated grids were the same, with a total of 33 points sampled per tree (n) (Figure 1). Finally, two categories were established: outside of the crown cover (OC, $n = 22$) and inside of the crown cover (IC, $n = 11$). The tree crown projection was obtained with the help of a plumb line, marking the vertical projection of the crown margins over the grid. Regarding orientation, all the points belonging to the A and B quadrats were considered as North points (N) and the ones located in the C and D quadrats, as South points (S).

2.3. Processing and Analysis of Soil Samples

In spring (April) 2010, soil samples were taken at each point of the grid ($n_{total} = 132$). The soil surface layer and the decomposing organic matter layer were removed from a square of 30×30 cm centred on the grid point, and a homogeneous soil sample of approximately 1 kg was extracted from the mineral horizon, over the point, to a depth of 20 cm, with the help of a spade [13]. Each sample was placed in a sealed plastic bag to prevent loss of moisture, properly labelled and preserved in an ice-cooler (in the absence of light, at approximately 4 °C) during its transfer to the laboratory. The soil samples were air-dried at room temperature for 48 h and then processed by hand, eliminating the rough fraction after mechanical milling and crushing. Prior to aliquot separation, the fine fraction was homogenised and passed through a 2-mm-Ø sieve. In total, 12 variables (11 edaphic variables plus cfu abundance) were evaluated for each soil sample. First, one aliquot of approximately 100 g was separated for quantification of *P. cinnamomi* cfu, and other aliquot of approximately 500 g, was sent to Innoagral laboratories, Grupo Hespérides Biotech S.L., (Sevilla, Spain) for the analysis of 10 soil physicochemical variables (% of clay, silt, and sand, pH, organic matter in soil, total nitrogen, carbon-nitrogen ratio and amount of phosphorus, calcium and potassium) (Supplementary Material, Table S2).

Finally, soil moisture (% volumetric content of water) was measured at a depth of 12 cm at all of the sampling points, with a Time Domain Reflectometry (TDR) sensor (Field Scout TDR 100, Spectrum Technologies, Inc. Aurora, IL, USA), at the same time as the collection of soil samples.

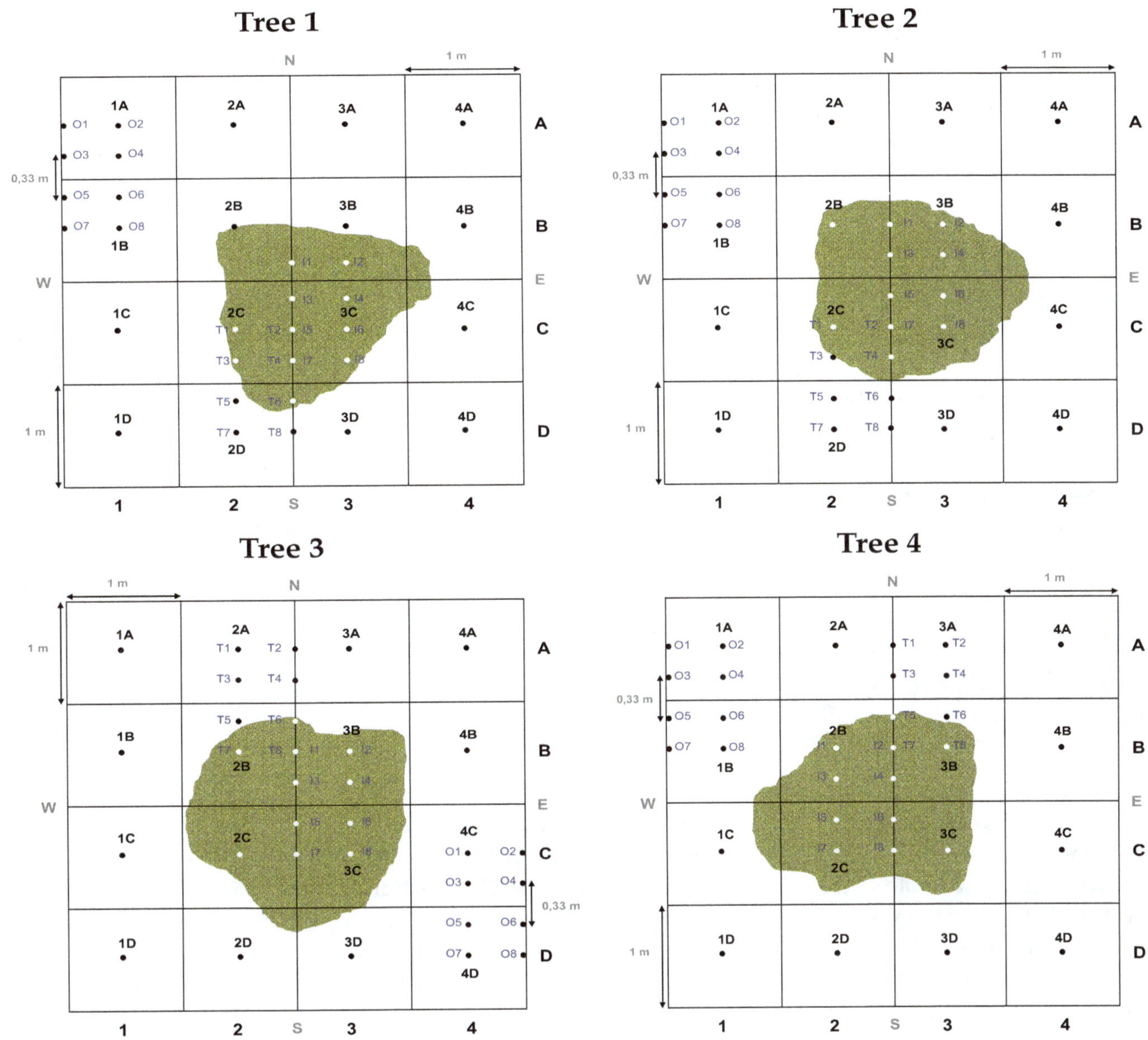

Figure 1. Soil sample design (according to Gallardo et al. [20] at two different scales: general grid (1 × 1 m) (G, $n = 16$) and specific position grid (0.5 × 0.33 m) corresponding to inside of the crown cover (I, $n = 8$), transition (T, $n = 8$), and outside of the crown cover (O, $n = 8$) for each of the four sampled trees. The black points belong to the group outside of the crown cover (OC, $n = 22$) and the white points belong to the group inside of the crown cover (IC, $n = 11$). The letters N, E, S and W, indicate the cardinal points.

2.4. Quantification of Cfu

Ten grams of each homogenised aliquot of soil for cfu analysis, previously dried at room temperature, was suspended in 100 mL 0.2% agar solution, as described in Romero et al. [27]. The solution was gently shaken and 1 mL of the soil suspension was plated with a 1000 µl pipette on 10 Petri plates for each soil sample, containing NARPH selective medium [14]. The soil suspension was carefully distributed over the surface of the selective medium with the help of an inoculation loop.

The plates were incubated at room temperature in darkness for 24 h, after which the soil suspension was removed with sterile water. Colonies growing on each plate were counted after an additional 72 h of incubation after washing.

The hyphal bodies that had grown in the culture medium were quantified visually by light contrast with a 10 × 1 magnifying glass (Nikon SMZ800, Nikon Corp., Tokyo, Japan) and a millimetre mesh. Some growing structures were identified through the observation of aleatory chosen NARPH dishes under microscope (Motic BA310E, Motic Instruments Co., Ltd., Chengdu, China) to ensure that no other organisms different from *P. cinnamomi* were counted. This species was easy identified through the characteristic aseptate coralloid hyphae with clustered hyphal swellings. The sum of the cfu for the 10 replicates per sample was expressed as cfu/g of dry soil. Due to the positive identification of *P. cinnamomi* on the soil, through molecular methods, and the results of the checking of aleatorily chosen NARPH plates, all the cfu counted were considered directly related with *P. cinnamomi* cfu abundance.

2.5. Statistical Analysis

All the variables were examined for normality (Shapiro-Wilk test, $P < 0.05$) and homoscedasticity (Levene test, $P < 0.05$). When the data did not fit to a normal distribution, the variable was subjected to a square root or logarithmic transformation. Once the normality and homoscedasticity requirements were met, the variables N, OM, C/N, percentage of silt and sand, pH and Ca were analysed using one-way analysis of variance (ANOVA), considering crown position and orientation as independent factors. In those cases where the variables were significant, Tukey's test for multiple comparisons of means was used to check for differences [33]. In the case of the non-normal variables (cfu, clay percentage, P, K and moisture), a Kruskal-Wallis (H) mean comparison test and the Mann-Whitney U test were applied for pairs of independent groups ($P < 0.05$). When Mann-Whitney U test was used for mean comparisons on the I-O-T grids, the α threshold to reject the null hypothesis was corrected according to Bonferroni ($P < 0.0167$) [34]. Specific correlations between the soil variables and cfu were determined using a non-parametric Spearman rho coefficient (ρ) at a significance level of 5% ($P < 0.05$), including soil variables that did not follow a normal distribution. Statistical analysis was performed using "R" version 3.3.1. [35].

The spatial patterns of cfu were characterised using Spatial Analysis by Distance Indices (SADIE), implemented in the program "SADIEShell v2.0" and the aggregation index (I_a) and clustering index (ν) were calculated per plot [36,37]. The I_a provides information on the overall spatial pattern of each environmental variable. According to Quero et al., [38] the spatial pattern is aggregated if $I_a > 1$, random if I_a is close to one and regular if $I_a < 1$. The index ν measures the degree of clustering of the data into patches (mean ν_i: areas of high values of the target variable) and gaps (mean ν_j: areas of low values). Then, ν was contoured by kriging in a two-dimensional map showing their spatial distribution of patches ($\nu \geq 1.5$) and gaps ($\nu \leq -1.5$) using Surfer 10.1 (Golden Software, CO, USA). An independent SADIE analysis was performed for each variable and tree. Afterwards the mean values of I_a for all the variables in the four trees were calculated and were also represented using the bilinear interpolation method in Surfer 10.1.

Finally, the relationship of the soil variables with the inoculum concentration and crown cover was assessed through a generalized linear mixed model (GLMM). This methodology allows the analysis of non-normal data that involve fixed or random effects [39]. The GLMM was implemented through the "lme4" package. The cfu variable was modelled through Poisson distribution with log transformation. The independent variables were previously filtered using a Variance Inflation Factor (VIF) threshold of 10, and the tree and crown position were selected as random effects [40]. Despite the relevance of crown position as the main factor of this study explaining spatial distribution of several variables, this factor did not present significant influence over the cfu variable when used in the model as fixed effect. Different model configurations were tested, providing the use of crown position and tree as random effects the best result. Autocorrelation of the output model was evaluated through the analysis of model residuals and correlation matrix of fixed effects. The model was selected based on the lowest value of Akaike's Information Criterion (AIC), which indicates the optimal fit, and the influence of effects was tested through a likelihood ratio test [41]. Comparison between Random Effects influence was performed through ANOVA linear model deleting each effect with glmerControl optimization type "bobyqa" (package "lmer4") and the influence of single fixed factors through automatic model reduction and Chi-squared test.

3. Results

3.1. Spatial Distribution of the Cfu

The cfu values of soil samples ranged from 0 to 211 cfu g^{-1}, for all four trees, showing significant correlation with the tree defoliation level ($\rho = 0.986$, $P < 0.05$). Moreover, cfu showed significant differences according to the crown cover factor (I, T and O; $H_{I\text{-}O} = 20.886$, $P_{I\text{-}O} < 0.001$; $H_{O\text{-}T} = 20.491$, $P_{O\text{-}T} < 0.001$; $H_{I\text{-}T} = 7.549$, $P_{I\text{-}T} < 0.01$). The I-grid showed a significant greater concentration of cfu than O grid in all cases ($U_{I\text{-}O} = 174.5$, $P_{I\text{-}O} < 0.001$), and the T-grid presented more variability in its results depending on the tree (Table 1). Moreover, a significantly higher cfu value in the IC grid was observed with respect to OC ($H_{IC\text{-}OC} = 27.4$; $U_{IC\text{-}OC} = 882.5$, $P_{IC\text{-}OC} < 0.001$).

Table 1. Colony forming units of each tree (cfu $sample^{-1}$, mean ± standard error) according to sample grid and sample position with respect to tree crown cover. Mean cfu: mean value of cfu considering all the samples; *n*: number of soil samples per grid under each tree; I: inside crown intensive grid; T: transition intensive grid; O: outside crown intensive grid; IC: all samples inside crown cover; OC: all samples outside crown cover. Different lowercase letters in superscript indicate significant differences between position with respect to crown cover (Mann-Whitney U Test, $P < 0.05$ for IC-OC comparisons and $P < 0.0167$ for I-O-T comparisons). No comparisons were made between means corresponding to different factors (I, O and T with IC and OC).

		Position Respect to Crown Cover				
Tree	**Mean cfu** $n = 33$	**I** $n = 8$	**T** $n = 8$	**O** $n = 8$	**IC** $n = 11$	**OC** $n = 22$
1	12 ± 3	22 ± 5 ^{a}	6 ± 4 ^{b}	5 ± 2 ^{b}	23 ± 7 ^{a}	6 ± 2 ^{b}
2	13 ± 5	34 ± 13 ^{a}	4 ± 3 ^{b}	3 ± 2 ^{b}	26 ± 10 ^{a}	7 ± 6 ^{b}
3	19 ± 8	38 ± 20 ^{a}	39 ± 26 ^{a}	0 ± 0 ^{b}	44 ± 21 ^{a}	5 ± 4 ^{b}
4	44 ± 10	119 ± 26 ^{a}	49 ± 11 ^{b}	16 ± 6 ^{c}	100 ± 22 ^{a}	16 ± 4 ^{b}
Mean	22 ± 4	53 ± 11 ^{a}	24 ± 8 ^{b}	6 ± 2 ^{b}	48 ± 9 ^{a}	9 ± 2 ^{b}

The highest value of cfu in the T-grid occurred in trees with higher level of defoliation (trees 3 and 4, 35 and 70%, respectively). As in the I-grid, the cfu value of IC samples was directly proportional to the defoliation level, being significantly higher for tree 4. Moreover, cfu was significantly higher in points located on the north side of the grid ($H_{cfu} = 8.422$, $P < 0.01$).

3.2. Spatial Distribution of Edaphic Variables

Significant differences were found between the overall positions for all variables, except for N, C/N, silt and moisture (Table 2). Clay, P and K were also significantly different between IC and OC ($H_{clay} = 6.3$, $P_{clay} < 0.05$; $H_P = 4.6$, $P_P < 0.05$; and $H_K = 4.2$, $P_K < 0.0167$), but regarding the concentrated grids (I, T and O positions) only OM, Clay and P presented differences. OM, clay and phosphorus were higher at points of the I grid in a significant extent to O grid ($F_{I\text{-}O\,(OM)} = 5.5$, $P_{I\text{-}O\,(OM)} < 0.0167$; $U_{I\text{-}O\,(clay)} = 318.0$, $P_{I\text{-}O\,(clay)} < 0.0167$; $U_{I\text{-}O\,(P)} = 327.0$, $P_{I\text{-}O\,(P)} < 0.0167$).

The Kruskal-Wallis test showed significant differences for clay, P and K relative to the IC and OC grids ($H_{clay} = 6.304$, $P_{clay} < 0.05$; $H_P = 4.564$, $P_P < 0.05$; $H_K = 4.232$, $P_K < 0.05$). Similarly, to the cfu results (Table 1), P and K had significantly higher concentrations under the crown cover, while clay had significantly lower percentages (Table 2). Regarding orientation, OM ($F_{OM} = 6.049$, $P < 0.05$), pH

($F_{pH} = 5.543$, $P < 0.05$), Ca ($F_{Ca} = 14.792$, $P < 0.05$) and K ($H_K = 2746$, $P < 0.01$) presented significant differences between North and South sides, with higher values in North side, except in the case of OM.

Moreover, the analysis of the Spearman's bivariate correlations showed a significant and positive relationship between cfu and sand ($\rho = 0.242$, $P < 0.01$), pH ($\rho = 0.319$, $P < 0.001$), Ca ($\rho = 0.374$, $P < 0.001$) and K ($\rho = 0.352$, $P < 0.001$), and a significant and negative one with clay ($\rho = -0.398$, $P < 0.001$).

Table 2. Edaphic variables. Mean values (mean ± standard error) according to sample grid and position respect to tree crown cover. *n*: total number of samples. I: inside crown grid; T: transition grid; O: outside crown grid. IC: samples inside crown cover; OC: samples outside crown cover; (ρ): Spearman correlation coefficient between cfu ($n = 132$) and physicochemical soil parameters. Different letters in superscript indicate significant differences with respect to crown cover (Tukey test for normal distributed variables, $P < 0.05$ and Mann-Whitney U Test for non-normal distributed variables, $P < 0.05$ for IC-OC comparisons and $P < 0.0167$ for I-O-T comparisons †). No comparisons were made between means corresponding to different factors (I, O and T with IC and OC).

Variable	Position with Respect to Crown Cover					
	I $n = 32$	T $n = 32$	O $n = 32$	IC $n = 44$	OC $n = 88$	Corr. coef. (ρ)
N (%)	0.19 ± 0.01^{a}	0.18 ± 0.01^{a}	0.18 ± 0.01^{a}	0.19 ± 0.01^{a}	0.17 ± 0.01^{a}	−0.045
OM (%)	1.91 ± 0.11^{b}	1.47 ± 0.12^{b}	1.77 ± 0.11^{a}	1.78 ± 0.08^{a}	1.61 ± 0.07^{a}	−0.008
C/N	6.67 ± 0.52^{a}	5.13 ± 0.49^{a}	6.25 ± 0.54^{a}	6.09 ± 0.39^{a}	5.98 ± 0.36^{a}	0.020
Clay † (%)	27.17 ± 1.01^{b}	$28.21 \pm 0.90^{a,b}$	26.85 ± 1.33^{a}	$25.43 \pm 0.84^{a,*}$	$28.66 \pm 0.67^{b,*}$	−0.398 ***
Silt (%)	37.28 ± 1.36^{a}	38.86 ± 1.16^{a}	37.68 ± 1.28^{a}	38.49 ± 1.23^{a}	37.25 ± 0.69^{a}	−0.026
Sand (%)	35.54 ± 2.03^{a}	32.93 ± 1.34^{a}	35.48 ± 2.17^{a}	36.08 ± 1.72^{a}	34.09 ± 1.07^{a}	0.242 **
P † (mg/kg)	32.45 ± 9.05^{b}	7.41 ± 3.38^{b}	19.58 ± 6.94^{a}	$27.07 \pm 6.55^{a,*}$	$12.19 \pm 3.31^{b,*}$	0.109
pH	5.13 ± 0.06^{a}	5.17 ± 0.04^{a}	5.20 ± 0.05^{a}	5.13 ± 0.04^{a}	5.19 ± 0.03^{a}	0.319 ***
Ca (meq/100 g)	0.70 ± 0.04^{a}	0.59 ± 0.03^{a}	0.73 ± 0.05^{a}	0.64 ± 0.03^{a}	0.67 ± 0.03^{a}	0.374 ***
K † (meq/100 g)	0.50 ± 0.13^{a}	0.20 ± 0.02^{a}	0.22 ± 0.05^{a}	$0.41 \pm 0.09^{a,*}$	$0.19 \pm 0.02^{b,*}$	0.352 ***
Moisture † (%)	15.04 ± 1.24^{a}	15.75 ± 0.65^{a}	15.24 ± 0.77^{a}	15.25 ± 0.84^{a}	15.30 ± 0.46^{a}	0.066

*: Significant at $P < 0.05$; **: Significant at $P < 0.01$ ***: Significant at $P < 0.001$.

3.3. *Spatial Analysis and Location of Edaphic Variables and Relationship with Cfu*

All variables showed spatial aggregation ($I_a > 1$) in at least one tree (Figure 2). The aggregation was significant ($P < 0.05$) for clay in trees 2, 3 and 4; OM, P and silt in trees 2 and 4; Ca in trees 3 and 4; the soil moisture for trees 2 and 3; and K and sand only in trees 1 and 4, respectively. In all other options, the variables tend to be random ($I_a \approx 1$, $P > 0.05$, Figure 2). Overall, nine of the twelve variables analysed with SADIE presented significant spatial aggregation patterns and up to six did so for trees 2 and 4, with the latter the most defoliated.

The clustering indices (ν) showed the presence of patches ($\nu_i \geq 1.5$) and/or gaps ($\nu_j \leq -1.5$) for all variables with $I_a > 1$ (Figure 2; Supplementary Material, Figure S1). Clay showed significant patches ($P < 0.05$) for trees 2 and 3 and gaps also for tree 3 under the crown cover, tending to cluster in tree 4. Organic matter showed clustering patches for trees 1, 2 and 4, as well as gaps for tree 2. Phosphorus and silt showed significant patches and gaps for trees 2 and 4, and Ca for trees 3 and 4, tending also to gaps in these trees. Soil moisture showed significant patches for trees 2 and 3 as well as gaps for tree 2, and K showed significant patches and gaps only for tree 1.

The remaining edaphic variables (N, C/N, sand, pH, Ca, K and moisture) did not show any significant aggregation patterns ($I_a \approx 1$) or significant clustering of patches and gaps, for all trees.

On the other hand, cfu showed a local aggregated distribution pattern (Ia > 1) that was not strong enough to contribute significantly ($P > 0.05$) to the overall aggregation of the influence of the crown area (Figure 2), but these clustering patches ($\nu > 1.5$) were found under the crown cover of all trees as well as in the immediate transition zone (Figure 3).

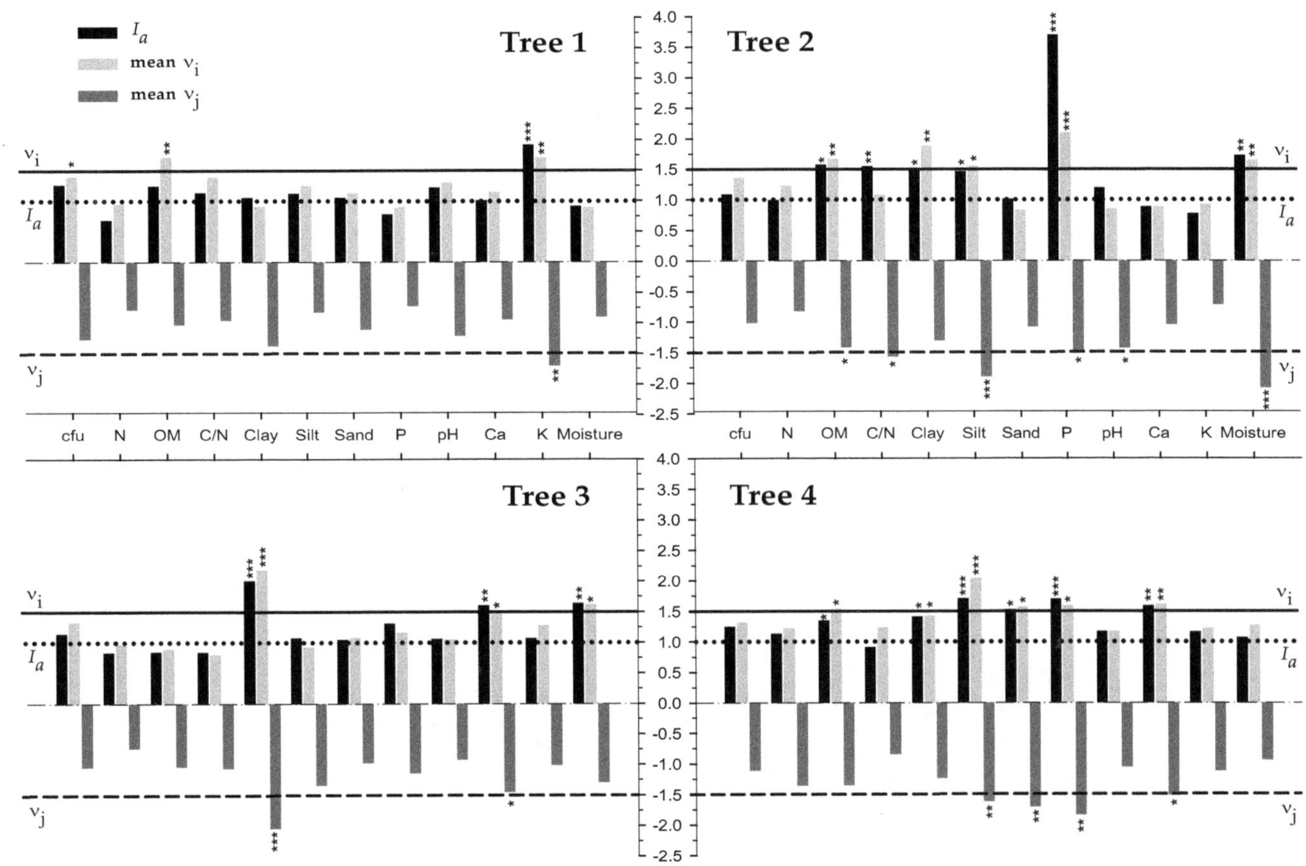

Figure 2. Aggregation indices (I_a) and clustering indices (ν) for the study of 12 variables in the four sampled oaks. The horizontal dotted line $I_a = 1$ indicates the limit of the type of distribution pattern of the variable ($I_a > 1$ aggregated, $I_a < 1$ regular and $I_a = 1$ random). The horizontal continuous line (mean $\nu_i \geq 1.5$) indicates the limit for which the index ν is grouped into patches, i.e., higher values of a given variable. The horizontal discontinuous line (mean $\nu_j \leq -1.5$) indicates the limit for which the index ν is grouped into gaps, i.e., lower values of a given variable. * - $P < 0.05$; ** - $P < 0.01$; *** - $P < 0.001$.

The variables that presented the most aggregated mean pattern were P, OM, K and clay (Figure 4). Phosphorous and OM tended to accumulate under crown cover as well as in the transition zone, with this tendency being obvious in trees 2-4, and to a lesser extent in tree 1. Potassium showed a tendency to be clustered under crown cover, being more evident in trees 3 and 4, which had higher levels of defoliation (35 and 70%, respectively).

Clay showed a dominance of gaps under the crown cover, which is clearer in trees 2 and 4, with patches and gaps in tree 2 ($\nu_i = 1.884$) and 3 ($\nu_i = 2.187$; $\nu_j = -2.045$) and smaller gaps with a tendency to randomness in tree 1. Potassium showed, under the crown, a dominance of spots in trees 3 and 4 and a tendency to be in gaps in tree 1 ($\nu_j = -1.698$) and 2. For P there was, under the crown, a dominance of spots in trees 2 ($\nu_i = 2.098$), 3 and 4 ($\nu_i = 1.592$), and a tendency to randomness in tree 1. Organic matter showed a dominance of patches under the crown in tree 1 ($\nu_i = 1.721$) and 2 ($\nu_i = 1.669$), small patches and gaps in tree 4 ($\nu_i = 1.532$) and a tendency to randomness in tree 3.

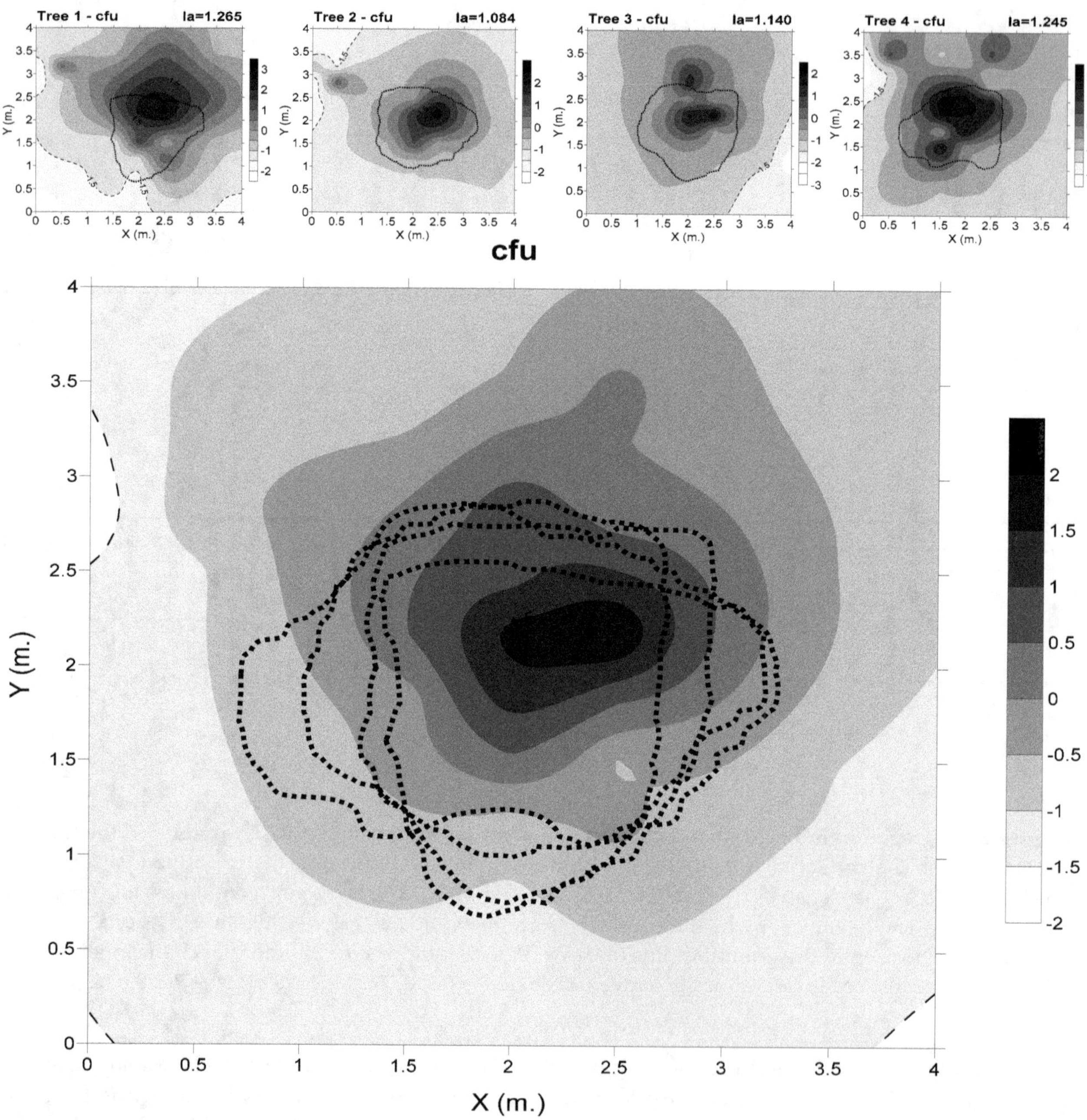

Figure 3. Maps of clustering indices (ν) of the cfu for the four sampled trees (top) and the mean for all the trees (bottom). The dark areas show cfu clustering patches (ν > 1.5) delimited by a continuous line, and the light areas show clustering gaps (ν < −1.5) delimited by a discontinuous line. The dotted lines represent the crown cover of each tree. (I_a): General aggregation index. Legend (ν) is unitless.

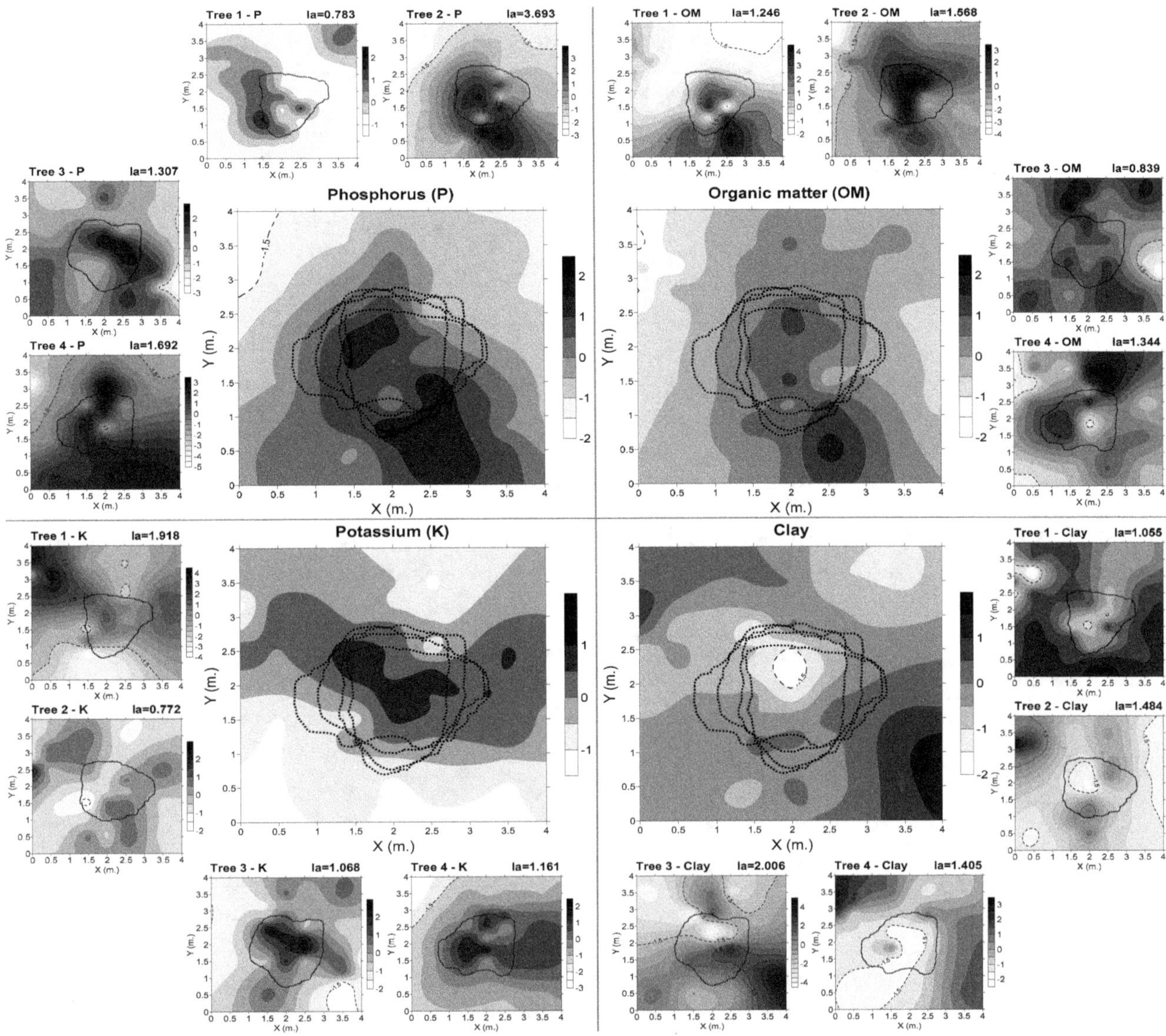

Figure 4. Maps of clustering indices (ν) of the edaphic variables whose aggregation index (I_a) presents statistical significance in at least two trees (Figure 2) and significant differences in mean value between grids regarding crown cover (Table 2). By columns: P (phosphorus), OM (organic matter), silt and clay. By rows, trees 1 to 4 (projected to 4 × 4 m surface). The darker areas show clustering patches of edaphic variables ($\nu > 1.5$) are delimited by a continuous line, and the light areas show edaphic variables clustering gaps ($\nu < -1.5$) are delimited by a discontinuous line. The dotted lines represent the crown cover of each tree. In the upper right corner of each sampling unit, the overall general aggregation pattern (I_a) is indicated. Legend (ν) is unitless.

The remaining edaphic variables (N, C/N, silt, sand, pH, Ca, and moisture) showed randomness in spatial distribution of patches and gaps for the most trees (Supplementary Material, Figure S1).

3.4. Generalized Linear Mixed Model for Cfu

Eight out the 11 edaphic variables (OM, C/N, clay, silt, P, Ca, K and moisture) were selected to test different mixed models using the tree and the position under the crown (IC/OC) as random effects. Finally, the generalised mixed model that fitted best to the cfu distribution was the one constructed by the variables silt, P, K and moisture as fixed effects, considering both tree and position as random effects, which presented model variances of 1.205 and 1.229, respectively (dimensionless). The likelihood ratio test showed also the significant influence of both random effects in the overall model (Table 3).

Table 3. Generalized Linear Mixed Model effects Chi-squared test. First row shows the optimal model Akaike Information Criteria (AIC). Df: Degrees of freedom. *P*: Significance level.

Single Effects Influence		Df	AIC	*P*
None			2703.8	
Random effects	*Tree*	6	4876.2	<0.001
	Position	6	4103.0	<0.001
Fixed effects	*Silt*	1	2704.8	0.082
	K	1	2754.4	<0.001
	P	1	2933.8	<0.001
	Moisture	1	3864.5	<0.001

Although the deletion of silt only produced marginal differences in the resulting model (Table 3), this parameter was finally included in the output model because of the lower AIC compared with the model constructed without this variable. Moreover, when the four selected variables, including silt were considered as fixed effects, they contributed significantly ($P < 0.05$) to the model (Table 4).

Table 4. Generalized linear mixed model adjusted by maximum likelihood between colony forming units (cfu, $n = 132$) and the explanatory variables (silt, phosphorus, potassium and moisture). *P*: Significance level.

	Estimate	Std. Error	z Value	*P*
(Intercept)	2.501	0.957	2.612	<0.01
Silt	0.044	0.026	1.737	<0.05
P	0.803	0.051	15.877	<0.001
K	−0.188	0.027	−6.983	<0.001
Moisture	1.081	0.032	33.881	<0.001

No significant correlation was found between the fixed effects, and the residuals showed a normal distribution (D = 0.122, $P = 0.099$), showing therefore the absence of autocorrelation between fixed factors. The final model showed also a strong Intraclass Correlation Coefficient (Adjusted ICC = 0.975) indicating that clustering by position under the crown accounted for a high proportion of the total variability, and data within each cluster were well correlated.

4. Discussion

This work shows the existence of small-scale spatial patterns in the distribution of cfu of *P. cinnamomi* under holm oak trees, related to texture, P, pH, Ca, K and moisture distribution in soil. Our results are consistent with those obtained in other studies, where the canopy cover of woody plants has been shown to have an important effect on soil properties [20,23,42,43] and these, in turn, on soil microorganisms [22,25]; in this case, on the frequency of *P. cinnamomi* [3]. Additionally, our results show a significant linear relationship between the values of the cfu and the edaphic variables (texture, P, pH, Ca, K and moisture), which correlate with the presence in soil of these oomycetes [44].

Although crown defoliation is usually the main factor assessed when studying holm oak health status in relation to root rot [45,46], this factor was not considered in our work. This analysis would have required different experimental design, considering a plot with a wider range of crown defoliation and health status, a healthy plot with similar soil conditions as a control, and more biological repetitions for each defoliation level. Such increase in the number of repetitions would have been a big challenge regarding the sampling intensity established under each tree in our study. In this work the plot was chosen by its homogeneous conditions in order to determine if soil microsite changes driven by crown cover influence were related with cfu distribution. Thus, we assumed that our chosen plots had similar disease conditions. Moreover, SADIE or spatial analysis use in few plots can be assessed because of an implicitly high sample size [47]. Notwithstanding, and despite the relevance of our findings, it is important to indicate that the low number of trees used in our study would limit the transferability of the results.

4.1. *The Abundance of Cfu and Soil Parameters are Influenced by the Canopy Cover*

The soil samples were collected in spring (April) 2010, in an area severely affected by root rot in Andalusia [48], and showed cfu values similar or slightly superior to those obtained in comparable work in the Iberian Peninsula, with inoculum concentrations ranging between 3 and 130 cfu g^{-1} [1,49,50]. In this sense, the observed defoliation level in the absence of other biotic agents (pests and diseases) is considered to be caused mainly by the root pathogen *P. cinnamomi* [27,44].

The sampled trees were located in a 15 year-old plantation at low density ($\approx$ 312 trees ha^{-1}) with scattered shrub cover (*Cistus ladanifer* L.). The low presence of herbaceous and shrub species allows the existence of a homogeneous root system with abundant fine roots, which in the case of the holm oak rhizosphere, is allocated mainly in the area under the canopy cover [19]. In previous studies, it has been found a significant relationship between soil biological components and the rhizosphere under the crown [25,26].

Moreover, there was a tendency for the highest concentration of cfu to group under the crown on the northern side, independently of the different slopes present in the studied trees. This aggregation of cfu might be related with the microclimate that the tree crown's shade generates on the ground, driven by differences in nutrients and organic matter. Other authors have reported similar aggregation patterns in soils under holm and cork oaks [20,23,42]. Furthermore, the greater presence of fine roots under the crown cover [51,52] agrees with the fine root distribution observed in several Mediterranean oak species [53], and should be considered another relevant driver influencing this cfu distribution pattern. Woody plant cover influences the edaphic properties, as well as the related plant and faunal communities [20,23,54], establishing an important and direct relationship between the fertility and presence of soilborne pathogens and thus influencing the health status of the tree [25,55]. Moreover, other soil characteristics such as texture, porosity, fertility, organic matter, cultivation methods and the presence of host plants are other factors involved in the presence of resistant structures in the soil [56]. In our work, some outliers detected in the cfu values outside the crown were closely related to the presence of clusters of *Cistus ladanifer* (data not shown). Those data were not eliminated from the analysis because it might be considered that the effects of the presence of the shrub clusters on the studied variables should be similar to the effect of the presence of the trees [57].

The spatial analysis showed that 9 out of the 12 variables analysed with SADIE presented spatial aggregation ($I_a > 1$, $P < 0.05$) in some of the studied trees, and 11 out of the 12 variables did so for the most defoliated tree (#4), but only 6 with significant probability. Results on the spatial distribution of nutrients in the current study agreed with those observed by Gallardo [20] and Andivia et al. [23], where all variables showed spatial heterogeneity. Nitrogen and OM tended to be clumped under the crown cover for all trees, except tree 4 in the case of N. The N cycle is closely related to the processes of organic matter in soils, although it seems to be little influenced by the presence of roots and mycorrhizae [19]. The soil content of P is also linked to soil biological and chemical processes, and forms clusters under the crown cover in three (trees 1, 2 and 4) of the four trees studied. Roots and mycorrhizae play an important role in the mineralisation of the P in organic matter through the activity of phosphatases [58]; thus, the probable extension of roots beyond the crown cover could explain P distribution within and around the crown [59,60]. In this sense, soil N and P concentrations depend on the rate of mineralisation and their uptake by roots and microorganisms [20,61], and on the interactions with soil mineral components in the case of P [61,62]. Potassium should show less spatial dependence on the tree crown than the rest of the essential elements as it is not linked to the organic soil components; however, it showed aggregation of clusters in trees 3 and 4. These contrasting results among trees may be related with other processes such as the cortical runoff of K leaching from leaves and other aboveground plant organs [20,61,63].

4.2. Concentrarion of Cfu was Influenced by the Spatial Distribution of Edaphic Variables

The significant correlations between the concentration of cfu and the edaphic variables, clay, P, OM and K, were related to differences in general values and/or significant aggregation patterns with respect to crown cover. These results provide evidence of the influence of crown cover on soil conditions and cfu. The lifecycle of *P. cinnamomi* and other oomycetes is driven by soil conditions, persists in the soil as resistance structures when the conditions are unfavourable, and proliferates through zoospores when soil temperature and moisture are adequate [55]. Therefore, the influence of crown cover on soil conditions might have an indirect influence on the number of cfu.

For all trees, the concentration of clay was lower under the crown cover. Locations under the tree cover showed higher cfu levels accompanied with lower clay and higher sand percentages. Other studies have reported loam medium texture under crown cover and clay-loam medium texture outside [64]. Although the differences in soil texture were significant for the crown cover, the effect of those differences on cfu concentration would be lower than the effect of the presence of fine roots, which are supposed to be concentrated under the crown. The root rot of *Q. ilex* affects mostly the fine roots, the most serious effects being on roots growing in relatively dry soils, eventually undergo short flooding due to extreme precipitation events [9,31,64]. Therefore, it might be considered that the presence of fine roots and soil moisture are more influential environmental drivers for the presence of microorganisms compared with changes in soil texture.

In our study pH and Ca showed a positive correlation with cfu. High concentrations of Ca and high pH values in the soil are related on the literature with the inhibition of the pathogen growth and the accumulation of survival structures of oomycetes [28,31,65]. However, the overall values we found for pH and Ca did not differ regarding crown position and presented small variations among all the samples. All the soil samples analysed were acidic ($5.1 < pH < 5.5$) and the Ca^{2+} values ranged from 0.56 to 0.78 meq 100 g^{-1} (from 0.28 to 0.4 mMol Ca each 100 g of soil) (Table 2). The pH was very close to the interval established as the optimum for the development and infection of several soilborne pathogen oomycetes in laboratory tests, including *P. cinnamomi* (optimum pH between 5.5 and 6.0) [66], and the concentration of Ca was 100 fold lower than the necessary to inhibit the mycelium growth of *P. cinnamomi* on in vitro tests [28]. Therefore, Ca and pH should not be considered in our case as factors influencing the distribution of cfu, since the range of Ca and pH values could not produce differences in the behaviour of the pathogen.

The cfu values were predicted to a significant extent by the GLMM using silt and moisture as variables, together with P and K, demonstrating that there is an influence of soil conditions on cfu abundance apart from tree crown influence. The output model might be considered site-specific because it is dependent on a set of microsite characteristics. However, the results suggest that there exists a significant influence of the soil variables contributing to the suitability of the microsite environment for *P. cinnamomi*.

The levels of P and K were significantly different due to crown position, presenting higher values under the tree crown. However, when the influence of crown position was taken out of the analysis, considering it as a random effect in the mixed model, both P and K concentration explained to a significant extent the cfu values. Both P and K were present in higher concentrations under the crown, following the same trend as cfu. Other studies showed that N, P, K, OM and the biomass of microorganisms tend to concentrate under the crown [21,67]. It could be considered that the influence of these soil parameters on the number of cfu is related to the suitability of the microenvironment for oomycetes to complete their lifecycle. Some oomycetes species, including several pathogenic *Phytophthora* spp., have the ability to either survive or complete their lifecycle as saprobes [19,68,69], despite their poor ability to compete with other saprophytic organisms [70]. Therefore, higher levels of organic matter and nutrients might lead to an increase of oomycete resistance structures under a saprophytic environment.

The cfu distribution seems to be favoured mainly by the grouping of silt patches. Silt is considered to equilibrate the clay-sand trend of the soil. Texture and porosity are directly influenced by the silt percentage in intermediate textured soils. In our study, the crown cover did not significantly influence soil moisture or aggregation patterns. However, the mixed model showed that crown cover clearly influenced the cfu abundance. The same occurred with silt, when it was considered as fixed effect in the model. Soils with higher silt percentage retain more water [64,71], and with higher water content oomycetes are more readily able to produce sporangia and release zoospores [72,73]. In contrast, as these soils become drier, oomycetes produce resistant survival structures.

5. Conclusions

The distribution of *P. cinnamomi* cfu in soils associated with *Q. ilex* was not random in the soil but showed distribution patterns predictable to some degree, influenced by the crown cover, orientation and the levels of soil moisture and fertility. Our results could be useful to increase the sampling efficiency of the field surveys. Soil sampling searching for *P. cinnamomi* in holm oak dehesas would be oriented to those areas most likely to contain the pathogen identified in our work, allowing a greater number of trees to be sampled. This work also highlights the dynamics of soil properties in the presence of tree cover. Clear differences and aggregated spatial patterns in key soil elements were where shown to be influenced by canopy cover. The cfu tend to concentrate in the North side, probably influenced by the shadow of the crown cover, and in zones with more organic matter, nutrients and well-textured soils. In our case, the influence of the texture was driven mainly by silt concentration due to the low variation of clay and sand in the studied area.

Due to the homogeneity of environmental conditions in the selected plot, the output GLMM must be considered site-specific, but we demonstrate that it is a useful tool to study the influence of soil parameters in the distribution of microbial community due to the elimination of random effects, mainly the influence of canopy cover. The spatial analysis of the biotic and abiotic factors involved in oak root rot processes can be an effective management tool predicting favourable areas regarding spatial heterogeneity, quantification and distribution for those parameters that limit the development of *P. cinnamomi*, thus favouring oak establishment and survival in afforestation practices. However, further research is needed to assess the abundance of *Phytophthora* spp. and other oomycetes in soils with more heterogeneous conditions, in order to clarify whether generalized models can be used to predict cfu amounts, particularly in Mediterranean dehesa and montados ecosystems and in oak afforestation.

Supplementary Materials: Figure S1: Maps of clustering indices (ν) of the edaphic variables. By columns: N (nitrogen), C/N (carbon/nitrogen ratio), silt, sand, pH, Ca (calcium) and moisture. By rows, trees 1 to 4 (projected to 4×4 m surface). Darker areas show clustering spots of edaphic variables ($\nu > 1.5$) are delimited by a continuous line, and the light areas show edaphic variables clustering gaps ($\nu < -1.5$) are delimited by a discontinuous line. The dotted line represents the crown cover of each tree. In the upper right corner of each sampling unit, the general aggregation pattern (I_a) i s indicated. Legend has no units. Table S1: Visual symptomatology and morphological parameters of the four trees selected for this study. Table S2: Soil Physicochemical parameters and analytical technique used.

Author Contributions: R.S.-C. and R.M.N.-C. designed the study. R.S.-C. and F.J.R.-G. performed the laboratory analysis. R.S.-C., R.M.N.-C. and J.L.Q. conducted the statistical analysis. R.M.N.-C. and J.L.Q. obtained the funds for the research. R.S.-C., R.M.N.-C., F.J.R.-G., and J.L.Q. participated in the writing and editing of the manuscript. R.S.-C. and R.M.N.-C. contributed equally in this work. All authors have read and agreed to the published version of the manuscript.

Acknowledgments: This project was funded through the ESPECTRAMED (CGL2017-86161-R) project. Thanks, are given to the University of Cordoba and Spanish Ministry of Economy and Competitiveness for financial support. We also acknowledge the institutional support of the University of Cordoba-Campus de Excelencia CEIA3, the "Agencia de Medio Ambiente y Agua" ("Consejería de Medioambiente y Ordenación del Territorio"-Junta de Andalucía) for their help, and especially to Campo Baldio S.L. for providing access to the field plot and for data support. We also thank Joaquín Duque Lazo for his valuable assistance during statistical analysis.

References

1. Jiménez, J.J.; Serrano, M.S.; Vicente, M.; Fernández, P.; Trapero Casas, A.; Sánchez Hernández, E. Nuevas especies de *Pythium* que causan podredumbre radical de *Quercus* en España y Portugal. *Bol. Sanid. Veg. Plagas* **2008**, *34*, 549–562.
2. Camilo-Alves, C.d.S.e.P.; Clara, M.I.E.d.; Ribeiro, N.M.C.d.A. Decline of Mediterranean oak trees and its association with *Phytophthora cinnamomi*: A review. *Eur. J. For. Res.* **2013**, *132*, 411–432. [CrossRef]
3. Ruiz-Gómez, F.J.; Navarro Cerrillo, R.M.; Lara Gómez, M.A.; Sánchez-Cuesta, R. Aislamiento e identificación de oomicetos en focos de podredumbre radical de Andalucía y Extremadura. *Cuad. Soc. Esp. Cienc. For.* **2016**, *43*, 363–376. [CrossRef]
4. González-Alonso, C. Analysis of the Oak Decline in Spain-La Seca. Bachelor's Thesis, Forest Management, Swedish University of Agricultural Sciences, Uppsala, Sweden, 2008.
5. Vettraino, A.M.; Barzanti, G.P.; Bianco, M.C.; Ragazzi, A.; Capretti, P.; Paoletti, E.; Luisi, N.; Anselmi, N.; Vannini, A. Occurrence of *Phytophthora* species in oak stands in Italy and their association with declining oak trees. *For. Pathol.* **2002**, *32*, 19–28. [CrossRef]
6. Pérez-Sierra, A.; López-García, C.; León, M.; García-Jiménez, J.; Abad-Campos, P.; Jung, T. Previously unrecorded low-temperature *Phytophthora* species associated with *Quercus* decline in a Mediterranean forest in eastern Spain. *For. Pathol.* **2013**, *43*, 331–339. [CrossRef]
7. Navarro Cerrillo, R.M.; Fernández Rebollo, P.; Trapero, A.; Caetano, P.; Romero, M.A.; Sánchez, M.E.; Fernández Cancio, A.; Sánchez, I.; López Pantoja, G. *Los Procesos de Decaimiento de Encinas y Alcornoques*; Dirección General de Gestión del Medio Natural. Consejería de Medio Ambiente; Junta de Andalucía: Sevilla, Spain, 2004; ISBN 84-95785-89-7.
8. Trapero, A.; Sánchez, M.E.; Pérez de Algaba, A.; Romero, M.A.; Navarro, N.; Varo, R.; Gutiérrez, J. Enfermedades de especies forestales en Andalucía. *Agricultura* **2000**, *821*, 822–824.
9. Oßwald, W.; Fleischmann, F.; Rigling, D.; Coelho, A.C.; Cravador, A.; Diez, J.; Dalio, R.J.; Horta Jung, M.; Pfanz, H.; Robin, C.; et al. Strategies of attack and defence in woody plant—*Phytophthora* interactions. *For. Pathol.* **2014**, *44*, 169–190. [CrossRef]
10. Ruiz Gómez, F.; Pérez-de-Luque, A.; Sánchez-Cuesta, R.; Quero, J.; Navarro Cerrillo, R. Differences in the response to acute drought and *Phytophthora cinnamomi* Rands Infection in *Quercus ilex* L. seedlings. *Forests* **2018**, *9*, 634. [CrossRef]
11. Jung, T.; Orlikowski, L.; Henricot, B.; Abad-Campos, P.; Aday, A.G.; Aguín Casal, O.; Bakonyi, J.; Cacciola, S.O.; Cech, T.; Chavarriaga, D.; et al. Widespread Phytophthora infestations in European nurseries put forest, semi-natural and horticultural ecosystems at high risk of *Phytophthora* diseases. *For. Pathol.* **2016**, *46*, 134–163. [CrossRef]
12. Marais, L.J.; Menge, J.A.; Bender, G.S.; Faber, B. *Phytophthora Root Rot*; AvoResearch California Avocado Commission Publishing: Irvine, CA, USA, 2002; Volume 2, pp. 3–6.
13. Zamora Rojas, E.; Andicoberry de los reyes, S.; Sánchez Clemente, M.E. Anexo A.1. IV. El decaimiento y la podredumbre radical en las dehesas andaluzas. In *Ecosistemas de Dehesa: Desarrollo de Políticas y Herramientas Para la Gestión y Conservación de la Biodiversidad (Life Biodehesa Project)*; Consejería de Medio Ambiente y Ordenación del Territorio: Sevilla, Spain, 2014.
14. Hüberli, D.; Tommerup, I.C.; Hardy, G.E.S.J. False-negative isolations or absence of lesions may cause mis-diagnosis of diseased plants infected with *Phytophthora cinnamomi*. *Australas. Plant Pathol.* **2000**, *29*, 164–169. [CrossRef]
15. Ashton, M.S.; Larson, B.C. Germination and seedling growth of *Quercus* (section *Erythrobalanus*) across openings in a mixed-deciduous forest of southern New England, USA. *For. Ecol. Manag.* **1996**, *80*, 81–94. [CrossRef]
16. Mauer, O.; Houškova, K.; Mikita, T. The root system of pedunculate oak (*Quercus robur* L.) at the margins of regenerated stands. *J. For. Sci.* **2017**, *63*, 22–33.
17. Aponte, C.; Matías, L.; González-Rodríguez, V.; Castro, J.; García, L.V.; Villar, R.; Marañón, T. Soil nutrients and microbial biomass in three contrasting Mediterranean forests. *Plant Soil* **2014**, *380*, 57–72. [CrossRef]
18. Carranca, C.; Castro, I.V.; Figueiredo, N.; Redondo, R.; Rodrigues, A.R.F.; Saraiva, I.; Maricato, R.; Madeira, M.A.V. Influence of tree canopy on N2 fixation by pasture legumes and soil rhizobial abundance in Mediterranean oak woodlands. *Sci. Total Environ.* **2015**, *506–507*, 86–94. [CrossRef]

19. Ruiz Gómez, F.J.R.; Navarro-Cerrillo, R.M.; Pérez-de-Luque, A.; Oβwald, W.; Vannini, A.; Morales-Rodríguez, C. Assessment of functional and structural changes of soil fungal and oomycete communities in holm oak declined dehesas through metabarcoding analysis. *Sci. Rep.* **2019**, *9*, 5315. [CrossRef]
20. Gallardo, A. Effect of tree canopy on the spatial distribution of soil nutrients in a Mediterranean Dehesa. *Pedobiologia* **2003**, *47*, 117–125. [CrossRef]
21. Cubera, E.; Moreno, G. Effect of land-use on soil water dynamic in dehesas of Central–Western Spain. *Catena* **2007**, *71*, 298–308. [CrossRef]
22. Quero, J.L.; Maestre, F.T.; Ochoa, V.; García-Gómez, M.; Delgado-Baquerizo, M. On the importance of shrub encroachment by sprouters, climate, species richness and anthropic factors for ecosystem multifunctionality in semi-arid Mediterranean ecosystems. *Ecosystems* **2013**, *16*, 1248–1261. [CrossRef]
23. Andivia, E.; Fernández, M.; Alejano, R.; Vázquez-Piqué, J. Tree patch distribution drives spatial heterogeneity of soil traits in cork oak woodlands. *Ann. For. Sci.* **2015**, *72*, 549–559. [CrossRef]
24. Cappai, C.; Kemanian, A.R.; Lagomarsino, A.; Roggero, P.P.; Lai, R.; Agnelli, A.E.; Seddaiu, G. Small-scale spatial variation of soil organic matter pools generated by cork oak trees in Mediterranean agro-silvo-pastoral systems. *Geoderma* **2017**, *304*, 59–67. [CrossRef]
25. Gómez-Aparicio, L.; Ibáñez, B.; Serrano, M.S.; De Vita, P.; Ávila, J.M.; Pérez-Ramos, I.M.; García, L.V.; Sánchez, M.E.; Marañón, T. Spatial patterns of soil pathogens in declining Mediterranean forests: Implications for tree species regeneration. *New Phytol.* **2012**, *194*, 1014–1024. [CrossRef] [PubMed]
26. Ibáñez, B.; Gómez-Aparicio, L.; Ávila, J.M.; Pérez-Ramos, I.M.; García, L.V.; Marañón, T. Impact of tree decline on spatial patterns of seedling-mycorrhiza interactions: Implications for regeneration dynamics in Mediterranean forests. *For. Ecol. Manag.* **2015**, *353*, 1–9. [CrossRef]
27. Romero, M.A.; Sánchez, J.E.; Jiménez, J.J.; Belbahri, L.; Trapero, A.; Lefort, F.; Sánchez, M.E. New *Pythium* taxa causing root rot on Mediterranean *Quercus* species in South-West Spain and Portugal. *J. Phytopathol.* **2007**, *155*, 289–295. [CrossRef]
28. Serrano, M.S.; De Vita, P.; Fernández-Rebollo, P.; Sánchez-Hernández, M.E. Calcium fertilizers induce soil suppressiveness to *Phytophthora cinnamomi* root rot of *Quercus ilex*. *Eur. J. Plant Pathol.* **2012**, *132*, 271–279. [CrossRef]
29. Vadell, E.; de-Miguel, S.; Pemán, J. Large-scale reforestation and afforestation policy in Spain: A historical review of its underlying ecological, socioeconomic and political dynamics. *Land Use Policy* **2016**, *55*, 37–48. [CrossRef]
30. Lehtijärvi, A.; Aday Kaya, A.G.; Woodward, S.; Jung, T.; Doğmuş Lehtijärvi, H.T. Oomycota species associated with deciduous and coniferous seedlings in forest tree nurseries of Western Turkey. *For. Pathol.* **2017**, *47*, e12363. [CrossRef]
31. Burgess, T.I.; Scott, J.K.; Mcdougall, K.L.; Stukely, M.J.C.; Crane, C.; Dunstan, W.A.; Brigg, F.; Andjic, V.; White, D.; Rudman, T.; et al. Current and projected global distribution of *Phytophthora cinnamomi*, one of the world's worst plant pathogens. *Glob. Chang. Biol.* **2017**, *23*, 1661–1674. [CrossRef]
32. Eichhorn, J.; Roskams, P.; Potočić, N.; Timmermann, V.; Ferretti, M.; Mues, V.; Szepesi, A.; Durrant, D.; Seletković, I.; Schroeck, H.-W.; et al. Part IV: Visual assessment of crown condition and damaging agents. In *UNECE ICP Forests Programme Co-ordinating Centre. Manual on Methods and Criteria for Harmonized Sampling, Assessment, Monitoring and Analysis of the Effects of Air Pollution on Forests*; Thünen Institute of Forest Ecosystems: Eberswalde, Germany, 2017; p. 54. ISBN 978-3-86576-162-0.
33. Jeffers, S.N.; Martin, S.B. Comparison of two media selective for *Phytophthora* and *Pythium* species. *Plant Dis.* **1986**, *70*, 1038–1043. [CrossRef]
34. Rohlf, F.J.; Sokal, R.R. *Statistical Tables*; W. H. Freeman: San Francisco, CA, USA, 1981.
35. R Core Team. *R: A Language and Environment for Statistical Computing*; R Foundation for Statistical Computing: Vienna, Austria, 2013.
36. Quero, J.L. La heterogeneidad en ecología: Herramientas de cuantificación y aplicaciones para la restauración. *Acta Granatense* **2006**, *4*, 107–114.
37. Winder, L.; Alexander, C.; Griffiths, G.; Holland, J.; Woolley, C.; Perry, J. Twenty years and counting with SADIE: Spatial Analysis by Distance Indices software and review of its adoption and use. *Rethink. Ecol.* **2019**, *4*, 1–16. [CrossRef]

38. Perry, J.N.; Dixon, P.M. A new method to measure spatial association for ecological count data. *Écoscience* **2002**, *9*, 133–141. [CrossRef]
39. Bates, D.; Mächler, M.; Bolker, B.; Walker, S. Fitting Linear Mixed-Effects Models using lme4. *arXiv* **2014**, arXiv:1406.5823.
40. Bolker, B.M.; Brooks, M.E.; Clark, C.J.; Geange, S.W.; Poulsen, J.R.; Stevens, M.H.H.; White, J.S.S. Generalized linear mixed models: A practical guide for ecology and evolution. *Trends Ecol. Evol.* **2009**, *24*, 127–135. [CrossRef] [PubMed]
41. Schielzeth, H. Simple means to improve the interpretability of regression coefficients: Interpretation of regression coefficients. *Methods Ecol. Evol.* **2010**, *1*, 103–113. [CrossRef]
42. Gea-Izquierdo, G.; Allen-Díaz, B.; San Miguel, A.; Canellas, I. How do trees affect spatio-temporal heterogeneity of nutrient cycling in Mediterranean annual grasslands? *Ann. For. Sci.* **2010**, *67*, 112. [CrossRef]
43. Simón, N.; Montes, F.; Díaz-Pinés, E.; Benavides, R.; Roig, S.; Rubio, A. Spatial distribution of the soil organic carbon pool in a Holm oak dehesa in Spain. *Plant Soil* **2013**, *366*, 537–549. [CrossRef]
44. Sánchez, M.; Caetano, P.; Ferraz, J.; Trapero, A. *Phytophthora* disease of *Quercus ilex* in south-western Spain. *For. Pathol.* **2002**, *32*, 5–18. [CrossRef]
45. González, M.; Romero, M.; Ramo, C.; Serrano, M.S.; Sánchez, M.E. *Control of Phytophthora Root Rot on Mediteranean Quercus spp. Using Fosetyl-Al Trunk Injections. M*; Firenze University Press: Firenze, Italy, 2017.
46. Navarro-Cerrillo, R.M.; Varo-Martínez, M.Á.; Acosta, C.; Palacios Rodriguez, G.; Sánchez-Cuesta, R.; Ruiz Gómez, F.J. Integration of WorldView-2 and airborne laser scanning data to classify defoliation levels in *Quercus ilex* L. dehesas affected by root rot mortality: Management implications. *For. Ecol. Manag.* **2019**, *451*, 117564. [CrossRef]
47. Maestre, F.T.; Quero, J.L. Análisis espacial mediante índices de distancia (SADIE). In *Introducción al Análisis Espacial de Datos en Ecología y Ciencias Ambientales: Métodos y Aplicaciones*; Universidad Rey Juan Carlos: Madrid, Spain, 2008; pp. 637–648. ISBN 978-84-9849-308-5.
48. Sánchez, M.E.; Caetano, P.; Ferranz, J.; Trapero Casas, A. El decaimiento y muerte de encinas en tres dehesas de la provincia de Huelva. *Bol. Sanid. Veg. Plagas* **2000**, *26*, 447–464.
49. Rodríguez, M.; Sánchez, M.E.; Trapero, A. Desarrollo de un método eficaz para la cuantificación de *Phytophthora cinnamomi* en muestras de suelo. In Proceedings of the XII Congreso de la Sociedad Española de Fitopatología, Girona, Spain, 26 September–1 October 2004.
50. Serrano, M.S.; Leal, R.; Vita, P.D.; Fernández-Rebollo, P.; Sánchez, M.E. Control of *Phytophthora cinnamomi* by soil application of calcium fertilizers in a Dehesa ecosystem in Spain. *Integr. Prot. Oak For. IOBC-WPRS Bull.* **2014**, *101*, 139–143.
51. Canadell, J.; Rodà, F. Root biomass of *Quercus ilex* in a montane Mediterranean forest. *Can. J. For. Res.* **1991**, *21*, 1771–1778. [CrossRef]
52. Silva, J.S.; Rego, F.C. Root to shoot relationships in Mediterranean woody plants from Central Portugal. *Biologia (Bratislava)* **2004**, *59*, 109–115.
53. Lopez, B.; Sabate, S.; Gracia, C. Fine roots dynamics in a Mediterranean forest: Effects of drought and stem density. *Tree Physiol.* **1998**, *18*, 601–606. [CrossRef] [PubMed]
54. Eldridge, D.J.; Bowker, M.A.; Maestre, F.T.; Roger, E.; Reynolds, J.F.; Whitford, W.G. Impacts of shrub encroachment on ecosystem structure and functioning: Towards a global synthesis. *Ecol. Lett.* **2011**, *14*, 709–722. [CrossRef] [PubMed]
55. Corcobado, T. Influencia de *Phytophthora cinnamomi* Rands en el Decaimiento de *Quercus ilex* L. y su Relación con las Propiedades del Suelo y las Ectomicorrizas. Ph.D. Thesis, Universidad de Extremadura, Plasencia, Spain, 2013.
56. García Moreno, A.M.; Fernández Rebollo, P.; Ortiz Berrocal, F.; Carbonero Muñoz, M.D. *Podredumbre Radical, Descripción y Control Aplicado a Los Ecosistemas de Dehesa*; Instituto de Investigación y Formación Agraria y Pesquera, Consejería de Agricultura, Pesca y Desarrollo Rural, Junta de Andalucía: Córdoba, Spain, 2016.
57. Moreira, A.C.; Martins, J.M.S. Influence of site factors on the impact of *Phytophthora cinnamomi* in cork oak stands in Portugal. *For. Pathol.* **2005**, *35*, 145–162. [CrossRef]
58. Satyaprakash, M.; Nikitha, T.; Reddi, E.U.B.; Sadhana, B.; Vani, S.S. Phosphorous and phosphate solubilising bacteria and their role in plant nutrition. *Int. J. Curr. Microbiol. Appl. Sci.* **2017**, *6*, 2133–2144.

59. Rodà, F.; Retana, J.; Gracia, C.A.; Bellot, J. *Ecology of Mediterranean Evergreen Oak Forests*; Springer Science & Business Media: Berlin/Heidelberg, Germany, 1999; Volume 137, ISBN 3-540-65019-9.
60. Shen, J.; Yuan, L.; Zhang, J.; Li, H.; Bai, Z.; Chen, X.; Zhang, W.; Zhang, F. Phosphorus dynamics: From soil to plant. *Plant Physiol.* **2011**, *156*, 997–1005. [CrossRef]
61. Sardans, J.; Peñuelas, J. Plant-soil interactions in Mediterranean forest and shrublands: Impacts of climatic change. *Plant Soil* **2013**, *365*, 1–33. [CrossRef]
62. Delgado-Baquerizo, M.; Maestre, F.T.; Gallardo, A.; Bowker, M.A.; Wallenstein, M.D.; Quero, J.L.; Ochoa, V.; Gozalo, B.; García-Gómez, M.; Soliveres, S.; et al. Decoupling of soil nutrient cycles as a function of aridity in global drylands. *Nature* **2013**, *502*, 672–676. [CrossRef]
63. Aber, J.A.; Melillo, J.M. *Terrestrial Ecosystems*, 2nd ed.; Brooks/Cole Publishing: Pacific Grove, CA, USA, 2001; ISBN 978-0-12-041755-1.
64. Brown, R.B. *Soil Texture*; Soil and Water Science Department, Florida Cooperative Extension Service, Institute of Food and Agriculture Sciences, University of Florida: Gainesville, FL, USA, 2003.
65. Serrano, M.S.; Fernández-Rebollo, P.; De Vita, P.; Sánchez, M.E. Calcium mineral nutrition increases the tolerance of *Quercus ilex* to *Phytophthora* root disease affecting oak rangeland ecosystems in Spain. *Agrofor. Syst.* **2013**, *87*, 173–179. [CrossRef]
66. Zentmyer, G.A. *Phytophthora Cinnamomi and the Diseases It Causes*; Monograph, American Phytopathological Society: Riverside, CA, USA, 1980; ISBN 0-89054-030-6.
67. Gallardo, A.; Rodríguez-Saucedo, J.J.; Covelo, F.; Fernández-Alés, R. Soil nitrogen heterogeneity in a Dehesa ecosystem. *Plant Soil* **2000**, *222*, 71–82. [CrossRef]
68. Hardham, A.R. Phytophthora cinnamomi. *Mol. Plant Pathol.* **2005**, *6*, 589–604. [CrossRef] [PubMed]
69. Cerri, M.; Sapkota, R.; Coppi, A.; Ferri, V.; Foggi, B.; Gigante, D.; Lastrucci, L.; Selvaggi, R.; Venanzoni, R.; Nicolaisen, M.; et al. Oomycete communities associated with reed die-back syndrome. *Front. Plant Sci.* **2017**, *8*, 1550. [CrossRef] [PubMed]
70. Dunstan, W.A.; Rudman, T.; Shearer, B.L.; Moore, N.A.; Paap, T.; Calver, M.C.; Dell, B.; Hardy, G.E.S.J. Containment and spot eradication of a highly destructive, invasive plant pathogen (*Phytophthora cinnamomi*) in natural ecosystems. *Biol. Invasions* **2010**, *12*, 913–925. [CrossRef]
71. Garrido Valero, M.S. *Interpretación de Análisis de Suelos*; Ministerio de Agricultura, Pesca y Alimentación: Madrid, Spain, 1994; ISBN 84-341-0810-0.
72. Tecon, R.; Or, D. Biophysical processes supporting the diversity of microbial life in soil. *FEMS Microbiol. Rev.* **2017**, *41*, 599–623. [CrossRef]
73. Jeon, S.; Krasnow, C.S.; Kirby, C.K.; Granke, L.L.; Hausbeck, M.K.; Zhang, W. Transport and retention of *Phytophthora capsici* zoospores in saturated porous media. *Environ. Sci. Technol.* **2016**, *50*, 9270–9278. [CrossRef]

11

Isolation and Pathogenicity of *Phytophthora* Species from Poplar Plantations in Serbia

Ivan Milenković [1,2,*], Nenad Keča [2], Dragan Karadžić [2], Zlatan Radulović [3], Justyna A. Nowakowska [4], Tomasz Oszako [5], Katarzyna Sikora [6], Tamara Corcobado [1] and Thomas Jung [1,7]

1 Phytophthora Research Centre, Mendel University in Brno, Zemědělská 1, 61300 Brno, Czech Republic; tamicorsa@hotmail.com (T.C.); dr.t.jung@t-online.de (T.J.)
2 Faculty of Forestry, University of Belgrade, Kneza Višeslava 1, 11030 Belgrade, Serbia; nenad.keca@sfb.bg.ac.rs (N.K.); dragan.karadzic@sfb.bg.ac.rs (D.K.)
3 Institute of Forestry, Belgrade, Kneza Višeslava 3, 11030 Belgrade, Serbia; zlatan.radulovic@forest.org.rs
4 Faculty of Biology and Environmental Sciences, Cardinal Stefan Wyszynski University in Warsaw, Wóycickiego 1/3 Street, 01-938 Warsaw, Poland; j.nowakowska@uksw.edu.pl
5 Faculty of Forestry, Białystok University of Technology, Marszałka J. Piłsudskiego 1A, 17-200 Hajnówka, Poland; T.Oszako@ibles.waw.pl
6 Forest Research Institute-IBL, Braci Leśnej 3, 05-090 Raszyn, Poland; K.Sikora@ibles.waw.pl
7 Phytophthora Research and Consultancy, Am Rain 9, 83131 Nussdorf, Germany
* Correspondence: ivan.milenkovic@sfb.bg.ac.rs.

Abstract: During a survey in three declining and three healthy poplar plantations in Serbia, six different *Phytophthora* species were obtained. *Phytophthora plurivora* was the most common, followed by *P. pini*, *P. polonica*, *P. lacustris*, *P. cactorum*, and *P. gonapodyides*. Pathogenicity of all isolated species to four-month and one-year-old cuttings of *Populus* hybrid clones I-214 and Pánnonia, respectively, was tested using both a soil infestation and stem inoculation test. Isolates of *P. polonica*, *P.* × *cambivora*, *P. cryptogea*, and *P.* × *serendipita* from other host plants were included as a comparison. In the soil infestation test, the most aggressive species to clone I-214 were *P. plurivora*, *P.* × *serendipita*, and *P. pini*. On clone Pánnonia, *P. gonapodyides* and *P. pini* were the most aggressive, both causing 100% mortality, followed by *P. cactorum*, *P.* × *cambivora*, and *P. polonica*. In the underbark inoculation test, the susceptibility of both poplar clones to the different *Phytophthora* species was largely similar, as in the soil infestation test, with the exception of *P. polonica*, which proved to be only weakly pathogenic to poplar bark. The most aggressive species to clone I-214 was *P. pini*, while on clone Pánnonia, the longest lesions and highest disease incidence were caused by *P. gonapodyides*. *Phytophthora cactorum* and *P. plurivora* were pathogenic to both clones, whereas *P.* × *cambivora* showed only weak pathogenicity. The implications of these findings and possible pathways of dispersion of the pathogens are discussed.

Keywords: soilborne pathogens; pathways; *Populus*; *Phytophthora plurivora*; *Phytophthora pini*; pathogenicity tests

---

1. Introduction

Phytophthora species are fungus-like organisms belonging to the kingdom Chromista (Stramenopiles) within the SAR (Stramenopiles, Alveolata, Rhizaria) super group [1]. These pathogens can infect numerous woody host plants in natural ecosystems, nurseries, and plantings [2–5]. The wide distribution of these damaging pathogens is mainly a consequence of the increasing international trade in living plants resulting in the introduction of non-native *Phytophthora* spp. into previously unaffected regions on infected nursery stock [3–8]. Therefore, forests and plantations established via the planting of nursery stock are at high risk of *Phytophthora* diseases [4,5].

Poplar plantations are the most widespread, artificially established broadleaved stands in Serbia. In the alluvial plains along the large rivers in Serbia, natural diverse stands of *Populus alba* L. and *P. nigra* L. are also common and of high ecological and economic importance. Due to the high productivity rate of the different hybrid clones, the area of poplar plantations in Serbia has increased considerably [9], reaching 48,000 ha or 2.1% of the total forest area [10]. In the area of Public Enterprise (PE) "Vojvodinašume", poplar plantations along the Sava, Tisa, and Danube rivers are of particular importance. Wet soil conditions and seasonal floodings create favourable conditions for the spread, infection, and survival of *Phytophthora* species. In riparian poplar plantations along the Sava River and in central Serbia, decline and dieback symptoms indicative of *Phytophthora* root infections have been recorded during the previous decade. Apart from individual reports of *P. cactorum* (Lebert and Cohn) Schröter on white poplar in the Czech Republic and in Hungary [4,11], and of *P.* × *cambivora* (Petri) Buisman on poplar trees in Croatia [12], there are no studies of *Phytophthora* species distribution in poplar stands in Europe. In a preliminary investigation of the fungal and oomycete community associated with poplars in Serbia, *P. cactorum* and *P. plurivora* Jung and Burgess were recorded in the rhizosphere of poplar trees [13].

The present study aimed to (1) determine the occurrence and diversity of *Phytophthora* species in poplar plantations in Serbia; and (2) test the aggressiveness of the isolated *Phytophthora* spp. to two poplar clones widely used in Serbia.

2. Material and Methods

2.1. Studied Sites

Four sites were selected near Srem in the northern province of Vojvodina. The 5–33 year old plantations were located on alluvial forest sites with humogley soils along the Sava River and belonged to Public Enterprise (PE) "Vojvodinašume" (Table 1). In stands 1 and 2, poplars showed increased crown transparency and yellowing of leaves (Figure 1a,b), while stands 3 and 4 were healthy. Two additional sites were located in central Serbia, in the forests of PE "Srbijašume" (stand 5; 25 years old) and in a private property in the village Brzeće, near the Kopaonik mountain (stand 6; 2 years old), respectively (Table 1). The latter two stands showed severe dieback and mortality (Figure 1c,d). Due to their vicinity to local rivers, both stands were growing on wet alluvial soils. All six plantations were established via the planting of nursery stock. In four plantations, poplar clone *P.* × *euramericana* I-214, a hybrid between *P. nigra* and the North American *P. deltoides* Bartram ex. Marshall, was used, while the other two plantations contained *P. deltoides* (Table 1).

Table 1. Isolation of *Phytophthora* species from rhizosphere soil samples in poplar plantations in Serbia.

Stand, Location (River)	Tree No.	*Populus* Species/Clone	Age	Disease Symptoms	Sample	*Phytophthora* Species (No. Isolates)	GenBank Accession Numbers
Stand No. 1, Klenak 44°45′48″ N 19°48′18″ E 86 m asl (Sava River)	1	*Populus deltoides*	10	No symptoms	Rhizosphere soil	*P. plurivora* *P. cactorum*	
	2	*P. deltoides*	10	No symptoms	Wet rhizosphere soil	*P. lacustris*	
	3	*P. deltoides*	10	No symptoms	Wet rhizosphere soil	*P. gonapodyides*	
	4	*P. deltoides*	10	No symptoms	Rhizosphere soil	*P. plurivora* *P. cactorum*	
	5	*P. deltoides*	10	No symptoms	Wet rhizosphere soil	*P. gonapodyides*	
	6	*P. deltoides*	10	No symptoms	Rhizosphere soil	-	

Table 1. *Cont.*

Stand, Location (River)	Tree No.	*Populus* Species/Clone	Age	Disease Symptoms	Sample	*Phytophthora* Species (No. Isolates)	GenBank Accession Numbers
Stand No. 2, Kupinovo, Kupinski Kut 44°40′01″ N 19°59′34″ E 76 m asl (Sava River)	7	*Populus × euramericana* clone I-214	31	No symptoms	Wet rhizosphere soil	*P. gonapodyides* *P. cactorum*	
	8	I-214	31	No symptoms	Rhizosphere soil	*P. plurivora*	
	9	I-214	31	No symptoms	Rhizosphere soil	-	
	10	I-214	31	No symptoms	Rhizosphere soil	*P. pini* (2) *P. plurivora* *P. lacustris*	KF234654 KF234736
	11	I-214)	31	No symptoms	Rhizosphere soil	*P. lacustris* *P. plurivora*	
	12	I-214	31	No symptoms	Rhizosphere soil	*P. pini* *P. plurivora* (2)	KF234655 KF234737
	13	I-214	31	High crown transparency	Rhizosphere soil	*P. plurivora* *P. pini*	
Stand No. 3, Kupinovo, Kupinski Kut 44°39′52″ N 19°59′43″ E 80 m asl (Sava River)	14	I-214	33	No symptoms	Rhizosphere soil	*P. plurivora* *P. pini* (3)	KF234740 KF234656
	15	I-214	33	No symptoms	Wet rhizosphere soil	*P. lacustris*	
	16	I-214	33	No symptoms	Rhizosphere soil	*P. plurivora* *P. cactorum*	JX276094
	17	I-214	33	No symptoms	Rhizosphere soil	*P. pini* *P. pini*	KF234660 KF234657
	18	I-214	33	No symptoms	Rhizosphere soil	*P. plurivora* (2) *P. pini*	KF234658
Stand No. 4, Kupinovo, Jasenska Belilo44°43′02″ N 20°06′20″ E 91 m asl (Sava River)	19	*P. deltoides*	5	No symptoms	Rhizosphere soil	-	
	20	*P. deltoides*	5	Yellowing of leaves	Rhizosphere soil	*P. plurivora* *P. polonica*	KF234729 KF234759
	21	*P. deltoides*	5	Yellowing of leaves	Rhizosphere soil	-	
	22	*P. deltoides*	5	Yellowing of leaves	Rhizosphere soil	*P. polonica* (2)	
	23	*P. deltoides*	5	High crown transparency	Rhizosphere soil	*P. plurivora* *P. polonica*	KF234727
	24	*P. deltoides*	5	Crown transparency	Rhizosphere soil	*P. plurivora* *P. polonica*	KF234760
	25	*P. deltoides*	5	Yellowing of leaves	Rhizosphere soil	*P. plurivora*	KF234728
Stand No. 5, Veliki Jastrebac-Blace43°21′19″ N 21°15′36″ E 492 m asl (stream Popovačka reka)	26	I-214	25	Dieback	Wet rhizosphere soil	*P. gonapodyides* *P. plurivora*	
	27	I-214	25	Crown transparency and dieback	Wet rhizosphere soil	*P. gonapodyides*	
Stand No. 6 Brus, Brzeće 43°18′07″ N 20°53′08″ E 1011 m asl (stream Bela reka)	28	I-214	2	Dieback, root necroses	Rhizosphere soil	*P. gonapodyides*	
No. of positive samples						24	
No. of obtained isolates						46	
No. of sequenced isolates						15	

Figure 1. Disease symptoms in poplar plantations in Serbia: (**a**,**b**) increased crown transparency and yellowing of leaves on five-year old trees of *Populus deltoides* in stand No. 4; (**c**,**d**) severe mortality and dieback of 25-years old trees of *Populus* × *euramericana* clone I-214 in stand No. 5.

2.2. Sampling, Isolation, and Morphological Identification of Phytophthora Spp.

Sampling and isolation methodology were performed according to [14,15]. In the four plantations in the Srem forest area, 25 poplar trees were randomly sampled in May 2011 and May/June 2012, while in the two stands in central Serbia, three trees were sampled in the spring and summer 2017. Three to four soil monoliths were taken from the rhizosphere of each tree, mixed, and ca. 3–4 L of soil per tree was taken to the laboratory. Both symptomatic and asymptomatic trees were sampled. Each soil sample was thoroughly mixed and a subsample of ca. 200 mL used for the isolation test using young leaves of *Quercus robur* L. and *Fagus sylvatica* L., as baits. The baiting test was performed at 20 °C and natural light. After the appearance of the first necrotic spots, baiting leaves were examined for the presence of *Phytophthora* sporangia under the light microscope. Small pieces from the necroses were then plated onto selective PARPNH agar [14–16]. First hyphae from plated leaves were subcultured onto V8-agar (V8A) and carrot juice agar (CA) (800 mL distilled water, 200 mL carrot or vegetable juice (Biotta®, Tägerwilen, Switzerland), 18 g agar (Torlak, Belgrade, Serbia) and 3 g $CaCO_3$; [15]), and stored at 20 °C for further examinations.

For classical species identification, the morphology of obtained isolates was examined at × 400 magnifications using a light microscope (CETI®MAGNUM-T/Trinocular Microscope, Oxon, UK). Structures were measured using a camera (Si3000®, Medline Scientific, Oxon, UK) and the XliCap® (Xl Imaging Ltd., Swansea, UK) imaging software. Sporangia were produced by flooding agar discs of young CA colonies for 24–48 h in non-sterile soil extract according to [14]. Gametangia were studied from four-week-old CA cultures, incubated at 20–22 °C in the dark. Self-sterile isolates were paired with known tester strains of *P. cryptogea* Pethybridge and Lafferty (A1 mating type: BBA 65909; A2 mating type: BBA 63651) to clarify whether they were sterile or heterothallic, and to which mating type heterothallic isolates belong. Colony growth patterns were examined after the growth of isolates for one week at 20 °C in the dark on four different agar media, including CA, V8A, malt-extract-agar

(MEA; 48 g/L malt-extract-agar; Merck KGaA, Darmstadt, Germany), and potato-dextrose-agar (PDA; 39 g/L potato-dextrose-agar; Merck KGaA, Darmstadt, Germany). Morphological features and colony growth patterns were compared with descriptions in the literature [2,17–22].

2.3. Molecular identiflcation of Phytophthora Spp.

The ITS1-5.8S-ITS2 rDNA region of 15 selected isolates representative of all morphotypes was sequenced after direct PCR using Phire™ Plant Direct PCR Kits (Thermo Fisher Scientific Inc., Waltham, MA, USA). Mycelium from three to five days old V8A colonies was scraped without agar using a sterile needle and placed in 2 mL Eppendorf tubes, with 30 μL of previously added Dilution Buffer (Thermo Fisher Scientific Inc., Waltham, MA, USA). PCR reactions were performed with a 20 μL volume, containing 10 μL 1 × Phire Plant PCR Buffer-a, 1 μL 0.5 μM of primers ITS4/ITS6 [23,24], 0.4 μL Phire Hot Start II DNA Polymerase, 0.5 μL of Dilution Buffer with diluted young hypha, and water (mQ) up to 20 μL, according to the manufacturer's recommendation. Reactions were performed in a PTC-200™ machine (MJ Research Inc., Waltham, MA, USA) with three-step PCR protocol, according to the manufacturer's recommendations.

The PCR program was 5 min at 98 °C followed by 40 cycles of 5 s at 98 °C, 5 s at 55 °C, and 50 s at 72 °C. The presence and size of PCR products was confirmed by analyzing 1 μL of product by electrophoresis in 1% TAE-agarose gel, stained with GelRed™ Nulceid Acid Dye (Biotium, Inc., Fremont, CA, USA), with FastRuler MR DNA Ladder (Thermo Fisher Scientific, Waltham, MA, USA) as the molecular mass standard of DNA. For sequencing, 20 μL PCR product was purified with the CleanUp Kit (A&A Biotechnology, Gdynia, Poland), following the manufacturer's protocol, and sequenced with ABI 3730xl DNA Analyzer (Applied Biosystems, Foster City, CA, USA).

Obtained sequences were aligned using the ClustalW algorithm of the BioEdit program subjected to an NCBI BLAST search (http://www.ncbi.nlm.nih.gov/BLAST/). Sequence analysis was done with MEGA6 software [25]. Isolates were assigned to a *Phytophthora* species when sequence identities were above a 99% cut-off in respect to those of extype isolates or key isolates. All ITS sequences obtained in this study were submitted to GenBank (Table 1).

2.4. Soil Infestation Test

One-year old cuttings of *P.* × *euramericana* clones I-214 and Pánnonia, commonly used in Serbian forestry, were rooted and grown for four months in the laboratory at 22–25 °C and natural daylight in 15 l plastic containers containing an autoclaved mixture of peat (Florabella, AGRO-FertiCrop d.o.o., Subotica, Serbia) and perlite (4–6 mm; Agro Perlit Extra, Termika, Zrenjanin, Serbia) with a 3:1 volume ratio. Isolates of all *Phytophthora* species recovered from the rhizosphere of poplar trees in this study, and, for comparisons, each isolate of *P. polonica* Belbahri et al. from *Q. robur*, *P. cryptogea* from *Q. petraea* (Matt.) Liebl., *P.* × *cambivora* from *F. sylvatica*, and *P.* × *serendipita* Man in't Veld et al. from *Pyrus pyraster* (L.) Burgsd. were included (Table 2).

Twelve plants per clone and per treatment and control were used (Table 2). Inocula of 12 isolates from nine different *Phytophthora* species were prepared using fine vermiculite, millet seeds, and V8 juice [14]. The substrate in the controls received a sterile mixture of fine vermiculite, millet seeds, and V8 juice [14]. After the inoculation, plants were immediately flooded for 72 h, and flooding was repeated at three-week intervals.

After each flooding period, the water was removed and sterilized with bleach. During the second and third flooding, leaves of *Prunus laurocerasus* L. and *Q. robur* were floated on the water surface in order to check the ability of the *Phytophthora* strains to produce sporangia and cause infections.

Table 2. Pathogenicity of nine *Phytophthora* species to *Populus* × *euramericana* clones I-214 and Pánnonia in the soil infestation test after 10 weeks.

Poplar Clone	*Phytophthora* Species (Isolates)	No. of Inoculated Plants	No. of Dead Plants	No. of Plants with Bark Necroses	No. of Plants with 100% Root Rot	Dry Weight of Small Roots (g)	Re-Isolation Frequency (%)	
							Necroses	Fine Roots
I-214	Control	12	0	0	0	3.591	0	0
	P. cactorum (JX276094)	12	8	2^{2}	9	2.073	100	100
	P. cryptogea (KF234765)	12	7	1 [a]; 2 [b]	9	2.199	100	100
	P. gonapodyides (2011/Pop.04)	12	9	1 [a]; 1 [b]	10	1.506	83.3	100
	P. lacustris (2011/Pop.Blato.03)	12	9	2 [a]; 1 [b]	10	1.59	66.67	100
	P. pini (KF234655)	12	10	2 [a]	9	1.155	100	100
	P. pini (KF234658)	12	10	3 [a]	10	1.143	100	100
	P. plurivora (KF234737)	12	7	4 [a];1 [b]	10	1.092	100	100
	P. plurivora (KF234740)	12	11	3 [a]	11	1.26	100	83.3
	P. polonica (JX276065)	12	9	2 [a]; 1 [b]	9	1.542	100	100
	P. polonica (KF234760)	12	6	4 [a]; 1 [b]	8	1.773	100	100
	P. × *cambivora* (JX276088)	12	5	4 [a]	5	2.112	100	100
	P. × *serendipita* (KM272262)	12	11	2 [a]	10	2.199	100	100
Pánnonia	Control	12	0	0	0	4.014	0	0
	P. cactorum (JX276094)	12	10	1 [a]; 1 [b]	4	1.245	100	100
	P. cryptogea (KF234765)	12	9	1 [b]	10	1.239	100	100
	P. gonapodyides (2011/Pop.04)	12	12	0	12	1.02	100	100
	P. lacustris (2011/Pop.Blato.03)	12	8	2 [a]; 1 [b]	9	1.209	50	100
	P. pini (KF234655)	12	9	2 [b]	11	1.509	100	100
	P. pini (KF234658)	12	12	0	12	1.077	100	100
	P. plurivora (KF234737)	12	7	3 [a]; 1 [b]	8	1.158	100	100
	P. plurivora (KF234740)	12	7	3 [a]; 2 [b]	9	1.357	100	100
	P. polonica (JX276065)	12	10	2 [a]	10	1.626	100	100
	P. polonica (KF234760)	12	10	1 [b]	10	1.434	100	100
	P. × *cambivora* (JX276088)	12	10	1 [b]	9	1.359	100	91.67
	P. × *serendipita* (KM272262)	12	9	3 [a]	11	1.131	100	100

[a] girdling necroses; [b] longitudinal necroses.

The experiment was performed in the laboratory at 22–25 °C and natural daylight. After ten weeks and three flooding periods, all plants were carefully excavated and the roots washed. Photos were taken and re-isolations were made by plating small pieces from the edges of necrotic lesions and segments of fine roots from each infested and control plant onto PARPNH. Then, all roots were harvested from the cuttings, dried at 65 °C for 72 h, and their dry weight measured using a fine scale (Exacta 300 EB, Tehtnica, Železniki, Slovenia).

2.5. Underbark Inoculation Test

One-year-old cuttings of *P.* × *euramericana* clones I-214 and Pánnonia were grown for one year in an autoclaved mixture of peat (Florabella, AGRO-FertiCrop d.o.o., Subotica, Serbia) and perlite (4–6 mm; Agro Perlit Extra, Termika, Serbia) with a 2:1 volume ratio. After sterilising the bark at a 10–15 cm distance from the collar with 70% ethanol, the cuttings were inoculated under the bark using a sterile 7-mm metal cork borer. Same-sized plugs cut from the edges of three to five -day old CA colonies, were used as inocula. In total, ten isolates of seven different *Phytophthora* species were used (Table 3), including isolates of all *Phytophthora* species recovered from the rhizosphere of poplar trees in this study, and, as comparisons, one isolate each of *P. polonica* from *Q. robur* and *P.* × *cambivora* from *F. sylvatica*. Control plants were inoculated with sterile CA plugs. The agar plugs were covered with the removed piece of bark and sterile moistened cotton, and sealed with Parafilm (Merck KGaA, Darmstadt, Germany). The plants were incubated in the laboratory at 22–25 °C and natural daylight, and checked weekly for the appearance of symptoms. After 11 weeks, the experiment was finished and necroses lengths were recorded after removal of the outer bark. Reisolations were made from all the inoculated and control plants by plating small pieces from the upper and lower margins of necrotic lesions or, in the absence of necroses, from the margins of the inoculation places onto PARPNH. Measuring of necrosis length was performed using a precise ruler, while necrosis width was measured using a flexible measurement tape.

Table 3. Results of the underbark inoculation test with *Populus* × *euramericana* clones I-214 and Pánnonia and seven *Phytophthora* spp. after 11 weeks.

Poplar Clone	*Phytophthora* Species (Isolates)	Number of Inoculated Plants	Stem Diameter ($\bar{x}$ ± SE (mm))	Plant Height ($\bar{x}$ ± SE (mm))	Number of Plants with Lesions (Bleeding)	Number of Plants with Secondary Shoots at Necroses Margins	Number of Plants with Dieback	Reisolation Frequency (%)
I-214	Control	12	10.6 ± 0.29	103.6 ± 3.16	0	0	0	0
	P. cactorum (JX276094)	12	9.3 ± 0.26	95.7 ± 3.47	12 (9)	2	0	100
	P. gonapodyides (2011/Pop.04)	12	9.7 ± 0.21	93.7 ± 2.74	12 (2)	0	0	83.3
	P. lacustris (2011/Pop.Blato.03)	12	9.1 ± 0.36	90.1 ± 3.72	12 (3)	0	0	83.3
	P. pini (KF234655)	12	11.2 ± 0.49	103.8 ± 3.58	12 (8)	5	5	100
	P. pini (KF234658)	12	10.6 ± 0.31	100.2 ± 2.16	12 (11)	10	1	100
	P. plurivora (KF234737)	12	9.3 ± 0.34	89 ± 3.02	12 (6)	0	1	100
	P. plurivora (KF234740)	12	10.1 ± 0.3	100.5 ± 3.49	12 (9)	4	1	100
	P. polonica (JX276065)	12	9.3 ± 0.22	94.1 ± 2.57	7 (0)	0	0	91.67
	P. polonica (KF234760)	12	10.1 ± 0.37	95.7 ± 2.15	9 (0)	0	0	100
	P. × *cambivora* (JX276088)	12	9.8 ± 0.51	100.6 ± 3.87	10 (0)	1	0	58.3

Table 3. *Cont.*

Poplar Clone	*Phytophthora* Species (Isolates)	Number of Inoculated Plants	Stem Diameter (x⁻ ± SE (mm))	Plant Height (x⁻ ± SE (mm))	Number of Plants with Lesions (Bleeding)	Number of Plants with Secondary Shoots at Necroses Margins	Number of Plants with Dieback	Reisolation Frequency (%)
Pánnonia	Control	12	9.4 ± 0.56	110.7 ± 5.74	0	0	0	0
	P. cactorum (JX276094)	12	10.9 ± 0.64	111.2 ± 6.11	12 (2)	3	2	100
	P. gonapodyides (2011/Pop.04)	12	9.9 ± 0.4	118 ± 5.48	12 (2)	0	7	91.67
	P. lacustris (2011/Pop.Blato.03)	12	7.9 ± 0.28	74.8 ± 2.27	12 (3)	5	0	66.67
	P. pini (KF234655)	12	9.9 ± 0.42	103.7 ± 4.49	12 (6)	0	1	100
	P. pini (KF234658)	12	9.8 ± 0.55	113 ± 7.39	12 (4)	2	1	100
	P. plurivora (KF234737)	12	8.4 ± 0.29	98.6 ± 6.06	12 (7)	0	0	100
	P. plurivora (KF234740)	12	9.9 ± 0.56	114.1 ± 6.01	12 (5)	0	1	100
	P. polonica (JX276065)	12	9.2 ± 0.25	103.6 ± 3.57	11 (4)	2	0	100
	P. polonica (KF234760)	12	10.3 ± 0.54	104.4 ± 6.45	10 (0)	0	0	91.67
	P. × *cambivora* (JX276088)	12	10.5 ± 0.53	104.2 ± 7.28	8 (0)	2	0	83.3

2.6. Statistical Analyses

Based on the length and width, the surface of each necrosis was calculated using the mathematical formula for elliptic surfaces. For each treatment, means (x⁻) and standard errors (±SE) of necrosis length, width, and surface were calculated. Analysis of variance was performed using a Generalized Linear Model (GLM), with a significance level of $\alpha = 0.05$. Testing of significance of differences in mean necrosis length, width, and surface between different treatments was performed using Tukey's HSD post hoc test ($\alpha = 0.05$). Statistical procedures were performed with the RStudio software version 1.1.383 (Integrated Development for R. RStudio, Inc., Boston, MA, USA).

3. Results

3.1. Phytophthora Species in Poplar Plantations

In total, 45 isolates of six different *Phytophthora* species were isolated from 24 of 28 soil samples (86%) (Table 1). The most common was *P. plurivora* (15 samples = 53.6%), followed by *Phytophthora pini* Leonian (7 samples = 25%), *Phytophthora gonapodyides* (H.E. Petersen) Buisman (6 samples = 21.4%), and *P. polonica*, *Phytophthora lacustris* Nechwatal et al. and *P. cactorum* (each 4 samples = 14.3%) (Table 1). In the mainly healthy stands 1 and 3, all samples were taken underneath non-symptomatic trees. Additionally, in stand 2, five of the six samples originated from non-symptomatic trees. In contrast, six of the seven samples in stand 4 were taken from trees with high crown transparency and yellowing of leaves. In the rhizosphere of each four declining trees, *P. plurivora* and *P. polonica* were found (Table 1). The three samples taken in stands 5 and 6 originated from trees showing severe dieback. *Phytophthora gonapodyides* was isolated from all three samples, while *P. plurivora* was present in one sample (Table 1). Co-occurrence of two different *Phytophthora* species was found in 11 of the 25 samples, while in one sample, three *Phytophthora* spp. co-occurred (Table 1).

The morphology of all isolates of *P. cactorum*, *P. pini*, *P. plurivora*, and *P. polonica* conformed with the original descriptions [2,18–21]. In accordance with [22], *P. lacustris* and *P. gonapodyides* did not form gametangia in pure cultures or in the mating tests. This is the first report of *P. lacustris*, *P. gonapodyides*, *P. pini*, and *P. polonica* on poplars in Serbia.

3.2. Soil Infestation Test

Five weeks after the inoculation (pi), the first symptoms like slight yellowing and atrophy of leaves were noticed in clone I-214 in the substrate infested by *P. plurivora*. Eight weeks pi, the first dieback of plants occurred in both clones in the substrate infested by *P. plurivora*, *P. pini*, and *P. gonapodyides*. Ten weeks pi, dieback also started in clone I-214 infested by *P. cactorum* and in clone Pánnonia in the substrate infested by the *P. polonica* strain originally isolated from oak. These diebacks were followed by premature shedding of leaves and total plant collapse. In addition, in the treatment Pánnonia/*P. plurivora* (KF234737), collar necroses were observed on two plants (Figure 2o). Ten weeks pi and after three flooding periods, numerous plants started to decline in most of the treatments (Table 2), and the experiment was assessed.

Figure 2. Representative cuttings of poplar clone I-214 after ten weeks in soil infested with nine *Phytophthora* spp.: (**a,b**) *P. cactorum*; (**c,d**) *P. cryptogea*; (**e**) *P. gonapodyides*; (**f,g**) *P. lacustris*; (**h,i**) *P. pini* KF234655; (**j,k**) *P. pini* KF234658; (**l,m**) *P. plurivora* KF234737; (**n,o**) *P. plurivora* KF234740; (**p,q**) *P. polonica* JX276065; (**r,s**) *P. polonica* KF234760; (**t,u**) *P.* × *serendipita*; (**v,w**) *P.* × *cambivora*; (**x,z**) control.

The most aggressive species to poplar clone I-214 were *P. plurivora* (KF234740) and *P.* × *serendipita*, each causing the decline of 11 poplar plants (91.67%), followed by both strains of *P. pini* (10 declining plants = 83.3%) (Table 2). In several treatments, necrotic bark lesions were recorded (Figure 2). Most *Phytophthora* species caused root rot and the loss of fine roots (Figure 2). The reduction of total root dry weight compared to the control was most severe for the two isolates of *P. plurivora* (30.4% and 35.1%, respectively) and the two isolates of *P. pini* (31.8% and 32.2%, respectively). Both pathogens caused 100% root rot in 75–91.2% of plants (Table 2). In the case of clone Pánnonia, the most aggressive species were *P. gonapodyides* and one isolate of *P. pini* (KF234658), which both caused 100% mortality, followed by *P. cactorum*, *P. × cambivora* and *P. polonica* each killing 10 plants (83.3%) (Table 2). Root rot and loss of fine roots were recorded in all treatments (Figure 2). The highest reduction of total root dry weight was caused by *P. gonapodyides* (25.4% compared to the control) and one isolate of *P. pini* (26.8%) (Table 2). These two species also killed all plants (Table 2). Additionally, all other *Phytophthora* species/isolates caused severe reductions of total root dry weight (28.2–40.5% compared to the control) and 100% root rot in 33.3–91.7% of plants (Table 2). In addition, several *Phytophthora* species caused girdling and longitudinal bark lesions (Figure 3).

Figure 3. Representative cuttings of poplar clone Pánnonia after ten weeks in soil infested with nine *Phytophthora* spp.: (**a**,**b**) *P. cactorum*; (**c**,**d**) *P. cryptogea*; (**e**,**f**) *P. gonapodyides*; (**g**,**h**) *P. lacustris*; (**i**,**j**) *P. pini* KF234655; (**k**,**l**) *P. pini* KF234658; (**m**,**o**) *P. plurivora* KF234737; (**p**,**r**) *P. plurivora* KF234740; (**s**) *P. polonica* JX276065; (**t**) *P. polonica* KF234760; (**u**,**v**) *P. × cambivora*; (**w**) *P. × serendipita*; (**x**,**y**) control.

3.3. Underbark Inoculation Test

Recorded symptoms and numbers of plants with dieback are shown in Table 3 and Figures 4 and 5. Four weeks after inoculation (pi), the first discrete bark necroses were observed, soon followed by the oozing of exudates. In clone I-214, these symptoms appeared first in the treatments with both *P. pini* isolates and one isolate of *P. plurivora* (KF234737). In clone Pánnonia, the first bleeding lesions were caused by *P. cactorum*. Six weeks pi, intensive bleeding lesions on both clones were recorded in the treatments with *P. pini*, *P. plurivora*, and *P. cactorum*. In addition, in clone I-214, small necroses were caused by *P. lacustris*, while in clone Pánnonia, bleeding cankers started to appear on individual plants infected by *P. lacustris*, *P. gonapodyides*, and the *P. polonica* isolate from oak (JX276065). Eight weeks pi, symptoms had progressed and numerous secondary shoots started to develop at the margins of necrotic zones (Figure 5b,c). In addition, the first plants with dieback were observed in the I-214/*P. pini* (KF234655) and the Pánnonia/*P. gonapodyides* treatments. Eleven weeks pi, more plants of both clones showed dieback and other symptoms, and the experiment was finished. Reisolations were successful in all treatments except of the controls, although with slightly lower reisolation rates in some cases (Table 3).

Figure 4. Representative symptoms caused by seven *Phytophthora* spp. on cuttings of poplar clone I-214 in the underbark inoculation test: (**a**,**d**) bleeding bark lesions after eight weeks: (**a**,**b**) *P. pini*; (**c**) *P. cactorum*, with beginning formation of secondary shoot; (**d**) *P. plurivora*, with beginning formation of secondary shoot; (**e**) cuttings inoculated with *P. pini* (KF234655) after 11 weeks; (**f**,**r**) necrotic lesions of the inner bark after 11 weeks: (**f**,**g**) *P. cactorum*; (**h**) *P.* × *cambivora*; (**i**) *P. gonapodyides*; (**j**) *P. lacustris*; (**k**) *P. polonica* JX276065; (**l**) *P. polonica* KF234760; (**m**) *P. pini* KF234658; (**n**) *P. pini* KF234655; (**o**,**p**) *P. plurivora* KF234740; (**q**) *P. plurivora* KF234737; (**r**) control.

Figure 5. Representative symptoms caused by seven *Phytophthora* spp. on cuttings of poplar clone Pánnonia in the underbark inoculation test: (**a**,**b**) bleeding bark lesions after eight weeks: (**a**) *P. cactorum*; (**b**) *P. lacustris*, with formation of secondary shoot. (**c**,**d**) necrotic lesions of outer bark after 11 weeks: (**c**) *P. plurivora*, with beginning formation of secondary shoot; (**d**) *P. gonapodyides*. (**e**) declining and healthy plants inoculated with *P. pini* after eight weeks. (**f**–**p**) necrotic lesions of the inner bark after 11 weeks: (**f**) *P. cactorum*; (**g**) *P.* × *cambivora*; (**h**) *P. lacustris*; (**i**) *P. gonapodyides*; (**j**) *P. polonica* JX276065; (**k**) *P. polonica* KF234760; (**l**) *P. pini* KF234658; (**m**) *P. pini* KF234655; (**n**) *P. plurivora* KF234737; (**o**) *P. plurivora* KF234740; (**p**) control.

The statistical analyses showed that in clone I214, most *Phytophthora* species significantly influenced most of the tested parameters. Exceptions were found for *P. gonapodyides* and *P. polonica* and lesion lengths; *P. gonapodyides*, *P. lacustris*, *P. polonica* and *P.* × *cambivora* and lesion widths; and *P. gonapodyides*, *P. lacustris*, and *P. polonica* and the surface areas of the lesions (Table 4). In the case of clone Pánnonia, only in plants infected with *P. polonica* (JX276065) and *P.* × *cambivora* did the lesion lengths differ significantly from the control. In contrast, plants infected with *P. gonapodyides*, *P. pini* (KF234655), and *P. polonica* (KF234760) showed significantly higher lesion widths compared to the control plants (Table 4). Lesions caused by these three *Phytophthora* spp. and *P. cactorum* also showed significantly higher surface areas compared to the control (Table 4). Results of Tukey's test are shown in Table 5.

Table 4. Estimates, *t* values, *p* values, and Residual deviances from GLM for tested parameters in the underbark inoculation trial with poplars. In clone I-214, degrees of freedom (df) for Residual deviance was 114; in clone Pánnonia df for Residual, deviance was 109. Significant effects are marked in bold type.

Poplar Clone Host	Treatment	Length			Width			Surface		
		Estimate	*t* Value	(*p*)	Estimate	*t* Value	(*p*)	Estimate	*t* Value	(*p*)
I214	Control	2.32	19.44	**0.000**	1.95	33.29	**0.000**	4.03	21.78	**0.000**
	P. cactorum	1.66	10.04	**0.000**	0.63	7.70	**0.000**	2.35	9.17	**0.000**
	P. gonapodyides	0.11	0.67	**0.506**	-0.07	-0.84	0.405	0.05	0.18	0.855
	P. lacustris	0.45	2.71	**0.008**	0.01	0.14	0.890	0.49	1.92	0.057
	P. pini (KF234655)	1.72	9.96	**0.000**	0.67	7.91	**0.000**	2.47	9.21	**0.000**
	P. pini (KF234658)	1.46	8.65	**0.000**	0.56	6.69	**0.000**	2.05	7.82	**0.000**
	P. plurivora (KF234737)	0.89	5.24	**0.000**	0.32	3.89	**0.000**	1.25	4.78	**0.000**
	P. plurivora (KF234740)	1.45	8.59	**0.000**	0.57	6.98	**0.000**	2.0	7.62	**0.000**
	P. polonica (JX276065)	-0.03	-0.18	0.861	0.01	0.07	0.947	-0.02	-0.09	0.933
	P. polonica (KF234760)	-0.02	-0.12	0.909	-0.08	-0.99	0.323	-0.1	-0.38	0.703
	P. × *cambivora*	0.50	2.96	**<0.010**	-0.05	-0.56	0.579	0.66	2.53	**0.013**
	Residual deviance		12.4			4.05			27.33	
Pánnonia	Control	2.29	14.83	**0.000**	1.99	22.85	**0.000**	4.05	15.02	**0.000**
	P. cactorum	1.07	4.76	**0.000**	0.19	1.48	0.141	1.32	3.38	**<0.010**
	P. gonapodyides	1.11	3.89	**0.000**	0.56	3.45	**<0.010**	2.16	4.33	**0.000**
	P. lacustris	0.57	2.62	**<0.010**	0.02	0.16	0.870	0.64	1.68	0.096
	P. pini (KF234655)	1.03	4.61	**0.000**	0.48	3.81	**0.000**	1.72	4.41	**0.000**
	P. pini (KF234658)	0.53	2.36	**0.020**	0.14	1.11	0.270	0.68	1.73	0.086
	P. plurivora (KF234737)	0.55	2.51	**0.014**	0.16	1.26	0.212	0.72	1.88	0.063
	P. plurivora (KF234740)	0.64	2.85	**<0.010**	0.16	-0.43	0.210	0.79	-0.02	**0.045**
	P. polonica (JX276065)	0.05	0.22	0.826	-0.05	2.66	0.670	-0.01	2.95	0.982
	P. polonica (KF234760)	0.50	2.26	**0.026**	0.33	1.18	**<0.010**	1.12	1.61	**<0.010**
	P. × *cambivora*	0.26	1.20	0.232	0.15	1.26	0.241	0.61	2.02	0.111
	Residual deviance		20.89			7.36			53.52	

The two isolates of *P. pini* were most aggressive to clone I-214, causing bleeding lesions in 66.7 and 91.3% of plants and dieback in 8.3% and 41.7% of plants, respectively (Table 3). Also, *P. cactorum* (bleeding lesions in 75% of plants) and the two isolates of *P. plurivora* (bleeding lesions in 50% and 75% of plants, respectively) showed considerable aggressiveness to clone I-214 (Table 3). These three *Phytophthora* species also caused the largest bark lesions with mean necrosis areas that were statistically significantly different from all other treatments and the control, ranging between 415.5 ± 60.2 and 668.1 ± 128.1 mm^2 (Table 5). *Phytophthora gonapodyides* and both isolates of *P. polonica* were non-pathogenic to the bark of clone I-214, while *P.* × *cambivora* was mildly pathogenic (Tables 3 and 5).

Table 5. Mean values, standard errors, and results of Tukey's post hoc tests for the bark necroses caused by seven *Phytophthora* spp. in the underbark inoculation test on poplar clones I-214 and Pánnonia after 11 weeks. Different letters behind values indicate significant differences ($\alpha = 0.05$).

Phytophthora Species (Isolates)	Poplar Clone I-214 (Mean ± SE+Tukey's Test)			Poplar Clone Pánnonia (Mean ± SE+Tukey's Test)		
	Necrosis Length (mm)	Necrosis Width (mm)	Necrosis Area (mm²)	Necrosis Length (mm)	Necrosis Width (mm)	Necrosis Area (mm²)
Control	10.2 ± 0.18f	7 ± 0.18F	56.5 ± 1.87f	9.9 ± 0.38d	7.3 ± 0.19c	57.5 ± 3.30d
P. cactorum	53.7 ± 5.61abcd	13.2 ± 1.13abcd	592.9 ± 104.52abcd	28.8 ± 5.14abc	8.9 ± 0.42abc	214.9 ± 47.02abcd
P. gonapodyides	11.4 ± 0.61f	6.6 ± 0.14f	59.2 ± 3.88f	30 ± 17.43 [a]abc	12.8 ± 3.61 [a]abc	496.4 ± 404.56 [a]abcd
P. lacustris	16 ± 1.69ef	7.1 ± 0.29f	92.4 ± 13.19ef	17.6 ± 1.54abcd	7.5 ± 0.46c	109.3 ± 16.97abcd
P. pini (KF234655)	57.2 ± 6.39abcd	13.8 ± 1.12abcd	668.8 ± 128.10abcd	27.8 ± 6.24abc	11.9 ± 1.50abc	321.5 ± 118.21abcd
P. pini (KF234658)	44 ± 3.69abcd	12.3 ± 0.52abcde	437.1 ± 59.78abcd	16.8 ± 0.96abcd	8.4 ± 0.54cbc	113.2 ± 11.83abcd
P. plurivora (KF234737)	24.7 ± 2.93ef	9.7 ± 0.46cde	197.2 ± 33.26cdef	17.2 ± 0.70abd	8.6 ± 0.40cbc	117.7 ± 10.72abcd
P. plurivora (KF234740)	43.5 ± 4.76abcd	11.6 ± 0.72abcde	415.5 ± 60.22abcde	18.8 ± 1.27abcd	8.6 ± 0.38cbc	126.8 ± 6.29abcd
P. polonica (JX276065)	9.9 ± 0.25f	7.1 ± 0.22f	55.3 ± 2.57f	10.4 ± 0.41d	7 ± 0.15c	57 ± 2.67d
P. polonica (KF234760)	10 ± 0.24f	6.5 ± 0.20f	51.2 ± 2.22f	16.3 ± 4.35abcd	10.2 ± 1.31cbc	177 ± 93.63abcd
P. × *cambivora*	16.8 ± 4.99ef	6.7 ± 0.59f	109.4 ± 51.31ef	12.9 ± 2.14ad	8.5 ± 1.20cbc	106.2 ± 39.73abcd

[a] Due to the mortality of seven plants only five plants had clearly visible necroses margins which were measured.

In contrast, on clone Pánnonia, *P. gonapodyides* was by far the most aggressive species, causing the dieback of 58.3% of plants and the largest bark lesions with a mean necrosis area of 496.37 ± 404.56 mm^2, followed by *P. cactorum* with dieback in 16.7% of plants and *P. pini* and *P. plurivora* each causing dieback in 8.3% of plants (Table 3). All *Phytophthora* species and isolates, except of the *P. polonica* isolate from oak (JX276065), caused bleeding bark lesions with sizes significantly different from the control (Figure 5; Table 5).

4. Discussion

Prior to this study, very little was known about the occurrence and role of *Phytophthora* species in poplar stands. *Phytophthora cactorum* had been isolated from bleeding cankers of *Populus alba* in the Czech Republic [11] and from the rhizosphere of young *P. alba* trees in Hungary [4], while *P.* × *cambivora* was recovered from poplar trees in Croatia [12]. A preliminary study demonstrated the presence of *P. cactorum* and *P. plurivora* in the rhizosphere of poplar trees in Serbia [13].

In the present work, a community of six *Phytophthora* species, *P. cactorum*, *P. gonapodyides*, *P. lacustris*, *P. pini*, *P. plurivora* and *P. polonica*, was found in the rhizosphere of 85% of sampled trees in six riparian poplar plantations in Serbia. While *P. gonapodyides* and *P. lacustris* might be native to Europe, the other four *Phytophthora* species are considered as introduced invasive pathogens in Europe [3–5,20]. The known host-*Phytophthora* associations of *P. cactorum* and *P. plurivora* and *Populus* spp. [4,11–13] were confirmed by this study. In both the soil infestation and the underbark inoculation test, *P. cactorum* and *P. plurivora* demonstrated high aggressiveness to the roots and bark of poplar clones I-214 and Pánnonia. In addition, this study also extended the knowledge on the distribution and host ranges of several *Phytophthora* species with *P. gonapodyides*, *P. lacustris*, *P. pini*, and *P. polonica* being the first records from poplar trees, and the latter two *Phytophthora* species also being the first records from Serbia and the Balkan region in general.

Phytophthora pini, like *P. plurivora*, belongs to the '*P. citricola* complex' within *Phytophthora* phylogenetic Clade 2 [21,25]. It was previously recorded on at least seven different plant species in North America and Europe, mainly ornamentals, but also *Pinus resinosa* Ait. and mature *F. sylvatica* trees [4,5,21,26–29]. However, its host range is most likely considerably wider since many plant diseases assigned in the past to *P. citricola s.l.* Sawada were most likely caused by *P. plurivora* and *P. pini* [20,21]. *Phytophthora pini* also occurs in water courses and irrigation reservoirs in North America [21]. The findings of this species in riparian poplar stands along the Sava River in Serbia are in accordance with its preference of wet sites. Like *P. plurivora*, in both the soil infestation and the underbark inoculation test, *P. pini* was highly aggressive to both tested poplar cultivars, and in particular to clone I-214. This is of high concern since I-214 is one of the most common clones used in poplar plantations in Serbia. Due to their high aggressiveness to poplars and other tree species, their wide host ranges, and homothallic production of oospores, acting as enduring survival structures [3,5,20,21], the presence of *P. pini* and *P. plurivora* most likely poses a serious risk to planted and natural stands of poplars and other tree species in Serbia.

Phytophthora polonica from Clade 9 [30] is also homothallic and was originally described from the rhizosphere of declining *Alnus glutinosa* (L.) Gaertn. trees in Poland [19]. Recently, this species was isolated from soils of declining *Juglans nigra* L. and *Q. robur* stands in Hungary and Poland [31,32], respectively. Furthermore, in Serbia, *P. polonica* was previously isolated in 2011 from declining *Q. robur* trees [33]. In previous pathogenicity tests, *P. polonica* proved only weakly pathogenic to *Alnus* shoots and was non-pathogenic to shoots of *Fraxinus angustifolia* Vahl and three *Quercus* spp. [19]. In the present study, *P. polonica* was isolated from the rhizosphere of symptomatic poplar trees and caused significant root rot, extensive loss of fine roots, and dieback on poplar clones I-214 and Pánnonia in the soil infestation test. In contrast, this pathogen was almost non-pathogenic to poplar bark in the underbark inoculation test. These results suggest the involvement of *P. polonica* in the complex of poplar dieback as a serious fine root pathogen which is not progressing into the suberised roots. In previous studies, a similar aetiology was demonstrated for *P. quercina* Jung and European oak decline [14,16,34–36].

Interestingly, in both pathogenicity tests, *P. gonapodyides* from Clade 6 [30] was the most aggressive species to clone Pánnonia, causing extensive root rot, fine root loss, bark lesions, and dieback. *Phytphthora gonapodyides* is considered as an opportunistic pathogen with a mainly aquatic lifestyle [37,38]. However, *P. gonapodyides* is also involved in the declines of oak and beech stands on mesic sites in Germany and the decline of *Quercus ilex* L. in xeric conditions in Spain [3,14–16,39]. Due to its ubiquitous presence in watercourses across Europe and its high aggressiveness to clones Pánnonia and I-214, this pathogen poses a significant threat to riparian poplar stands. *Phytophthora lacustris* is another Clade 6 species [30] commonly occurring in waterways and in riparian stands [22]. In the soil infestation test, *P. lacustris* caused extensive fine root damage, girdling and longitudinal bark lesions developing from the infected root system into the stem, and dieback on both poplar clones. In contrast, in the underbark inoculation trial, this species was only slightly pathogenic to both clones, indicating flooding as an indispensable requisite for successful infections in this aquatic *Phytophthora* species. As with *P. gonapodyides*, the ubiquitous presence of *P. lacustris* in European waterways might pose a risk to the health of riparian poplar stands in Serbia and elsewhere. *Phytophthora cactorum*, *P. lacustris* (referred to as *P.* taxon salixsoil Brasier et al.), and *P. plurivora* also proved to be pathogenic to the roots and bark of *Fraxinus excelsior*, another common species in riparian forest stands [40].

It is possible that the six isolated *Phytophthora* species were already present at the sites before the six poplar plantations sampled in this study were established. River water appears as the most likely natural pathway of introduction since all six plantations are located in the flood plains of the Sava River, experiencing in most years strong floodings [41], or smaller rivers in central Serbia, respectively. On a global scale, river water is ubiquitously infested with a wide array of *Phytophthora* species [37,42–49]. The importance of water as a source of *Phytophthora* inoculum was convincingly demonstrated by several studies [50–56]. Severe and long-during floodings such as those affecting wide regions of Serbia in the years 2014 and 2016 are, hence, of particular concern as they enable the spread of a diverse range of harmful *Phytophthora* species from large catchments to forests and plantations situated in the floodplains downstream. However, against the background of almost ubiquitous infestations of nursery stands in Europe with a total of more than 50 *Phytophthora* species [4,7,57], including

P. cactorum, *P. gonapodyides*, *P. lacustris*, *P. pini*, *P. plurivora*, and *P. polonica*, a possible introduction of these *Phytophthora* species to the Serbian poplar plantations with infested nursery stock cannot be ruled out. Unfortunately, it was not possible to identify and sample the nurseries from where the plants, used for the establishment of the *Phytophthora*-infested poplar plantations in Serbia, originated.

In conclusion, the presence of the six *Phytophthora* species in the sampled poplar stands poses a serious threat to poplars, in particular to plants in young plantations, and potentially also to other riparian tree species. Continued monitoring of the presence and diversity of *Phytophthora* species in riparian poplar stands and other hygrophilic stands and in water courses in Serbia is urgently required in order to assess the magnitude of the problem and develop appropriate management concepts for the disease. Since the riparian poplar plantations are seasonally flooded with *Phytophthora*-infested river water, the repeated introduction and spread of *Phytophthora* spp. into and within the poplar plantations cannot be prevented. Direct control of *Phytophthora* spp. with fungizides like metalaxyl is impossible since their application is not permitted in riparian stands. Furthermore, the use of potassium phosphite, which is globally the most efficient control measure for forest *Phytophthora* diseases [5,58], is no longer possible due to its recent registration as a fungizide in Europe. Therefore, the most promising management measure will be the use of less susceptible poplar clones or other less susceptible riparian tree species, which requires extensive host range testing with the six *Phytophthora* species and other poplar clones and riparian tree species.

5. Conclusions

(1) A community of six different *Phytophthora* species, *P. cactorum*, *P. gonapodyides*, *P. lacustris*, *P. pini*, *P, plurivora*, and *P. polonica*, was detected in each of the three symptomatic and healthy, riparian poplar plantations in Serbia.

(2) In both a soil infestation test and an underbark inoculation test, all six *Phytophthora* species proved their pathogenicity to four-month and one-year-old cuttings of poplar clones I-214 and Pánnonia, respectively.

(3) The results suggest the involvement of soilborne *Phytophthora* species as fine root and bark pathogens in the decline of poplar plantations. The presence of these *Phytophthora* species in riparian poplar plantations might also pose a serious risk to other riparian forest communities, in particular the natural stands of *Quercus robur* and *Fraxinus angustifolia*.

Author Contributions: Conceptualization, I.M. and T.J.; Funding acquisition, T.O.; Methodology, I.M., N.K., D.K., Z.R., J.A.N., K.S., T.C. and T.J.; Writing-original draft, I.M., N.K., and T.J.

Acknowledgments: We are grateful to Public Enterprise "Vojvodinašume" for material support during this work. The COST Actions FP0801 and TD1209, as well as Forest Research Institute-IBL Scholarship Grant, are acknowledged for the Short Term Scientific Missions of Ivan Milenković. Malgorzata Borys (IBL, Warsaw, Poland) and Rajka Domuzin (Institute of Forestry, Belgrade, Serbia) are appreciated for their excellent support during the laboratory works. Sanja Ćirić (Jeleč Dekor, Jagodina, Serbia) is acknowledged for providing the poplar cuttings for the pathogenicity tests, and Sabine Werres (JKI, Braunschweig, Germany) for providing the tester strains of *P. cryptogea*.

References

1. Beakes, G.W.; Thines, M.; Honda, D. Straminipile "Fungi" – Taxonomy. In *Encyclopedia of Life Sciences (eLS)*; John Wiley and Sons, Ltd.: Chichester, UK, 2015; pp. 1–9.
2. Erwin, D.C.; Ribeiro, O.K. *Phytophthora Diseases Worldwide*; American Phytopathological Society (APS) Press: St. Paul, MN, USA, 1996; p. 592. ISBN 0-89054-212-0.

3. Jung, T.; Vettraino, A.M.; Cech, T.; Vannini, A. The impact of invasive *Phytophthora* species on European forests. In *Phytophthora: A Global Perspective*; Lamour, K., Ed.; Plant Protection Series 2; CABI: Wallingford, UK, 2013; pp. 146–158. ISBN 978-1-78064-093-8.
4. Jung, T.; Orlikowski, L.; Henricot, B.; Abad-Campos, P.; Aday, A.G.; Aguín Casal, O.; Bakonyi, J.; Cacciola, S.O.; Cech, T.; Chavarriaga, D.; et al. Widespread *Phytophthora* infestations in European nurseries put forest, semi-natural and horticultural ecosystems at high risk of *Phytophthora* diseases. *For. Pathol.* **2016**, *46*, 134–163. [CrossRef]
5. Jung, T.; Pérez–Sierra, A.; Durán, A.; Horta Jung, M.; Balci, Y.; Scanu, B. Canker and decline diseases caused by soil- and airborne *Phytophthora* species in forests and woodlands. *Persoonia* **2018**, *40*, 182–220. [CrossRef]
6. Brasier, C.M. The biosecurity threat to the UK and global environment from international plant trade. *Plant Pathol.* **2008**, *57*, 792–808. [CrossRef]
7. Pérez-Sierra, A.; Jung, T. Phytophthora in woody ornamental nurseries. In *Phytophthora: A Global Perspective*; Lamour, K., Ed.; Plant Protection Series 2; CABI: Wallingford, UK, 2013; pp. 166–177. ISBN 978-1-78064-093-8.
8. Scott, P.M.; Burgess, T.I.; Hardy, G.E.St.J. Globalization and *Phytophthora*. In *Phytophthora: A Global Perspective*; Lamour, K., Ed.; Plant Protection Series 2; CABI: Wallingford, UK, 2013; pp. 226–232. ISBN 978-1-78064-093-8.
9. Keča, L.; Keča, N.; Pantić, D. Net present value and internal rate of return as indicators for assessment of cost-efficiency of poplar plantations: A Serbian case study. *Int. For. Rev.* **2012**, *14*, 145–156. [CrossRef]
10. Banković, S.; Medarević, M.; Pantić, D.; Petrović, N.; Šljukić, B.; Obradović, S. The growing stock of the Republic of Serbia—State and problems. *Bull. Fac. For.* **2009**, *100*, 7–29, (In Serbian with English summary). [CrossRef]
11. Cerny, K.; Strnadova, V.; Gregorova, B.; Holub, V.; Tomsovsky, M.; Mrazkova, M.; Gabrielova, S. *Phytophthora cactorum* causing bleeding canker of common beech, horse chestnut, and white poplar in the Czech Republic. *Plant Pathol.* **2009**, *58*, 394. [CrossRef]
12. Pernek, M.; Županić, M.; Diminić, D.; Cech, T. Vrste roda *Phytophthora* na bukvi i topolama u Hrvatskoj (*Phytophthora* species on beech and poplars in Croatia). *Šumar. List* **2011**, *135*, 130–137. (In Croatian)
13. Keča, N.; Milenković, I.; Keča, L. Mycological complex of poplars in Serbia. *J. For. Sci.* **2015**, *61*, 169–174. [CrossRef]
14. Jung, T.; Blaschke, H.; Neumann, P. Isolation, identification and pathogenicity of *Phytophthora* species from declining oak stands. *Eur. J. For. Path.* **1996**, *26*, 253–272. [CrossRef]
15. Jung, T. Beech decline in Central Europe driven by the interaction between *Phytophthora* infections and climatic extremes. *For. Pathol.* **2009**, *39*, 73–94. [CrossRef]
16. Jung, T.; Blaschke, H.; Oßwald, W. Involvement of soilborne *Phytophthora* species in Central European oak decline and the effect of site factors on the disease. *Plant Pathol.* **2000**, *49*, 706–718. [CrossRef]
17. Stamps, D.J.; Waterhouse, G.M.; Newhook, F.J.; Hall, G.S. *Revised Tabular Key to the Species of Phytophthora*; Mycological Papers 162; CAB International Mycological Institute: Kew, UK, 1990; pp. 1–28. ISBN 0851986838.
18. Jung, T.; Cooke, D.E.L; Blaschke, H.; Duncan, J.M.; Oßwald, W. *Phytophthora quercina* sp. nov., causing root rot of European oaks. *Mycol. Res.* **1999**, *103*, 785–798. [CrossRef]
19. Belbahri, L.; Moralejo, E.; Calmin, G.; Oszako, T.; Garcia, J.A.; Descals, E.; Lefort, F. *Phytophthora polonica*, a new species isolated from declining *Alnus glutinosa* stands in Poland. *FEMS Microbiol. Lett.* **2006**, *261*, 165–174. [CrossRef] [PubMed]
20. Jung, T.; Burgess, T.I. Re-evaluation of *Phytophthora citricola* isolates from multiple woody hosts in Europe and North America reveals a new species, *Phytophthora plurivora* sp. nov. *Persoonia* **2009**, *22*, 95–110. [CrossRef] [PubMed]
21. Hong, C.X.; Gallegly, M.E.; Richardson, P.A.; Kong, P. *Phytophthora pini* Leonian resurrected to distinct species status. *Mycologia* **2011**, *103*, 351–360. [CrossRef] [PubMed]
22. Nechwatal, J.; Bakonyi, J.; Cacciola, S.O.; Cooke, D.E.L.; Jung, T.; Nagy, Z.Á.; Vannini, A.; Vettraino, A.M.; Brasier, C.M. The morphology, behaviour and molecular phylogeny of *Phytophthora* taxon Salixsoil and its redesignation as *Phytophthora lacustris* sp. nov. *Plant Pathol.* **2013**, *62*, 355–369. [CrossRef]
23. White, T.J.; Bruns, T.; Lee, S.; Taylor, J. Amplification and direct sequencing of fungal ribosomal RNA genes for phylogenetics. In *PCR Protocols: A Guide to Methods and Applications*; Innis, M.A., Gelfand, D.H., Sninsky, J.J., White, T.J., Eds.; Academic Press: San Diego, CA, USA, 1990; pp. 315–322. ISBN 0123721806.
24. Cooke, D.E.L.; Drenth, A.; Duncan, J.M.; Wagels, G.; Brasier, C.M. A molecular phylogeny of *Phytophthora* and related oomycetes. *Fungal Genet. Biol.* **2000**, *30*, 17–32. [CrossRef] [PubMed]

25. Tamura, K.; Stecher, G.; Peterson, D.; Filipski, A.; Kumar, A. MEGA6: Molecular Evolutionary Genetics Analysis version 6.0. *Mol. Biol. Evol.* **2013**, *30*, 2725–2729. [CrossRef] [PubMed]
26. Jung, T.; Hudler, G.W.; Jensen-Tracy, S.L.; Griffiths, H.M.; Fleischmann, F.; Oßwald, W. Involvement of *Phytophthora* spp. in the decline of European beech in Europe and the USA. *Mycologist* **2005**, *19*, 159–166. [CrossRef]
27. Hong, C.X.; Gallegly, M.E.; Richardson, P.A.; Kong, P.; Moorman, G.W. *Phytophthora irrigata*, a new species isolated from irrigation reservoirs and rivers in Eastern United States of America. *FEMS Microbiol. Lett.* **2008**, *285*, 203–211. [CrossRef] [PubMed]
28. Weiland, G.E.; Nelson, A.H.; Hudler, G.W. Aggressiveness of *Phytophthora cactorum*, *P. citricola* I and *P. plurivora* from European beech. *Plant Dis.* **2010**, *94*, 1009–1014. [CrossRef]
29. Rytkönen, A.; Lilja, A.; Werres, S.; Sirkiä, S.; Hantula, J. Infectivity, survival and pathology of Finnish strains of *Phytophthora plurivora* and *Ph. pini* in Norway spruce. *Scand. J. For. Res.* **2013**, *28*, 307–318. [CrossRef]
30. Yang, X.; Tyler, B.M.; Hong, C. An expanded phylogeny for the genus *Phytophthora*. *IMA Fungus* **2017**, *8*, 355–384. [CrossRef] [PubMed]
31. Kovacs, J.; Lakatos, F.; Szabo, I. The role of *Phytophthora* species in the decline of Black walnut stands. In Proceedings of the International Scientific Conference on Sustainable Development and Ecological Footprint, Sopron, Hungary, 26–27 March 2012; University of West Hungary Press: Sopron, Hungary, 2012; pp. 1–5.
32. Jankowiak, R.; Stępniewska, H.; Bilański, P.; Kolařík, M. Occurrence of *Phytophthora plurivora* and other *Phytophthora* species in oak forests of southern Poland and their association with site conditions and the health status of trees. *Folia Microbiol.* **2014**, *59*, 531–542. [CrossRef] [PubMed]
33. Milenković, I.; Keča, N.; Karadžić, D.; Radulović, Z.; Milanović, S.; Tomšovsky, M.; Jung, T. Phytophthora diversity in natural ecosystems in Serbia and Montenegro. Unpublished work. 2018.
34. Jönsson, U. A conceptual model for the development of *Phytophthora* disease in *Quercus robur*. *New Phytol.* **2006**, *171*, 55–68. [CrossRef] [PubMed]
35. Jönsson, U.; Jung, T.; Rosengren, U.; Nihlgard, B.; Sonesson, K. Pathogenicity of Swedish isolates of *Phytophthora quercina* to *Quercus robur* in two different soil types. *New Phytol.* **2003**, *158*, 355–364. [CrossRef]
36. Jönsson, U.; Jung, T.; Sonesson, K.; Rosengren, U. Relationships between *Quercus robur* health, occurrence of *Phytophthora* species and site conditions in southern Sweden. *Plant Pathol.* **2005**, *54*, 502–511. [CrossRef]
37. Brasier, C.M.; Cooke, D.E.L.; Duncan, J.M.; Hansen, E.M. Multiple new phenotypic taxa from trees and riparian ecosystems in *Phytophthora gonapodyides—P. megasperma* ITS Clade 6, which tend to be high-temperature tolerant and either inbreeding or sterile. *Mycol. Res.* **2003**, *17*, 277–290. [CrossRef]
38. Jung, T.; Stukely, M.J.; Hardy, G.E.St.J.; White, D.; Paap, T.; Dunstan, W.A.; Burgess, T.I. Multiple new *Phytophthora* species from ITS Clade 6 associated with natural ecosystems in Australia: Evolutionary and ecological implications. *Persoonia* **2011**, *26*, 13–39. [CrossRef] [PubMed]
39. Corcobado, T.; Cubera, E.; Pérez-Sierra, A.; Jung, T.; Solla, A. First report of *Phytophthora gonapodyides* involved in the decline of *Quercus ilex* in xeric conditions in Spain. *New Dis. Rep.* **2010**, *22*, 33. [CrossRef]
40. Orlikowski, L.B.; Ptaszek, M.; Rodziewicz, A.; Nechwatal, J.; Thinggaard, K.; Jung, T. *Phytophthora* root and collar rot of mature *Fraxinus excelsior* in forest stands in Poland and Denmark. *For. Pathol.* **2011**, *41*, 510–519. [CrossRef]
41. Nikić, Z.; Letić, L.; Nikolić, V.; Filipović, V. Procedure for underground water calculation regime of Pedunculate oak habitat in Plain Srem (in Serbian with English summary). *Bull. Fac. For.* **2010**, *101*, 125–138. [CrossRef]
42. Huai, W.X.; Tian, G.; Hansen, E.M.; Zhao, W.X.; Goheen, E.M.; Grünwald, N.J.; Cheng, C. Identification of *Phytophthora* species baited and isolated from forest soil and streams in northwestern Yunnan province, China. *For. Pathol.* **2013**, *43*, 87–103. [CrossRef]
43. Hueberli, D.; Hardy, G.S.; White, D.; Williams, N.; Burgess, T.I. Fishing for *Phytophthora* from Western Australia' s waterways: A distribution and diversity survey. *Australas. Plant Pathol.* **2013**, *42*, 251–260. [CrossRef]
44. Hulvey, J.; Gobena, D.; Finley, L.; Lamour, K. 2010: Co-occurrence and genotypic distribution of *Phytophthora* species recovered from watersheds and plant nurseries of eastern Tennessee. *Mycologia* **2010**, *102*, 1127–1133. [CrossRef] [PubMed]

45. Jung, T.; Chang, T.T.; Bakonyi, J.; Seress, D.; Pérez-Sierra, A.; Yang, X.; Hong, C.; Scanu, B.; Fu, C.H.; Hsueh, K.L.; et al. Diversity of *Phytophthora* species in natural ecosystems of Taiwan and association with disease symptoms. *Plant Pathol.* **2017**, *66*, 194–211. [CrossRef]
46. Nagel, J.H.; Gryzenhout, M.; Slippers, B.; Wingfield, M.J.; Hardy, G.E.St.J.; Stukely, M.J.; Burgess, T.I. Characterization of *Phytophthora* hybrids from ITS clade 6 associated with riparian ecosystems in South Africa and Australia. *Fungal Biol.* **2013**, *117*, 329–347. [CrossRef] [PubMed]
47. Oh, E.; Gryzenhout, M.; Wingfield, B.D.; Wingfield, M.J.; Burgess, T.I. Surveys of soil and water reveal a goldmine of *Phytophthora* diversity in South African natural ecosystems. *IMA Fungus* **2013**, *4*, 123–131. [CrossRef] [PubMed]
48. Reeser, P.W.; Hansen, E.M.; Sutton, W.; Remigi, P.; Adams, G.C. *Phytophthora* species in forest streams in Oregon and Alaska. *Mycologia* **2011**, *103*, 22–35. [CrossRef] [PubMed]
49. Shrestha, S.K.; Zhou, Y.; Lamour, K. 2013: Oomycetes baited from streams in Tennessee 2010–2012. *Mycologia* **2013**, *105*, 1516–1523. [CrossRef] [PubMed]
50. Hong, C.X.; Richardson, P.A.; Kong, P. Pathogenicity to ornamental plants of some existing species and new taxa of *Phytophthora* from irrigation water. *Plant Dis.* **2008**, *92*, 1201–1207. [CrossRef]
51. Hong, C.X.; Gallegly, M.E.; Richardson, P.A.; Kong, P.; Moorman, G.W. *Phytophthora hydropathica*, a new pathogen identified from irrigation water, *Rhododendron catawbiense* and *Kalmia latifolia*. *Plant Pathol.* **2010**, *59*, 913–921. [CrossRef]
52. Jung, T.; Blaschke, M. 2004: *Phytophthora* root and collar rot of alders in Bavaria: Distribution, modes of spread and possible management strategies. *Plant Pathol.* **2004**, *53*, 197–208. [CrossRef]
53. Orlikowski, L.B.; Trzewik, A.; Orlikowska, T. Water as potential source of *Phytophthora citricola*. *J. Plant Prot. Res.* **2007**, *47*, 125–132.
54. Yang, X.; Hong, C.X. *Phytophthora virginiana* sp. nov., a high temperature tolerant species from irrigation water in Virginia. *Mycotaxon* **2013**, *126*, 167–176. [CrossRef]
55. Yang, X.; Copes, W.E.; Hong, C.X. *Phytophthora mississippiae* sp. nov., a new species recovered from irrigation reservoirs at a plant nursery in Mississippi. *J. Plant Pathol. Microbiol.* **2013**, *4*, 5. [CrossRef]
56. Yang, X.; Richardson, P.A.; Hong, C. *Phytophthora* ×*stagnum* nothosp. nov., a new hybrid from irrigation reservoirs at ornamental plant nurseries in Virginia. *PLoS ONE* **2014**, *9*, e103450. [CrossRef] [PubMed]
57. Moralejo, E.; Pérez-Sierra, A.; Alvarez, L.A.; Belbahri, L.; Lefort, F.; Descals, E. Multiple alien *Phytophthora* taxa discovered on diseased ornamental plants in Spain. *Plant Pathol.* **2009**, *58*, 100–110. [CrossRef]
58. Hardy, G.S.; Barrett, S.; Shearer, B.L. The future of phosphite as a fungicide to control the soilborne plant pathogen *Phytophthora cinnamomi* in natural ecosystems. *Australas. Plant Path.* **2001**, *30*, 133–139. [CrossRef]

12

A Survey in Natural Forest Ecosystems of Vietnam Reveals High Diversity of both New and Described *Phytophthora* Taxa Including *P. ramorum*

Thomas Jung [1,2,*], Bruno Scanu [3], Clive M. Brasier [4], Joan Webber [4], Ivan Milenković [1], Tamara Corcobado [1], Michal Tomšovský [1], Matěj Pánek [1,5], József Bakonyi [6], Cristiana Maia [7], Aneta Bačová [1], Milica Raco [1], Helen Rees [4,8], Ana Pérez-Sierra [4] and Marília Horta Jung [1,2]

1 Phytophthora Research Centre, Mendel University in Brno, 61300 Brno, Czech Republic; ivan.milenkovic@mendelu.cz (I.M.); tamara.sanchez@mendelu.cz (T.C.); tomsovsk@mendelu.cz (M.T.); panek@vurv.cz (M.P.); aneta.bacova@mendelu.cz (A.B.); milica.raco@mendelu.cz (M.R.); marilia.jung@mendelu.cz (M.H.J.)

2 Phytophthora Research and Consultancy, Am Rain 9, 83131 Nußdorf, Germany

3 Dipartimento di Agraria, Sezione di Patologia vegetale ed Entomologia (SPaVE), Università degli Studi di Sassari, Viale Italia 39, 07100 Sassari, Italy; bscanu@uniss.it

4 Forest Research, Alice Holt Lodge, Farnham, Surrey GU10 4LH, UK; clive.brasier@forestresearch.gov.uk (C.M.B.); joan.webber@forestresearch.gov.uk (J.W.); helen.rees@bristol.ac.uk (H.R.); ana.perez-sierra@forestresearch.gov.uk (A.P.-S.)

5 Crop Research Institute, Drnovská 507/73, 16106 Prague 6, Czech Republic

6 Plant Protection Institute, Centre for Agricultural Research, Herman Ottó út 15, 1022 Budapest, Hungary; bakonyi.jozsef@agrar.mta.hu

7 Centre of Marine Sciences (CCMAR), University of Algarve, 8005-139 Faro, Portugal; ccmaia@ualg.pt

8 School of Life Sciences, University of Bristol, 24 Tyndall Avenue, Bristol BS8 1TQ, UK

* Correspondence: thomas.jung@mendelu.cz.

Abstract: In 2016 and 2017, surveys of *Phytophthora* diversity were performed in 25 natural and semi-natural forest stands and 16 rivers in temperate and subtropical montane and tropical lowland regions of Vietnam. Using baiting assays from soil samples and rivers and direct isolations from naturally fallen leaves, 13 described species, five informally designated taxa and 21 previously unknown taxa of *Phytophthora* were isolated from 58 of the 91 soil samples (63.7%) taken from the rhizosphere of 52 of the 64 woody plant species sampled (81.3%) in 20 forest stands (83.7%), and from all rivers: *P. capensis*, *P. citricola* VII, VIII, IX, X and XI, *P.* sp. botryosa-like 2, *P.* sp. meadii-like 1 and 2, *P.* sp. tropicalis-like 2 and *P.* sp. multivesiculata-like 1 from *Phytophthora* major phylogenetic Clade 2; *P. castaneae* and *P. heveae* from Clade 5; *P. chlamydospora*, *P. gregata*, *P.* sp. bitahaiensis-like and *P.* sp. sylvatica-like 1, 2 and 3 from Clade 6; *P. cinnamomi (Pc)*, *P. parvispora*, *P. attenuata*, *P.* sp. attenuata-like 1, 2 and 3 and *P. ×heterohybrida* from Clade 7; *P. drechsleri*, *P. pseudocryptogea*, *P. ramorum (Pr)* and *P.* sp. kelmania from Clade 8, *P. macrochlamydospora*, *P.* sp. ×insolita-like, *P.* sp. ×kunnunara-like, *P.* sp. ×virginiana-like s.l. and three new taxa, *P.* sp. quininea-like, *P.* sp. ×Grenada 3-like and *P.* sp. ×Peru 4-like, from Clade 9; and *P.* sp. gallica-like 1 and 2 from Clade 10. The A1 and A2 mating types of both *Pc* and *Pr* co-occurred. The A2 mating type of *Pc* was associated with severe dieback of montane forests in northern Vietnam. Most other *Phytophthora* species, including *Pr*, were not associated with obvious disease symptoms. It is concluded that (1) Vietnam is within the center of origin of most *Phytophthora* taxa found including *Pc* and *Pr*, and (2) *Phytophthora* clades 2, 5, 6, 7, 8, 9, and 10 are native to Indochina.

Keywords: biosecurity; breeding systems; hybridization; *Phytophthora cinnamomi*; biogeography; center of origin

1. Introduction

The number of devastating declines of trees and other woody plants driven by introduced invasive *Phytophthora* species in natural ecosystems in Australia, Europe, and North America has increased exponentially since the 1960s [1–9]. Therefore, numerous surveys in natural and semi-natural ecosystems have been performed in the past two decades to assess *Phytophthora* diversity in these continents and in Africa, Asia, and South America [4,5,10–19]. As a result of these surveys and molecular re-evaluations of culture collections and several species complexes, the number of described species and informally designated taxa of *Phytophthora* has tripled since 1999 [2,18,20–28]. A conservative estimate predicted the existence of 200–600 unknown *Phytophthora* species in natural ecosystems of as yet unsurveyed regions of the world [26]. These are distributed among 12 major phylogenetic clades [23,28,29].

Accumulating circumstantial evidence suggests that Southeast and East Asia might be one center of origin of the genus. This included the common occurrence of both mating types of several heterothallic *Phytophthora* species, the occurrence of many *Phytophthora* diseases on mainly non-native horticultural trees and crops, and the apparent absence of *Phytophthora* diseases in natural ecosystems, despite the presence of species which cause severe forest dieback elsewhere [2,10,12,13,15,16,30–35]. In 2013, a survey in natural forests and streams of Taiwan demonstrated remarkably high diversity including ten described species and 17 previously unknown taxa of which nine were of hybrid origin. The results suggested that most of these taxa including the A1 mating type of *P. cinnamomi* were indigenous to Taiwan, whereas the A2 mating type of *P. cinnamomi* is introduced; that major *Phytophthora* phylogenetic clades 2, 5, 6, 7 and 9 are native to Southeast and Eastern Asia; and that interspecific hybridisation may have a major role in speciation and radiations in diverse natural ecosystems [10,22].

The high *Phytophthora* diversity in Taiwan probably reflects both the high floristic, geological, and climatic diversity of this island and repeated immigration of *Phytophthora* species from mainland Asia via temporary landbridges during glacial periods in the pleistocene followed by periods of separation and speciation during interglacials [10,22,36–39]. Similarly, due to its complex geology, geomorphology, and orographic climates and the repeated immigration of plant species from both northern latitudes and the numerous islands of Sundaland during glacial periods, Indochina is also a biodiversity hotspot, harbouring 20%–25% of the world' s plant species [39–41]. With a north–south extension of 1650 km and a west-east extension ranging between 50 and 600 km, Vietnam is located between 8°30′ and 23°30′ northern latitude and 102°10′ and 109°27′ eastern longitude in eastern Indochina along the South China Sea, covering approximately 330,000 km^2. In Vietnam, seven climatic regions are distinguished. In simple terms, northern Vietnam has a humid subtropical monsoon climate with cool winters and hot rainy summers in lowland areas and cold misty winters and warm rainy summers in montane regions. Southern Vietnam has a tropical monsoon climate with warm winters and hot summers and a pronounced rainy period between May and October due to the East Asian monsoon. However, regionally, temperature and precipitation patterns can vary considerably due to orographic influences. The geology and geomorphology of Vietnam are also highly complex. Due to this environmental heterogeneity, the flora of Vietnam is remarkably diverse, comprising more than 10,350 species and 2256 genera of vascular plants, of which 10% and 3%, respectively, are endemic [40]. This includes 245 and 211 native species of the Lauraceae and Fagaceae respectively, families known for the high susceptibility of their European and North American members to introduced *Phytophthora* species [5–7,42–44]. Therefore, as in Taiwan, a high diversity of unknown *Phytophthora* species might be expected in Vietnam. Further, due to their co-evolution with Vietnamese tree genera also present in Europe and North America, some of these might pose a threat to forests and natural ecosystems in the latter two continents.

In spring 2016 and 2017, in the frame of a collaborative research project between the Mendel University in Brno, Forest Research and the University of Sassari, a survey of *Phytophthora* diversity was performed in a diverse range of natural forest types and river systems across Vietnam. This paper reports on the results of this *Phytophthora* survey and the association of *Phytophthora* spp. with disease symptoms of forest trees in Vietnam, and discusses the potential threat posed by previously unknown *Phytophthora* spp. to European and North American forests.

2. Material and Methods

2.1. Sampling and Phytophthora Isolation

Twenty-five natural forest stands covering a wide range of tree species, climates, and landscapes across Vietnam were selected for sampling (Figures 1 and 2). The forest stands were located in northern Vietnam in Hoàng Liên National Park (NP) (12 stands) and on two neighboring mountains (two stands), in Ba Vì NP (three stands) and in Cuc Phuong NP (five stands), and in southern Vietnam in Bù Gia Mập NP, U Minh Hạ NP and Côn Đảo N P o n C ôn Đảo island (each one stand). In addition, 16 rivers and streams were sampled in northern Vietnam (Figures 1 and 2c). Soil sampling and isolation methodology were according to [4,10].

In total, 91 rhizosphere soil samples were taken from 142 mature specimens of 64 native tree and shrub species. Three 20 × 30 × 20 cm soil monoliths were taken around each tree, at a distance of 30–150 cm from the stem base and at a soil depth of 10–30 cm. Aliquots of ca. 2 litres of rhizosphere soil together with roots (diameter ≤5 mm) from all monoliths were bulked, and subsamples of ca. 200 mL were used for isolation tests. Isolations from soil samples were carried out at 18–20 °C in an airconditioned laboratory at natural light using 3- to 10-day-old leaflets of native tree species, mainly *Lithocarpus bacgangensis*, *L. corneus*, *Quercus glauca*, *Q. chapaensis*, *Q. gilva*, *Castanopsis indica* and *Chamaecyparis hodginsii*, and the introduced *Acacia mangium* as baits floated over flooded soil.

Brownish leaflets were examined at ×80 under a light microscope for presence of *Phytophthora* sporangia. Infected leaflets were blotted dry, necrotic lesions cut into small segments and plated onto selective PARPNH agar (V8-juice agar (V8A) amended with 10 µg/mL pimaricin, 200 µg/mL ampicillin, 10 µg/mL rifampicin, 25 µg/mL pentachloronitrobenzene (PCNB), 50 µg/mL nystatin and 50 µg/mL hymexazol).

In forest stand F07, the isolation of *Phytophthora* was also attempted from a bleeding bark lesion on a surface root of a mature *Castanopsis acuminatissima* (Figure 3e). Necrotic bark pieces were transported in distilled water to the lab and blotted dry on filter paper. Then, ca. 2 mm pieces were cut from the lesion margins and plated onto PARPNH agar.

In forest stand F11, freshly fallen leaves of a mature *Rhododendron arboreum* with necrotic lesions were collected from the forest floor close to forest stream R05 ca. 1 m above the waterline. The isolation of *Phytophthora* from these leaves was carried out as described below for leaves collected from rivers.

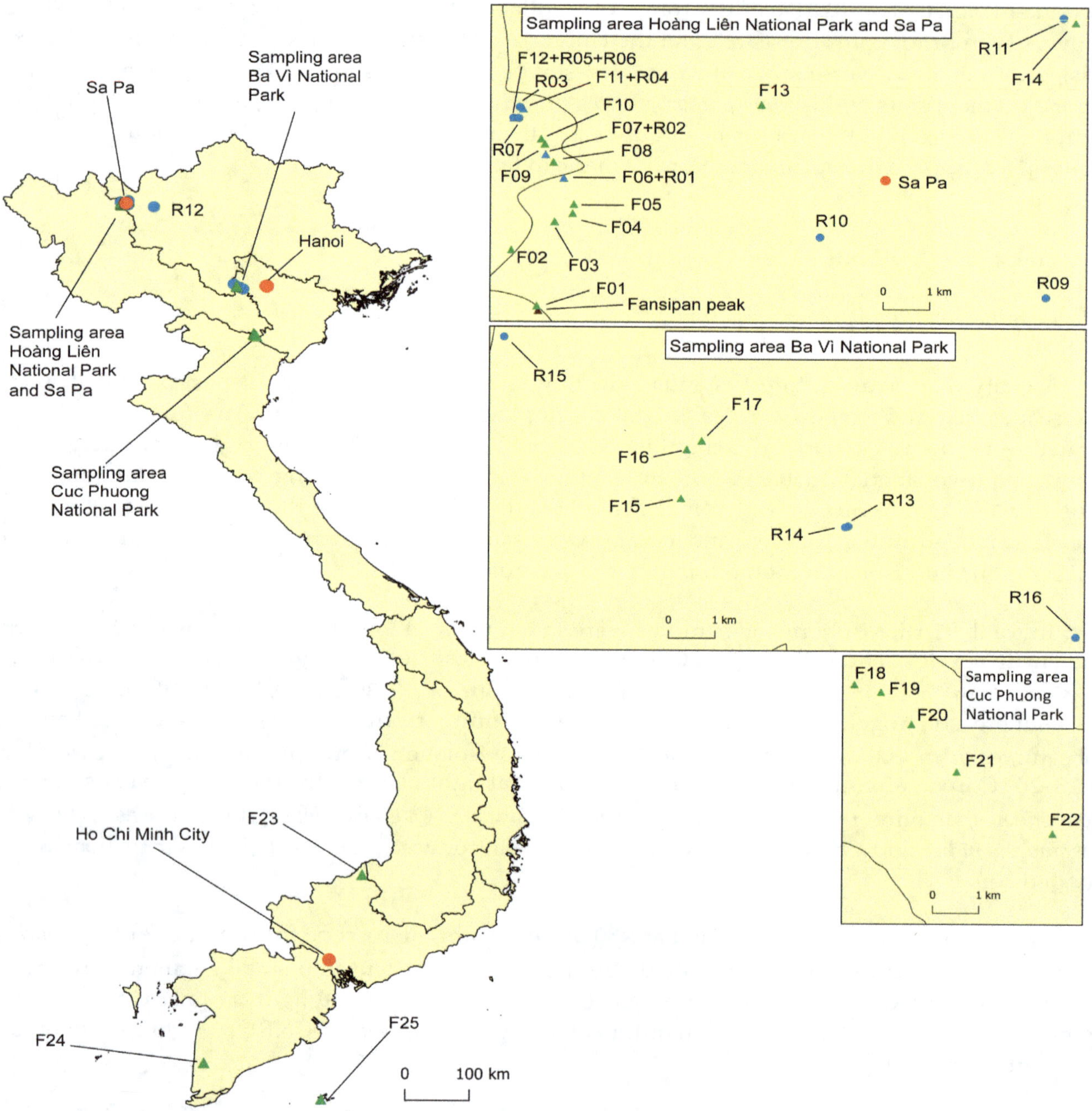

Figure 1. Location of the 25 forest sites (F01–F25; green triangles) and the 16 riparian sites (R01–R16; blue dots) included in the *Phytophthora* survey in Vietnam; blue triangles represent sites included in both the riparian and forest survey. For geographical coordinates and details of sites see Tables 1 and 2.

Figure 2. Representative forest stands and streams sampled in Vietnam; (**a**) Hoàng Liên National Park around the Fansipan mountain with diverse montane evergreen cloud forests and montane evergreen broadleaved forests; (**b**) diverse montane evergreen cloud forest F04 in Hoàng Liên National Park dominated by Fagaceae and Lauraceae species; (**c**) Cat Cat River (R10) running through a diverse montane evergreen forest in Hoàng Liên National Park; (**d**) montane *Chamaecyparis hodginsii—Quercus* forest on Sau Chua mountain; (**e**) montane *Alnus nepalensis* stand on Xin Chài mountain; (**f**) diverse, suptropical, humid evergreen forest F15 in Ba Vì National Park; (**g**) Cuc Phuong National Park with diverse, tropical, evergreen lowland rainforests growing on limestone; (**h**) diverse, tropical, evergreen lowland rainforest stand F20. For GPS coordinates see Tables 1 and 2; for location of sites see Figure 1.

Figure 3. Disease symptoms of mature native trees in natural forest stands in Vietnam associated with presence of *Phytophthora* species in the rhizosphere; (**a**–**f**) montane evergreen cloud forests in Hoàng Liên National Park; (**a**) crown thinning and dieback of *Quercus glauca* in forest stand F03 (2337 m a.s.l.; *P. cinnamomi* A2); (**b**) crown dieback and mortality of *Castanopsis acuminatissima* and *Neolitsea poilanei* in forest stand F05 (2249 m a.s.l.; *P. attenuata, P. castaneae, P. cinnamomi* A2); (**c**,**d**,**f**) severe crown dieback and mortality of *C. acuminatissima* in a swampy depression of forest stand F06 close to stream R01 (2083 m a.s.l.; *P. castaneae, P. cinnamomi* A2, *P. gregata*); (**f**) the white flowers and young leaves in the crowns of *C. acuminatissima* belong to the epiphytic *Rhododendron leptocladus*; (**e**) bark lesion with staining of the underlying cambium caused by *P. cinnamomi* A2 on a surface root of *C. acuminatissima* in forest stand F07; (**g**) mortality of *Dysoxylum juglans* in suptropical humid evergreen forest stand F15 in Ba Vì National Park (1108 m a.s.l.; *P.* sp. attenuata-like 3).

For the isolation of *Phytophthora* spp. from the 16 rivers and streams, an in-situ baiting technique was used [10,11]. Twelve of the 16 riparian baiting sites were located inside or downstream of natural forests (Figure 2c). At each site, 15–20 non-wounded young leaves of the native *C. indica*, *Citrus sinensis*, *L. bacgangensis*, *Q. glauca*, and, in some cases, *Carpinus* sp., *C. hodginsii*, *Cinnamomum iners*, *Dipterocarpus alatus*, *Prunus* sp., *Q. gilva* and *A. mangium* were placed as baits in a 25 × 30 cm raft, prepared using fly mesh and styrofoam, and the raft put to float at a place where water flow was calm. The rafts were collected after 2–3 days. In addition, in 2017 freshly fallen leaves of different tree species and flowers of *Rhododendron arboreum* and *R. leptocladus* were collected from forest streams R01, R02, R10 and R11. Baiting leaves and the collected fallen leaves and flowers were washed in distilled water and blotted dry on filter paper. Five to ten pieces (approximately 2 × 2 mm) were cut from the margins of each watersoaked or necrotic lesion of each leaf or flower, blotted on filter paper and plated onto PARPNH agar.

All Petri dishes with plated leaf, flower or bark pieces were incubated at 20 °C in the dark and repeatedly examined under the stereo microscope at ×20 for *Phytophthora*-like hyphae after 12–48 h. Pure cultures were obtained by transferring single hyphal tips from the edge of the colonies onto V8A. Stock cultures were maintained on carrot agar (CA) [45] at 10 °C in the dark.

2.2. Molecular Identification of Isolates

For all *Phytophthora* isolates obtained in this study mycelial DNA was extracted from one-week old V8A cultures. Total DNA was extracted using the Phire Plant Direct PCR Kit (Thermo Fisher Scientific Inc., Waltham, MA USA) following the manufacturer's instructions. DNA was stored at −20 °C until further use. For all isolates the region spanning the internal transcribed spacer (ITS1-5.8S-ITS2) region of the ribosomal DNA was amplified using primer-pairs ITS1/ITS4 or ITS6/ITS4 [29,46]. For representative isolates of several known and all putative new species the mitochondrial *cox1* gene was amplified with both primer-pairs COXF4N/COXR4N and FM84/FM83 [47,48]. The PCR reaction mixture and the amplification conditions for ITS and *cox1* were according to [29,47,48]. PCR consumables were provided by Thermo Fisher Scientific. PCR products were purified and sequenced by GATC Biotech (Konstanz, Germany) and by Source Bioscience (Nottinham, UK) in both directions with the primers used for PCR amplification.

Sequences were edited using Geneious (Version 11.1.2, Biomatters Ltd., Auckland, New Zealand). Heterozygous sites observed were labelled according to the IUPAC coding system. Consensus sequences were aligned using the CLUSTAL W algorithm. The consensus sequences were subjected to an NCBI BLAST search (http://www.ncbi.nlm.nih.gov/BLAST/) and to a blast search in a local database containing sequences of ex-type isolates or key isolates from published studies to identify the closest related sequences. Isolates were assigned to a species when sequence identities were above a 99% cut-off in respect to those of ex-type isolates or key isolates. ITS and *cox1* sequences from representative isolates of all known and all putative new *Phytophthora* species obtained in this study were deposited at GenBank and accession numbers are given in Supplementary Table S1.

2.3. Classical Identification of Isolates

Colony growth patterns of 7-d-old cultures grown at 20 °C in the dark on V8A, malt-extract agar (MEA; Oxoid Ltd., Basingstoke, UK) and PDA [21] and morphological characters of sporangia, oogonia, antheridia, chlamydospores, hyphal swellings, and aggregations were compared with isolates from known species and with species descriptions in the literature.

Sporangia production and microscopic examinations and measurements of morphological structures at ×400 were according to [21,22] using a compound microscope (Zeiss Axioimager.Z2, Carl Zeiss AG, Oberkochen, Germany), a digital camera (Zeiss Axiocam ICc5) and a biometric software (Zeiss ZEN). Self-sterile isolates were paired on both V8A and CA with known A1 and A2 mating type tester strains of *P. cinnamomi*, *P.* ×*cambivora* and *P.* ×*heterohybrida* (isolates with non-papillate sporangia) or *P. botryosa*, *P. colocasiae* and *P. meadii* (isolates with papillate sporangia). All pairings were examined after 4–6 weeks incubation at 20 °C in the dark in order to determine whether self-sterile isolates are heterothallic or sterile and to which mating type heterothallic isolates belong [21]. All isolates are preserved in the culture collections maintained at Mendel University and Forest Research.

3. Results

In total, 943 oomycete isolates, including 652 *Phytophthora* isolates and 291 isolates from other oomycete genera, were obtained from forest stands (Table 1) and river systems (Table 2) in Vietnam. The *Phytophthora* isolates belonged to 13 described species, five informally designated taxa and 21 previously unknown taxa. From the other oomycete genera, 122 isolates were identified to species level. They could be assigned to the recently described *Nothophytophthora vietnamensis* (26 isolates), *Phytopythium vexans* sensu lato (63 isolates from 14 partly highly different haplotypes), four other known species and three novel taxa of *Phytopythium* (16 isolates), two described species and six novel taxa of *Pythium* (17 isolates) and one novel taxon of *Elongisporangium*. The remaining 169 isolates, which were not identified to species level, belonged to *Phytopythium* (161 isolates), *Pythium* (7 isolates) and *Saprolegnia* (1 isolate), respectively. GenBank accession numbers of ITS sequences of representative isolates of all oomycete taxa and of *cox1* sequences of representative isolates of most *Phytophthora* taxa are given in Supplementary Table S1. Detailed descriptions of morphological characteristics, morphometric and temperature-growth data, and multigene phylogenies for all new *Phytophthora* species will be presented in separate publications.

3.1. Phytophthora Diversity in Natural and Semi-Natural Forest Stands

In 20 forest stands (80%), 20 *Phytophthora* taxa were isolated from 58 of the 91 soil samples (63.7%) taken from the rhizosphere of 52 of the 64 woody plant species sampled (81.3%); from the root lesion of *C. acuminatissima* in stand F07; and from all four freshly fallen *Rhododendron* leaves collected from the ground in stand F11: *P. attenuata*, *P. castaneae*, *P. chlamydospora*, *P. cinnamomi*, *P. gregata*, *P. heveae*, *P. parvispora*, *P. ramorum*, *P. citricola* VII, three new species related to *P. attenuata*, three new species from the '*Phytophthora citricola* complex', three new species related to *P. botryosa* and *P. meadii*, and one new species related to *P. multivesiculata* and another to *P. tropicalis*, respectively (Table 1). From 29 of the 35 *Phytophthora*-negative soil samples, several known and previously unknown *Phytopythium* or *Pythium* spp. were isolated (Table 1). The only forest site from which no oomycete isolates could be obtained was subalpine *Rhododendron* scrub at 2903 m altitude near the Fansipan peak (F01).

Phytophthora cinnamomi, Clade 7c, was isolated from 26 of 66 rhizosphere soil samples (39.4%) collected from 27 of the 50 tree and shrub species (54%) in 13 of the 17 mountainous forest stands sampled (76.5%), making it the most widespread and common *Phytophthora* species above 700 m altitude. The A2 mating type of *P. cinnamomi* was present in 11 forest stands with an altitudinal amplitude ranging from 713 to 2337 m above sea level (a.s.l.). In contrast, the A1 mating type was only found in four forest stands located between 1108 and 2636 m a.s.l. (Figure 1; Table 1). Both mating types co-occurred in one stand in Hoàng Liên NP and another in Ba Vì NP. Interestingly, in Hoàng Liên NP, the A1 mating type was present in the upper montane Rhododendron forest F02 at 2636 m a.s.l. and in the lower montane stands F11 and F12 at 1900 m a.s.l., but was not detected in the eight forest stands (F03–F10) sampled between 2337 and 2022 m a.s.l., all highly infested by the A2 mating type. The latter was also isolated from a bark lesion on a surface root of *C. acuminatissima* in stand F07. Two A2 isolates from stand F11 were able to produce oogonia in single culture on V8A (Table 1). Over all stands, the A1:A2 mating type ratio of the 151 *P. cinnamomi* isolates was 30.5:69.5, whereas in the two stands with co-occurrence of both mating types the A1:A2 ratio of the 44 isolates was 59.1:40.9. Among the 39 *P. cinnamomi* isolates for which ITS sequences were produced, 32 isolates belonged to the same haplotype as the ex-type isolate from Sumatra (CBS 142.22; GenBank accession no. KU899160) (Table S1). Six isolates, representing both mating types, from stands F02 and F05 in Hoàng Liên NP and stand F17 in Ba Vì NP were heterozygous at position 767 (K instead of G) (Table 1S) and shared the same haplotype with an isolate from a subtropical *Quercus* forest in Taiwan (TW213; GenBank accession no. KU682570). Another isolate from stand F17 shared the heterozygous position 767 and was also heterozygous at position 89 (Y instead of C) (Supplementary Table S1). The 15 isolates for which the *cox1* gene was sequenced belonged to eight different haplotype which differed over a 712 bp alignment from the ex-type isolate (KU899315) at 1–4 positions.

Phytophthora parvispora was exclusively found in stand F15 in Ba Vì NP where it co-occurred with both mating types of its closest relative *P. cinnamomi* (Table 1). Compared to the ex-type of *P. parvispora* (CBS 132772; KC478667), the three isolates had identical *cox1* sequences and differed in ITS by one heterozygous site at position 73 (Y instead of T) (Supplementary Table S1). In mating tests with A1 and A2 tester strains of *P. cinnamomi*, all isolates were sterile.

Phytophthora attenuata from Clade 7a and three previously unknown taxa closely related to *P. attenuata* were recovered from five forest stands in Hoàng Liên NP and Ba Vì NP (Table 1). The individual taxa from this '*P. attenuata* complex' differed in their altitudinal amplitude and geographical distribution (Figure 1; Table 1). *Phytophthora attenuata*, *P.* sp. attenuata-like 1 and *P.* sp. attenuata-like 2 were only found in Hoàng Liên NP. Most widespread was *P.* sp. attenuata-like 1 which was isolated from the rhizosphere of five tree species in three stands located between 2249 and 2636 m a.s.l., followed by *P. attenuata* (three tree species in two stands; 1910–2249 m a.s.l.) and *P.* sp. attenuata-like 2 (2 tree species in 1 stand; 1910 m a.s.l.). In contrast, *P.* sp. attenuata-like 3 was exclusively found between 713 and 1108 m altitude in two of the three forest stands sampled in Ba Vì NP where it was associated with six tree species (Table 1). The ITS and *cox1* sequences of *P. attenuata* isolates from Vietnam differed from the ex-type isolate (CBS 141199; GenBank nos. KU517154 and KU517148) and other isolates of *P. attenuata* from Taiwan at 0–1 and 0–5 positions. *Phytophthora* sp. attenuata-like 1, *P.* sp. attenuata-like 2 and *P.* sp. attenuata-like 3 showed differences to *P. attenuata* in ITS at 0–1, 1–2 and 2–3 positions, respectively, and in *cox1* at 6–8, 9–11 and 6–8 positions, respectively. The *cox1* sequences of the three new taxa differed from each other at 8–17 positions. Heterozygous sites were present in the ITS sequences of all isolates of *P.* sp. attenuata-like 2 (R at position 184) and most isolates of *P.* sp. attenuata-like 3 (Y in position 54; K in position 152). The ITS sequence of one isolate of *P.* sp. attenuata-like 1 from stand F05 contained seven heterozygous sites possibly suggesting hybrid origin.

Phytophthora castaneae from Clade 5 showed a similar altitudinal (1108–2242 m a.s.l.) and geographical distribution to *P. cinnamomi* (Figure 1; Table 1). It was isolated from the rhizosphere of 13 tree species from the genera *Castanopsis*, *Lithocarpus*, *Neolitsea*, *Meliosma*, *Illicium* and *Rhododendron* in seven stands in Hoàng Liên NP and Ba Vì NP, and *C. hodginsii* in stand F14 on Sau Chua mountain where it was the only *Phytophthora* species recovered (Table 1). The ITS sequences of all isolates from Hoàng Liên NP and several isolates from Ba Vì NP matched the ex-type of *P. castaneae* (ICMP 19434; GenBank no. KP295319). However, several isolates from Ba Vì NP had a unique polymorphism at position 54 (A or R instead of G) while all isolates from Sau Chua mountain were characterised by having a unique polymorphism at position 590 (A instead of G). The *cox1* sequences of 15 isolates from the seven stands constituted six haplotypes which differed from the ex-type isolate (KP295234) by 0–1 bp. Interestingly, all four tested isolates from Sau Chua mountain had a unique polymorphism at position 421 (A instead of G). Five of the six tested *P. castaneae* isolates from stand F15 in Ba Vì NP shared a T at position 369 with *P. heveae* isolates from the same stand and with the *P. heveae* ex-type (CBS296.29; GenBank nos. HQ643238 and KP295326) whereas *P. castaneae* isolates from the other stands and the *P. castaneae* ex-type have a C at this position. Compared to *P. castaneae*, the other Clade 5 species found in this survey, *P. heveae*, had a lower altitudinal amplitude. *Phytophthora heveae* was isolated from the rhizosphere of 10 tree species in the subtropical lower montane stands F15 and F16 in Ba Vì NP and in four tropical lowland rainforest stands in Cuc Phuong NP, Bù Gia Mập NP and Côn Đảo NP (Figure 1; Table 1). Both Clade 5 species only co-occurred in stand F15. The ITS sequences of all *P. heveae* isolates (Table S1) matched the ex-type of *P. heveae*. The *cox1* sequences of all isolates differed from the ex-type (GenBank no. KP295239) at position 536 (T instead of C). Isolates from Cuc Phuong NP and Bù Gia Mập NP had unique polymorphisms at positions 30 (C instead of A) and 390 (A instead of T), respectively. The morphology of all isolates of *P. castaneae* and *P. heveae* was in accordance with the original descriptions [2].

Table 1. Location, altitude, geological substrate and vegetation of 25 forest sites sampled in spring 2016 and 2017 in Vietnam, sampled tree species and *Phytophthora* and other oomycete taxa isolated.

Site no.	GPS Coordinates	Altitude (m a.s.l)	Location	Geological Substrate	Vegetation	Sampled Tree Species (no. of *Phytophthora*-Positive/ Sampled Trees)	*Phytophthora* and *Nothophytophthora* spp. (no. of Positive Samples) [a,b]
F01	N22 18.466 E103 46.480	2903	Fansipan, Hoàng Liên National Park (NP)	Triassic schists and sandstones	Subalpine *Rhododendron* scrub	*Rhododendron* spp. (0/3)	-
F02	N22 19.194 E103 46.177	2636	Fansipan, Hoàng Liên NP	Triassic schists and sandstones	Upper montane Rhododendron ('Elfin') cloud forest	*Rhododendron arboreum*, mix from 3 trees with dieback (DB) (1/1)	ATT1 (1), CIN A1 (1)
F03	N22 19.563 E103 46.679	2337	Hoàng Liên NP	Triassic schists and sandstones	Montane evergreen cloud forest	*Quercus glauca*, DB (2/2)	CIN A2 (2)
F04	N22 19.670 E103 46.885	2242	Hoàng Liên NP	Triassic schists and sandstones	Montane evergreen cloud forest	*Meliosma henryi* (1/1) *Betula alnoides* & *Elaeocarpus japonicus* (1/1) *Castanopsis acuminatissima*, mix from 2 trees, DB (1/1) *C. acuminatissima* with DB & *Acer campbellii* (1/1)	CAS (1) ATT1 (1), CIN A2 (1) [c] ATT1 (1), CIN A2 (1) VIE (1)
F05	N22 19.786 E103 46.899	2249	Hoàng Liên NP	Triassic schists and sandstones	Montane evergreen cloud forest	*Neolitsea poilanei*, DB (3/3) *C. acuminatissima* mix from 3 trees, DB (1/1) *Illicium griffithii* & *C. acuminatissima*, DB (1/1)	ATT1 (3), CAS (1), CIN A2 (3) ATT (1), CIN A2 (1) CAS (1)
F06	N22 20.127 E103 46.782	2083	Hoàng Liên NP	Triassic schists and sandstones	Montane evergreen cloud forest	*C. acuminatissima*, DB (2/2) *M. henryi* & *A. campbellii* (1/1) *M. henryi* & *Neolitsea merilliana* (1/1)	CIN A2 (2), GRE (1), CAS (1) [d] GRE (1) CIN A2 (1), MUV1 (1)
F07	N22 20.430 E103 46.574	2010	Hoàng Liên NP	Triassic schists and sandstones	Montane evergreen cloud forest	*Illicium tsaii* & *Rhododendron sinofalconeri* (1/1) *C. acuminatissima*, DB, necrotic root lesion (1/1)	CAS (1) CIN A2 (1)
F08	N22 20.331 E103 46.664	2066	Hoàng Liên NP	Triassic schists and sandstones	Montane evergreen cloud forest	*Casearia annamensis* (1/1) *Acer oblongum*, mix from 2 trees (0/1)	CIN A2 (1) - [e]
F09	N22 20.565 E103 46.565	2010	Hoàng Liên NP	Triassic schists and sandstones	Montane evergreen cloud forest	*C. acuminatissima* (0/1) *Q. glauca* (0/2)	- [e] -
F10	N22 20.632 E103 46.523	2022	Hoàng Liên NP	Triassic schists and sandstones	Montane, evergreen cloud forest	*Neolitsea polycarpa*, mix from 3 trees, DB (1/1) *N. polycarpa*, *Symplocos pseudobarberina* & *Beilschmiedia roxburghiana* (1/1)	CIN A2 (1) CIN A2 (1)
F11	N22 21.026 E103 46.315	1910	Hoàng Liên NP	Triassic schists and sandstones	Montane, evergreen broadleaved forest	*A. oblongum* & *Symplocos dryophila* (2/2) *C. acuminatissima* DB, *Ilex leseeneri* & *Eurya annamensis* (1/1) *R. arboreum* (1/1) [g]	ATT (1), ATT2 (1), CIN A1 (2), CIN A2 (1), CIN A2ho (1) [e,f] CIN A2 (1), CAS (1) CHL (1), RAM A1 (1)

Table 1. *Cont.*

Site no.	GPS Coordinates	Altitude (m a.s.l)	Location	Geological Substrate	Vegetation	Sampled Tree Species (no. of *Phytophthora*-Positive/ Sampled Trees)	*Phytophthora* and *Nothophytophthora* spp. (no. of Positive Samples) [a,b]
F12	N22 20.909 E103 46.199	1895	Hoàng Liên NP	Triassic schists and sandstones	Montane, evergreen broadleaved forest	*Acer oliverianum*, *Eryobotrya cavaleriei* & *Symplocos quillaminii* (1/1) *Q. glauca* (1/1)	CIN A1 (1) CIN A1 (1)
F13	N22 21.090 E103 49.092	1717	Xin Chài mountain	Triassic schists and sandstones	Montane *Alnus* forest on steep loamy slope	*Alnus nepalensis* (2/3)	CIT VII (1), MEA1 (1), ×TRO2 (1), VIE [e]
F14	N22 22.168 E103 52.758	1367	Sau Chua mountain	Triassic schists and sandstones	Montane *Chamaecyparis*-*Quercus* forest	*Chamaecyparis hodginsii* (7/9)	CAS (7) [c,e]
F15	N21 3.699 E105 21.733	1108	Ba Vì National Park (NP)	Triassic schists and sandstones and porphyrites	Suptropical humid evergreen forest	*Castanopsis chinensis* (2/2) *C. chinensis* & *Beilschmiedia fordii* (1/1) *Dysoxylum juglans*, DB (1/1) *Eberhardtia tonkinensis*, *Antidesma* sp. & *Jasminum* sp. (0/1) *Eurya japonica* & *Nephelium lappaceum* (1/1) *Lithocarpus bacgangensis* (1/1) *Lithocarpus pseudosundaicus* (0/1) *Machilus bonii* (1/1) *Magnolia annamensis* (1/1) *Q. glauca*, mix from 3 trees (1/1) *Vernicia montana* & *Antidesma* sp. (1/1)	ATT3 (1), CAS (1), CIN A1 (1) [e] CAS (1), HEV (1), PAR (1) [d] ATT3 (1) [d] - [d] CAS (1) [e] CAS (1) [d] - CIN A1 (1) [e] CIN A2 (1) [e] ATT3 (1), CIN A1 (1), HEV (1), PAR (1) [e] ATT3 (1), CIN A1 (1) [d]
F16	N21 04.455 E105 21.810	807	Ba Vì NP	Triassic schists and sandstones and porphyrites	Suptropical humid evergreen forest	*Caryodaphnosis baviensis* (0/2) *Lithocarpus bacgangensis* (1/1) *Meliosma arnottiana* (1/1) *Phoebe petelotii*, *Machilus thunbergii* & *Claoxylon indicum* (1/1)	- [e] HEV (1) [e] CIT IX (1) [e] CIN A2 (1) [e]
F17	N21 04.587 E105 22.016	713	Ba Vì NP	Triassic schists and sandstones and porphyrites	Suptropical humid evergreen forest	*Alsodaphne velutina* & *Litsea brevipetiolata* (1/1) *Bischofia javanica* & *Litsea monocephala* (0/1) *C. chinensis* (1/1) *Castanopsis tonkinensis* (1/1) *Q. glauca* (0/1)	- [e] - [e] ATT3 (1), CIN A2 (1) [d] ATT3 (1) [e] - [e]

Table 1. *Cont.*

Site no.	GPS Coordinates	Altitude (m a.s.l)	Location	Geological Substrate	Vegetation	Sampled Tree Species (no. of *Phytophthora*-Positive/ Sampled Trees)	*Phytophthora* and *Nothophytophthora* spp. (no. of Positive Samples) [a,b]
F18	N20 20.876 E105 35.793	392	Cuc Phuong National Park (NP)	Triassic limestones	Tropical evergreen lowland rainforest	*C. baviensis & Litsea robusta* (0/1) *Dracontomelum duppereanum*, mix from 2 trees (0/1) *Saraca dives*, mix from 2 trees (1/1)	- [e] - [e] HEV (1) [e]
F19	N20 20.779 E105 36.099	356	Cuc Phuong NP	Triassic limestones	Tropical evergreen lowland rainforest	*Allophylus cobbe*, mix from 2 trees (0/1) *D. duppereanum* & S. dives (0/2)	- [e] - [e,h]
F20	N20 20.366 E105 36.452	318	Cuc Phuong NP	Triassic limestones	Tropical evergreen lowland rainforest	*A. cobbe*, *Ficus* sp., *Merremia boisiana* & *Homalium* sp. (1/1) *S. dives* (0/2)	MEA2 (1) [e,i] - [e]
F21	N20 19.755 E105 36.979	267	Cuc Phuong NP	Triassic limestones	Tropical evergreen lowland rainforest	*Anogeissus acuminata* (0/1) *A. acuminata* & *Taxotrophis macrophylla* (1/2)	- [d or e] CIT X (1) [e,j]
F22	N20 18.963 E105 38.101	264	Cuc Phuong NP	Triassic limestones	Tropical evergreen lowland rainforest	*C. baviensis* (0/1) *C. baviensis* & *S. dives* (0/1) *S. dives*, mix from 2 trees (0/1)	- [e] - [e,j] - [e]
F23	N12 06.326 E107 09.396	417	Bù Gia Mập National Park	Quaternary alluvial sediments	Tropical evergreen lowland rainforest	*Dipterocarpus alatus*, *Ailanthus triphysa*, *Hopea odorata* & *Dalbergia oliveri* (1/1)	HEV (1) [e]
F24	N9 13.645 E104 57.330	4	U Minh Hạ National Park	Quaternary peat	Tropical lowland peat forest	*Melaleuca cajuputi* (0/3)	- [k]
F25	N8 40.621 E106 34.836	55	Côn Đảo National Park, Côn Lôn island	Rhyolite and diorite	Tropical evergreen lowland rainforest	*Chukrasia tabularis* (0/1) *A. triphysa*, *C. tabularis* (1/1) *Leucaena leucocephala*, *Canarium album* & *Hopea odorata* (1/1) *H. odorata*, *C. album*, *D. alatus* (1/1)	- [e,k] CIT XI [e,l] BOT2 [e,j] HEV [l, m]

[a] ATT = *P. attenuata*, ATT 1 = *P.* sp. attenuata-like 1, ATT 2 = *P.* sp. attenuata-like 2, ATT 3 = *P.* sp. attenuata-like 3, BOT2 = *P.* sp. botryosa-like 2, CAS = *P. castaneae*, CHL = *P. chlamydospora*, CIN = *P. cinnamomi*, CIT VII = *P. citricola* VII, CIT IX = *P. citricola* IX, CIT X = *P. citricola* X, CIT XI = *P. citricola* XI, GRE = *P. gregata*, HEV = *P. heveae*, MEA1 = *P.* sp. meadii-like 1, MEA2 = *P.* sp. meadii-like 2, MUV1= *P.* sp. multivesiculata-like 1, PAR = *P. parvispora*, RAM = *P. ramorum*; TRO2 = *P.* sp. tropicalis-like 2, VIE = *Nothophytophthora vietnamensis*. [b] Mating types: A1 = forming oogonia only in dual cultures with A2 tester strains; A2 = forming oogonia only in dual cultures with A1 tester strains; A2ho = forming oogonia in dual cultures with A1 tester strains and in ageing single cultures. [c] *Pythium senticosum* also isolated. [d] *Phytopythium* sp. also isolated. [e] *Phytopythium vexans* s.l. also isolated. [f] *Phytopythium* sp. 1 PB-2013 also isolated. [g] Fallen leaves collected from the ground. [h] *Pythium intermedium* also isolated. [i] *Pythium* sp. conidiophorum-like also isolated. [j] *Phytopythium chamaehyphon* also isolated. [k] *Phytopythium cucurbitacearum* also isolated. [l] *Phytopythium vexans* also isolated. [m] *Phytopythium* sp. Côn Đảo also isolated.

Table 2. Location and altitude of the 16 riparian sites sampled in spring 2016 and 2017 in Vietnam and *Phytophthora* and other oomycete taxa isolated.

Site no.	GPS Coordinates	Altitude (m a.s.l)	River; Province	Location of Catchment and Vegetation	Sampling Method [a]	*Phytophthora* and *Nothophytophthora* spp. [b,c]
R01	N22 20.127 E103 46.782	2083	Forest stream 1; Lào Cai	Hoàng Liên NP; subalpine and montane Rhododendron scrub and forests, montane broadleaved forests	Baiting raft Fallen leaves/flowers	CAP, ×HET A1, ×HET A1ho, MUV1, RAM A1 [d] CIT VII, RAM A1, SYL2, VIE
R02	N22 20.440 E103 46.576	2007	Forest stream 2; Lào Cai	Hoàng Liên NP; subalpine and montane Rhododendron scrub and forests, montane broadleaved forests	Baiting raft Fallen leaves/flowers	RAM A1, SYL2 GAL1, GAL2, MUV1, RAM A1, SYL2, VIE [e,f,g,h]
R03	N22 21.046 E103 46.273	1913	Forest stream 3, tributary of forest stream 5; Lào Cai	Hoàng Liên NP; montane broadleaved forests	Baiting raft	CHL, CIT VII, RAM A1, RAM A2, SYL2 [e]
R04	N22 21.029 E103 46.317	1904	Forest stream 4, tributary of forest stream 5; Lào Cai	Hoàng Liên NP; montane broadleaved forests	Baiting raft	CHL, ×HET A1, RAM A1, RAM A2, SYL2, SYL3 [i]
R05	N22 20.906 E103 46.197	1895	Forest stream 5, Gold river, downstream of R03, R04, R06-R08; Lào Cai	Hoàng Liên NP; montane broadleaved forests	Baiting raft	CHL, CIT VII, ×HET A1, RAM A1, SYL2 [e]
R06	N22 20.911 E103 46.199	1896	Forest stream 6, tributary of forest stream 5; Lào Cai	Hoàng Liên NP; montane broadleaved forests	Baiting raft	CHL, ×HET A1 [h,i]
R07	N22 20.902 E103 46.261	1912	Forest stream 7, tributary of the Gold river; Lào Cai	Hoàng Liên NP; montane broadleaved forests	Baiting raft	×HET A1, RAM A1
R08	N22 20.904 E103 46.259	1911	Forest stream 5, Gold river; Lào Cai	Hoàng Liên NP; montane broadleaved forests	Baiting raft	CIT VII, SYL2, SYL3
R09	N22 18.597 E103 52.426	1013	Muong Hoa River; Lào Cai	Hoàng Liên NP; subalpine and montane Rhododendron scrub and forests, montane broadleaved forests, rice fields	Baiting raft	KEL, PSC, ×KUN
R10	N22 19.372 E103 49.780	1193	Forest stream 9, Cat Cat River; Lào Cai	Hoàng Liên NP; montane broadleaved forests	Baiting raft Fallen leaves/flowers	CAP, CIT VII, CIT VIII, SYL1, SYL3 [e] CHL, PSC, QUI, SYL 1, SYL3, RAM A1, VIE [f,j,k]
R11	N22 22.230 E103 52.615	1308	Forest stream 8; tributary of Ngòi Duôi River; Lào Cai	Sau Chua mountain; *Chamaecyparis hodginsii* forest F24; broadleaved mountain forests and *Cunninghamia lanceolata* plantations	Baiting raft Fallen leaves/flowers	BIT, CIT VII, CIT IX, MAC, QUI, SYL3 CIT IX, KEL, RAM A1 [l]

Table 2. *Cont.*

Site no.	GPS Coordinates	Altitude (m a.s.l)	River; Province	Location of Catchment and Vegetation	Sampling Method [a]	*Phytophthora* and *Nothophytophthora* spp. [b,c]
R12	N22 16.787 E104 13.394	63	Red River (Sông Hồng); Lào Cai	Large catchment in N-Vietnam and Yunnan; subalpine and montane Rhododendron scrub and forests, montane broadleaved forests, forest plantations, rice fields, horticulture	Baiting raft	×KUN, ×PER4, ×VIR
R13	N21 03.275 E105 24.050	59	Stream 9; Hanoi	Ba Vi NP; subtropical evergreen forests, rice fields	Baiting raft	×INS, ×GRE3, ×KUN, ×PER4, ×VIR [e]
R14	N21 3.261 E105 24.012	60	Stream 10, tributary of stream 9; Hanoi	Ba Vi NP; subtropical evergreen forests, rice fields	Baiting raft	×KUN, ×PER 4 [i]
R15	N21 06.177 E105 19.267	26	Black River (Sông Đà); Hanoi	Large catchment in N-Vietnam and Yunnan; subalpine and montane Rhododendron scrub and forests, montane broadleaved forests, subtropical evergreen forests, forest plantations, rice fields, horticulture	Baiting raft	DRE A1, ×VIR [l,m]
R16	N21 01.576 E105 27.218	26	Stream 11; Hanoi	Forest plantations, rice fields, horticulture	Baiting raft	×PER4, ×VIR [n]

[a] Baiting rafts were collected in March–April 2016; fallen leaves were collected in March 2017. [b] BIT = *P.* sp. bitahaiensis-like, CAP = *P. capensis*, CHL = *P. chlamydospora*, CIT VII = *P. citricola* VII, CIT VIII = *P. citricola* VIII, CIT IX = *P. citricola* IX, DRE = *P. drechsleri*, GAL1 = *P.* sp. gallica-like 1, GAL2 = *P.* sp. gallica-like 2, KEL = *P.* sp. kelmania, MAC = *P. macrochlamydospora*, MUV1 = *P.* sp. multivesiculata-like 1, PSC = *P. pseudocryptogea*, QUI = *P.* sp. quininea-like, RAM = *P. ramorum*, SYL1 = *P.* sp. sylvatica-like 1, SYL2 = *P.* sp. sylvatica-like 2, SYL3 = *P.* sp. sylvatica-like 3, ×GRE3 = *P.* sp. ×Grenada 3-like, ×HET = *P. ×heterohybrida*, ×INS = *P.* sp. ×insolita-like, ×KUN = *P.* sp. ×kunnunara-like, ×PER4 = *P.* sp. ×Peru 4-like, ×VIR = *P.* sp. ×virginiana-like s.l., VIE = *Nothophytophthora vietnamensis*. [c] Mating types: A1, A2, A1ho (homothallic and stimulating oogonia formation in A2 tester strains). [d] *Elongisporangium* sp. Hoàng Liên also isolated. [e] Unidentified *Pythium* sp. also isolated. [f] *Phytopythium vexans* aff. also isolated. [g] *Pythium senticosum* also isolated. [h] *Pythium* sp. ×ZSF0056-like also isolated. [i] *Phytopythium* sp. 1 PB-2013 also isolated. [j] *Phytopythium litorale* also isolated. [k] *Pythium* sp. CAL_2011f also isolated. [l] *Pythium* sp. 1_MNS-2013 also isolated. [m] *Pythium* sp. 2_ROH-2015 also isolated. [n] *Phytopythium palingenes* also isolated.

In total, nine previously unknown *Phytophthora* species from four of the five subclades within Clade 2 were detected in forest stands. From Clade 2a, *P.* sp. meadii-like 1 was isolated from the montane *A. nepalensis* stand F13 at 1717 m a.s.l. on Xin Chài mountain, while *P.* sp. meadii-like 2 was found in the tropical lowland rainforest stand F20 in Cuc Phuong NP (Figure 1; Table 1). The ITS sequences of all isolates of *P.* sp. meadii-like 1 were identical except for one isolate with an extra T in position 11 and an A instead of a T in position 12 (Supplementary Table S1). This new taxon differed in the ITS from *P. meadii* (isolate P75; GenBank no. GU993903) at positions 137 and 632 which were shared with the ex-type isolate of *P. botryosa* (CBS586.69; GenBank no. HQ643151), and from *P. botryosa* at five positions (72, 152, 444, 460, 773) which were identical with *P. meadii*. In addition, all isolates of *P.* sp. meadii-like 1 had a unique deletion at position 146. The ITS sequences of *P.* sp. meadii-like 2 showed intraspecific variability at positions 11, 13, 22. Most isolates differed from *P.* sp. meadii-like 1, *P. meadii* isolate P75 and the ex-type isolate of *P. botryosa* by having four unique heterozygous positions (161, 444, 502, 713). In addition, *P.* sp. meadii-like 2 showed in the ITS the same differences to *P. meadii* and *P. botryosa* as *P.* sp. meadii-like 1. The ITS sequences of both new taxa showed differences to the ex-type isolate of the recently described *P. mekongensis* from southern Vietnam (CBS135136; GenBank no. KC875838) at eight positions (152, 155, 163, 165, 166, 175, 179, 750). A third new taxon from Clade 2a, *P.* sp. botryosa-like 2, was exclusively isolated from the tropical lowland rainforest stand F25 on Côn Đảo island (Figure 1; Table 1). The ITS sequences of all isolates were identical to each other and differed from *P. botryosa* and *P. meadii* at four (72, 137, 161, 460) and five positions (152, 161, 444, 632, 773), respectively. In a 610 bp alignment of *cox1*, *P.* sp. meadii-like 1, *P.* sp. meadii-like 2 and *P.* sp. botryosa-like 2 differed from *P. meadii* (isolate p75; GU945489) at 14, 13 and 12 positions, respectively, and from *P. botryosa* (HQ261256) at 10, 9, and 8 positions, respectively. According to sequence analyses, the closest relatives of *P.* sp. botryosa-like 2 were an isolate obtained in 1930 from *Cocos nucifera* in Sulavesi (CBS235.30) which differed in ITS (HQ643140) by five heterozygous positions and in *cox1* (HQ708214) at five positions, and an isolate of unknown origin which was obtained from a vanilla plant in 1928 (CBS238.28) and showed differences at five positions in both ITS (HQ643139) and *cox1* (HQ708213) of which three were heterozygous in ITS. All isolates of *P.* sp. botryosa-like 2, *P.* sp. meadii-like 1 and *P.* sp. meadii-like 2 produce caducous papillate sporangia with variable shapes and are heterothallic, exclusively belonging to mating type A1. Oospore abortion rates in mating tests with A2 tester strains of *P. meadii* and *P. botryosa* exceeded 95%.

From the montane *A. nepalensis* stand F13 a new *Phytophthora* species from Clade 2b was isolated which differed from the ex-type isolate of *P. tropicalis* (CBS434.91) in ITS (HQ643369) and *cox1* (HQ708417) at 5 and 7 positions, respectively, and is hence informally designated as *P.* sp. tropicalis-like 2. Similar to *P. tropicalis*, all isolates produce thickwalled chlamydospores and papillate sporangia. *Phytophthora* sp. tropicalis-like 2 differs from *P. tropicalis* [49] by producing sporangia which are only partially caducous and have shorter pedicels (24.7 ± 16.8 μm vs. > 50 μm) and shorter length/breath (l/b) ratio (1.8 ± 0.3 vs. 1.8–2.4).

Phytophthora citricola VII, informally designated from a mountain forest in Taiwan [10], and another three new taxa from the '*P. citricola* complex' in Clade 2c, informally designated here as *P. citricola* IX, *P. citricola* X and *P. citricola* XI, were isolated from the montane *A. nepalensis* stand F13, the subtropical evergreen forest stand F16 and the tropical lowland rainforest stands F21 and F25, respectively (Figure 1; Table 1). *Phytophthora citricola* VII, IX, X and XI differ from the authentic type of *P. citricola* s.s. (CBS295.29; ITS–FJ560913; *cox1*—KC855432) in the ITS (771 bp alignment) at 3, 3, 12 and 11 positions, and in *cox1* (1231 bp alignment) at 23, 19, 29, and 15 positions, respectively. Like other members of the '*P. citricola* complex', the four new species are homothallic forming smooth-walled oogonia with paragynous antheridia. The sporangia of *P. citricola* VII and IX resemble those produced by other species from Clade 2c in being semipapillate, persistent and with exclusively external proliferation. In contrast, *P. citricola* X and XI produce mainly papillate sporangia with both external and, infrequently, also internal extended and nested proliferation. In addition, *P. citricola* X is distinguished from all known related species by forming abundant catenulate hyphal swellings in water. *Phytophthora* citricola VII produces a high proportion of zoospores with a ring-like to oval coiling of both flagella ends.

From a swampy depression in the montane evergreen cloud forest F06 in Hoàng Liên NP, a previously unknown *Phytophthora* species from Clade 2e was isolated which is provisionally named as *P.* sp. multivesiculata-like 1. Its ITS and *cox1* sequences differ from the ex-type isolate of *P. multivesiculata* (CBS545.96; HQ643288 and HQ708340) at eight and 38 positions, respectively. The ITS sequences of two yet undescribed species, *Phytophthora* sp. aquatilis (GenBank no. FJ666126) and *Phytophthora* sp. Costa Rica 5 (KC479200), show differences to *P.* sp. multivesiculata-like 1 at 8 and 6 positions, respectively. Like *P. multivesiculata*, *P.* sp. multivesiculata-like 1 is homothallic with aplerotic oospores and produces in water numerous catenulate hyphal swellings and both nonpapillate and semipapillate sporangia with external and internal proliferation. However, it can easily be distinguished from *P. multivesiculata* [50] by forming considerably larger sporangia (on av. 57.2 × 32.8 vs. 45 × 33 μm), larger oogonia (45 vs. 41 μm) with highly variable shapes ranging from globose, excentric or elongated with long tapering bases to comma-shaped, and exclusively amphigynous antheridia.

Also in Hoàng Liên NP, *P. gregata* from Clade 6b was recovered from the rhizosphere of *Meliosma henryi* and *Neolitsea merilliana* in the montane evergreen cloud forest F06 while the other Clade 6b species *P. chlamydospora* and *P. ramorum* from Clade 8c were isolated from fallen leaves of *R. arboreum* collected from the forest ground in the montane, evergreen broadleaved forest F11 (Table 1).

Besides the recently described *Nothophytophthora vietnamensis* [51] which was isolated from the rhizosphere of *C. acuminatissima* and *Acer campbellii* in the montane evergreen cloud forest F04 and *A. nepalensis* in stand F13 on Xin Chài mountain, a range of *Pythium* and *Phytopythium* species including *Py. intermedium*, *Py. senticosum*, *Ph. chamaehyphon*, *Ph. cucurbitacearum*, *Ph.* sp. 1 PB-2013, 14 haplotypes from the *Ph. vexans* complex and two previously unknown taxa, informally designated as *Py.* sp. conidiophorum-like and *Ph.* sp. Côn Đảo, were obtained from 15 forest stands (Table 1 and Supplementary Table S1).

3.2. *Phytophthora* Diversity in Natural Forest Streams and Rivers

Using rafts with leaves of *C. indica*, *C. sinensis*, *L. bacgangensis*, *Q. glauca*, and, less frequently, *Carpinus* sp., *C. hodginsii*, *Cinnamomum iners*, *Dipterocarpus alatus*, *Prunus* sp., *Q. gilva* and *A. mangium* as in situ baits in all 16 rivers and streams tested, and freshly fallen leaves of different tree species and flowers of *R. arboreum* and *R. leptocladus* in four forest streams, seven known species (*P. capensis*, *P. chlamydospora*, *P. drechsleri*, *P. macrochlamydospora*, *P. pseudocryptogea*, *P. ramorum*, *P.* ×*heterohybrida*), five informally designated taxa (*P. citricola* VII, *P.* sp. kelmania, *P.* sp. ×insolita-like, *P.* sp. ×kunnunara-like, *P.* sp. ×virginiana-like s.l.) and 12 previously unknown taxa of *Phytophthora* were isolated (Table 2). The latter included *P.* sp. multivesiculata-like 1, two new species from the '*P. citricola* complex', three and one new species related to the Clade 6 taxa *P.* sp. sylvatica and *P.* sp. bitahaiensis, respectively, three new species from Clade 9 and two new species related to *P. gallica* from Clade 10.

The *Phytophthora* communities in the 11 montane streams above 1000 m a.s.l. with a temperate climate were dominated by species belonging to Clades 2, 6, 7, and 8 whereas from the five lowland rivers with subtropical to tropical climate almost exclusively *Phytophthora* species from Clade 9 were obtained (Figure 1; Table 2).

In montane streams, the most widespread species was *P. ramorum* which could be recovered from seven of the eight forest streams above 1890 m altitude in the Fansipan area and in 8–12 km distance to these sites from stream R11 originating from the *C. hodginsii* forest F24 at Sau Chua mountain in 1300 m altitude. Both mating types were obtained with the A1 mating type occurring in eight streams and the A2 in two streams. In the latter (streams R03, R04) both mating types co-occurred (Table 2). In the two streams (R01, R02) sampled in both 2016 and 2017 only mating type A1 was isolated. The 65 *P. ramorum* isolates exhibited five slightly different ITS genotypes. Eight isolates from four streams (R02, R04, R05, R10) were identical to the ex-type isolate from Germany, which belongs to the EU1 lineage (CBS101553; HQ643339). The most common genotype (46 isolates) differed from the ex-type by having a T instead of a Y at position 616, while three isolates from three streams (R01, R02, R10) had a C at this position. Four *P. ramorum* isolates from streams R02 and R03 were distinguished from the ex-type by being heterozygous at position 682 (R instead of G) while in one isolate from stream R03 both heterozygous positions occurred. For 43 isolates, representative for all eight streams, forest site F11 and both mating types, *cox1* was sequenced and compared to representative isolates of the four known *P. ramorum* lineages EU1, EU2, NA1, and NA2. In a 1240 bp long *cox1* alignment all but one of the Vietnamese *P. ramorum* isolates were identical and differed from four representative isolates of the North American NA2 lineage only at position 123. They differed from EU1 by 5 bp (positions 123, 808, 1141, 1156, 1202), EU2 by 5 bp (624, 966, 1035, 1156, 1240) and NA1 by 4 bp (123, 808, 1156, 1202), respectively. Isolate VN88 differed from the other 42 isolates by having a unique polymorphism at position 1228 (C instead of A). The morphological structures of all isolates were congruent with the original description of *P. ramorum* [52].

From Clade 2c, *P. capensis*, *P. citricola* VII, *P. citricola* VIII, and *P. citricola* IX were isolated from two, one, five, and one montane forest streams, respectively (Table 2). From stream R11, *P. capensis*, and *P. citricola* VII and VIII were obtained while in another two streams (R01, R10) two different species from the '*P. citricola* complex' co-occurred. Compared to the ex-type isolate of *P. capensis* (P1819; ITS—GU191232; *cox1*—GU191275) from South Africa, the ITS sequences of the Vietnamese isolates were identical but their *cox1* sequences were separated in a 598 bp alignment by 9 bp (1.5%). *Phytophthora citricola* VIII differed from the authentic type isolate of *P. citricola* s.s. (CBS295.29; ITS—FJ560913; *cox1*—KC855432) and from *P. citricola* VII, IX, X and XI in ITS (771 bp alignment) at 2, 2–3, 4, 12 and 11 positions, and in *cox1* (1231 bp alignment) at 33, 26, 35, 42, and 37 positions. Being homothallic with paragynous antheridia and producing semipapillate sporangia of variable shapes, *P. citricola* VIII morphologically resembles other species from Clade 2c.

Phytophthora sp. multivesiculata-like 1 from Clade 2e was present in streams R01 and R02 which originate from a catchment area around forest stand F06 where this new taxon was also found.

From Clade 6b, which contains numerous predominantly aquatic *Phytophthora* species, *P. chlamydospora*, three new species informally designated as *P.* sp. sylvatica-like 1, 2 and 3, and another new species designated as *P.* sp. bitahaiensis-like were recovered from 8 of the 11 mountain streams (Table 2). Most common were *P.* sp. sylvatica-like 2 (six streams), *P. chlamydospora* (5 streams) and *P.* sp. sylvatica-like 3 (four streams). In five streams more than one Clade 6b species were found. In the ITS (844 bp alignment) and *cox1* (861 bp alignment), *P.* sp. sylvatica-like 1 differs from its closest relative *P.* sp. forestsoil-like from Taiwan (KU682574) at 2 and 7 positions, respectively, while *P.* sp. sylvatica-like 2 and 3 show differences to *P.* sp. forestsoil-like in ITS at 12–13 and 11–12 positions, respectively, and in *cox1* at 29 and 27 positions. *Phytophthora* sp. sylvatica-like 1 differs from *P.* sp. sylvatica-like 2 and 3 in ITS by 9–11 and 7–8 bp, respectively, and in *cox1* by 44 and 46 bp, respectively. The latter two species can be distinguished in the ITS and *cox1* by differences at 2–4 and 11 positions, respectively. *Phytophthora* sp. bitahaiensis-like differs in the ITS from *P.* sp. bitahaiensis (isolate BHL1; KT183432) from a forest stream in Yunnan, China, at 4 positions. Unfortunately, for *P.* sp. bitahaiensis no *cox1* sequences are available. Similar to *P.* sp. forestsoil-like and many other aquatic Clade 6 species [10,21,53], *P.* sp. sylvatica-like 1, 2, and 3 and *P.* sp. bitahaiensis-like are sterile and form abundantly nonpapillate sporangia with internal nested and extended proliferation. All ten sequenced isolates from *P. chlamydospora* were identical in both the ITS and *cox1* and differed from the ex-type isolate of *P. chlamydospora* from the UK (P236; AF541900, MH136867) only by being in the ITS heterozygous at position 57 (R instead of G) while being identical to the ex-type in *cox1*.

From five mountain streams above 1890 m altitude, the recently described Clade 7a hybrid species *P. ×heterohybrida* was isolated. All 12 isolates differed in the ITS from the ex-type isolate from Taiwan (CBS141207; KU517151) at position 77 (Y instead T). In addition, three isolates from stream R01 were separated from the ex-type by being homozygous at position 428 (T instead of Y) and by having two unique heterozygous positions (656 and 748). In a 876 bp alignment of *cox1*, one isolate from stream R01 differed at three positions from the ex-type isolate (KU517145) and from all other Vietnamese and

Taiwanese isolates which were identical. Mating tests with A1 and A2 tester strains of *P.* ×*heterohybrida* from Taiwan showed that all isolates from Vietnam belonged to the A1 mating type. In addition, two isolates from stream R01 produced oogonia abundantly in single culture. The morphology of the ornamented oogonia, mostly two-celled amphigynous antheridia and nonpapillate sporangia of the Vietnamese isolates matched the original description of *P.* ×*heterohybrida* [22].

Phytophthora pseudocryptogea and *P.* sp. kelmania from Clade 8a were isolated from each two of the three lower montane streams R09-R11 (Table 2). The three isolates of *P. pseudocryptogea* from streams R09 and R10 differed in ITS (816 bp alignment) and *cox1* (672 bp alignment) from the Australian ex-type isolate (VHS16118; KP288376, KP288342) at three and one positions, respectively. The five isolates of *P.* sp. kelmania from streams R09 and R11 showed differences in ITS (841 bp alignment) and *cox1* (582 bp alignment) to isolate P10614 (HQ261691, HQ261438) from North America at 3, 4, and 5 positions, respectively. The morphology of all Vietnamese isolates of *P. pseudocryptogea* and *P.* sp. kelmania matched the descriptions in literature [54].

From forest stream R02 in 2007 m altitude two new species from Clade 10 were isolated from naturally fallen leaves. *Phytophthora* sp. gallica-like 1 and *P.* sp. gallica-like 2 were distantly related to *P. gallica* differing in a 885 bp alignment of the ITS from the ex-type isolate of the latter (CBS 111474 = GAL1; DQ286726) at 53 and 68 positions, respectively, while being separated from each other by 49 bp. Morphologically, both new species can easily be distinguished from the sterile *P. gallica* [55] by being homothallic forming smooth-walled oogonia with paragynous antheridia. *Phytophthora* sp. gallica-like 1 produces globose chlamydospores like *P. gallica* whereas *P.* sp. gallica-like 2 does not form chlamydospores.

From Clade 9a2, *P. macrochlamydospora* and a new species related to *P. quininea*, informally designated as *P.* sp. quininea-like, co-occurred in montane forest stream R11. Isolate VN1006 of *P. macrochlamydospora* differed in the ITS (816 bp alignment) and in *cox1* (670 bp alignment) from the Australian ex-type isolate (P10263; FJ801351, MH136923) at 2 and 3 positions, respectively. In accordance with the original description of *P. macrochlamydospora* [2], the Vietnamese isolate was sterile and produced large chlamydospores and semipapillate to non-papillate sporangia. The four isolates of *P.* sp. quininea-like were separated in the ITS and *cox1* from the ex-type isolate of *P. quininea* (CBS407.48 = P8488; HQ261660; AY564200 + HQ708386) from Peru by differences of 7 and 27 bp, respectively. Like *P. quininea*, *P.* sp. quininea-like produces non-papillate sporangia with internal and external proliferation, catenulate irregular hyphal swellings and thick-walled large chlamydospores. Interestingly, two of the four isolates were sterile while the other two isolates were homothallic like *P. quininea* [2] but can be distinguished from the latter by forming amphigynous instead of paragynous antheridia. Only one isolate of the Clade 9a1 hybrid taxon *P.* sp. ×kunnunara-like could be obtained from one lower montane stream (R09) in Hoàng Liên NP (Table 2).

In contrast to the montane streams, the *Phytophthora* communities in the five lowland rivers with subtropical to tropical climate were dominated by *Phytophthora* taxa from the high-temperature tolerant Clades 9a1 and 9a3. From Clade 9a1 potential hybrid isolates related to *P. virginiana* were obtained from the Red River, the Black River and two other streams (R13, R16) (Table 2). The potential hybrids differed in the ITS from the ex-type isolate of *P. virginiana* (46A2; KC295544) by having in total 10 heterozygous positions, with 1–8 heterozygous positions per isolate, which are partly not present in the three hybrid taxa *P.* sp. ×virginiana-like 1, 2, and 3 from Taiwan. Therefore, these Vietnamese isolates are informally designated as *P.* sp. ×virginiana-like sensu lato. *Phytophthora* sp. ×kunnunara-like was found in the Red River and the two streams originating from Ba Vì NP (R13, R14) (Table 2). Compared to *P.* sp. kunnunara from Western Australia, the ITS of the Vietnamese isolates had 10 heterozygous positions, with 1–8 heterozygous positions per isolate, which were only partly shared with Taiwanese isolates of *P.* sp. ×kunnunara-like (KU682602, KU682603). Another swarm of potential hybrid isolates was abundantly obtained from four of the five lowland streams. They differed in the ITS from *P.* sp. Peru 4 (KC479209) at 10 positions which were all heterozygous (1–6 per isolate) and are, hence, informally designated as *P.* sp. ×Peru 4-like. Finally, isolates of another new potential hybrid taxon, *P.* sp. ×Grenada 3-like were recovered from stream R13 which were distinguished in the ITS from *P.* sp. Grenada 3 (KC479208) by having five instead of one heterozygous position. All isolates of *P.* sp. ×Grenada 3-like, *P.* sp. ×kunnunara-like, *P.* sp. ×Peru 4-like and *P.* sp. ×virginiana-like were in culture fast-growing, self-sterile and produced intercalary or laterally globose, club-shaped to irregular hyphal swellings, mostly globose thin-walled chlamydospores and nonpapillate sporangia with internal nested and extended proliferation, typical features of aquatic Clade 9 species [56].

Phytophthora sp. ×insolita-like from Clade 9a3 was found in stream R13. Compared to the Taiwanese ex-type isolate of *P. insolita* (IMI288805; AF271222) the Vietnamese isolates showed differences in the ITS at eight positions of which four were heterozygous and mostly shared with Taiwanese isolates of *P.* sp. ×insolita-like (KU682601). Morphologically, all isolates from stream R13 were similar to both *P. insolita* and *P.* sp. ×insolita-like from Taiwan [2,10], producing in single culture smooth-walled oogonia without antheridia, thin-walled chlamydospores and non-papillate sporangia with internal nested and extended proliferation.

The only *Phytophthora* species recovered from a lowland stream (Black River) and not belonging to Clade 9 was *P. drechsleri* from Clade 8a. The isolates differed from the ex-type isolate of *P. drechsleri* (ATCC 46724 = 23J5; AF266798, MH620076) in the ITS at one position (136; Y instead of T) and in a 862 bp alignment of *cox1* at five positions. All three isolates belonged to the A1 mating type.

With 10, eight and six *Phytophthora* species, respectively, the montane forest streams R10, R11 and R02 harboured the highest diversity of *Phytophthora* species while the lowland rivers contained highly diverse assemblies of Clade 9 hybrids with almost all isolates being different from each other in the ITS.

Nothophytophthora vietnamensis was isolated from the montane forest streams R01, R02 and R10 in Hoàng Liên NP. In addition, the novel *Elongisporangium* sp. Hoàng Liên, *Phytopythium litorale*, *Ph. vexans* s.l., *Ph.* sp. 1 PB-2013, *Py. senticosum*, *Py.* sp. CAL_2011f, *Py.* sp. 1_MNS-2013, the previously unknown hybrid taxon *Py.* sp. ×ZSF0056-like and unidentified *Pythium* spp. were recovered from eight of the eleven mountain streams (Table 2 and Table S1). In four of the five lowland rivers, *Ph. palingenes*, *Ph.* sp. 1 PB-2013, *Py.* sp. 1_MNS-2013, *Py.* sp. 2_ROH-2015 and unidentified *Pythium* spp. were found (Table 2 and Table S1).

3.3. Association between Phytophthora Presence in the Rhizosphere and Disease Symptoms

In the 20 *Phytophthora*-inhabited forests sampled, the majority of the 52 tree species from which *Phytophthora* species were recovered appeared generally healthy (Figure 2a–f). Symptoms indicative of Phytophthora root diseases were almost exclusively found in eight montane forest stands and were mainly restricted to tree species belonging to the Ericaceae, Fagaceae and Lauraceae (Figures 3 and 4).

In Hoàng Liên NP, scattered dieback of *Rhododendron arboreum* trees with presence of *P. cinnamomi* A1 and *P.* sp. attenuata-like 1 in the rhizosphere was observed in the upper montane *Rhododendron* cloud forest at 2636 m altitude. In the montane evergreen cloud forest F03 at 2337 m altitude, groups of mature *Quercus glauca* trees showed severe thinning and dieback of the crowns (Figure 3a) which was associated with presence of *P. cinnamomi* A2 in the rhizosphere. In contrast, in forest stand F12 at 1895 m altitude, infested by *P. cinnamomi* A1, all *Q. glauca* trees appeared healthy. In six of the nine montane evergreen forest stands sampled between 1895 and 2242 m altitude, *C. acuminatissima* and sometimes also *Neolitsea poilanei* and *N. polycarpa* showed severe thinning and dieback of the crowns and mortality (Figure 3b–d,f). Disease incidence was particularly high in the swampy depression sampled in stand F06 (Figure 3c,d,f). In seven and five of the nine stands, *P. cinnamomi* A2 and *P. castaneae*, respectively, were recovered from rhizosphere soil samples while *P. cinnamomi* A1, *P. attenuata*, *P. gregata*, *P.* sp. attenuata-like 1, *P.* sp. attenuata-like 2 and *P.* sp. multivesiculata-like 1 were only infrequently found (Table 1). On a visual examination root samples from three declining trees each of *C. acuminatissima* and *N. polycarpa*, all infested by *P. cinnamomi* A2 and/or *P. castaneae*, exhibited severe losses of lateral and fine roots and open callussing lesions on coarse roots (Figure 4a–d). In stand F07, *P. cinnamomi* A2 was isolated from a bleeding bark lesion on a surface root of a declining *C. acuminatissima* tree (Figure 3e).

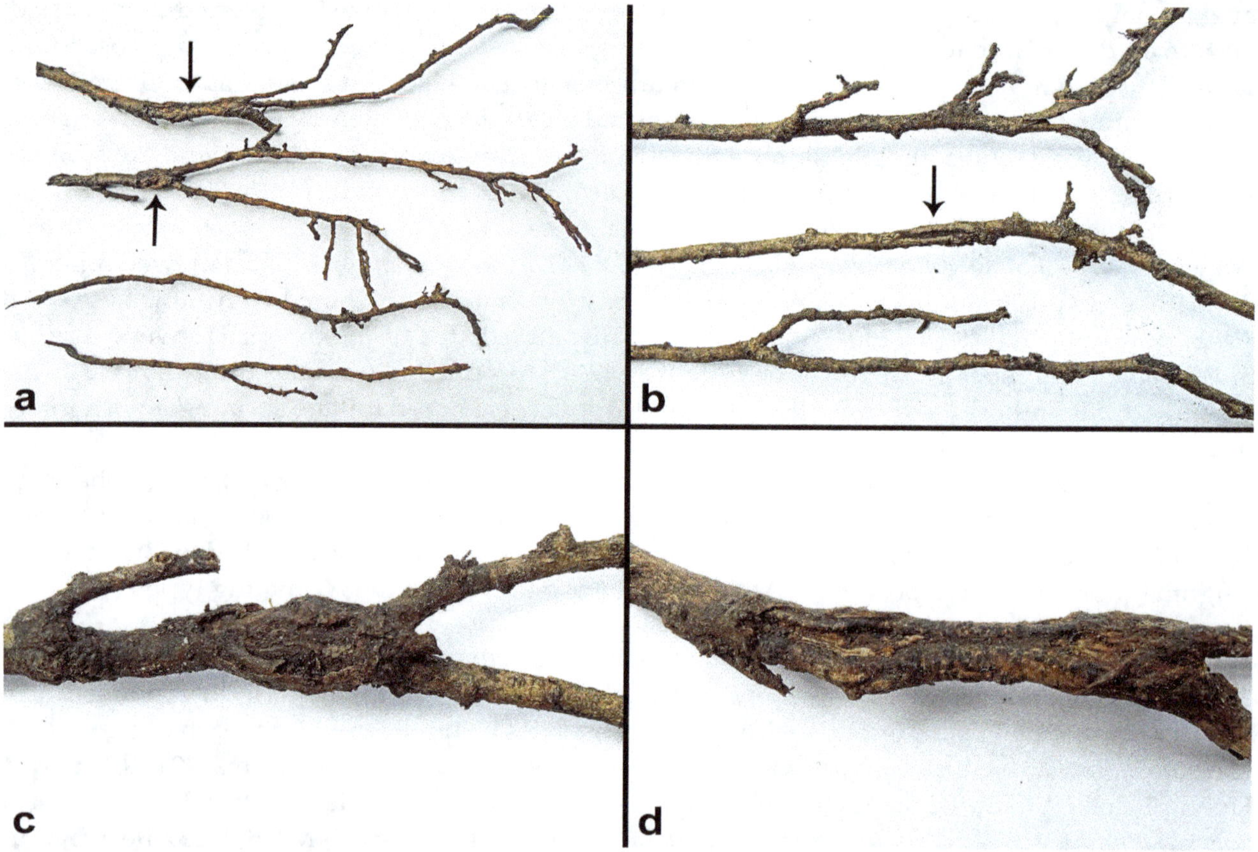

Figure 4. Symptoms on root systems of declining *Neolitsea polycarpa* trees in the montane evergreen cloud forest F10 in Hoàng Liên National Park associated with presence of *P. cinnamomi* A2 in the surrounding soil; (**a**,**b**) severe losses of lateral roots and fine roots and open callussing lesions on coarse roots (arrows); (**c**,**d**) detailed view of the open callussing lesions on the coarse roots from Figure 4a.

In contrast to the montane forests of Hoàng Liên NP, in the three submontane (700–1100 m a.s.l.) stands sampled in the subtropical, humid evergreen forests of Ba Vì NP all tree species, including several species from the Fagaceae genera *Castanopsis*, *Lithocarpus* and *Quercus* and the Lauraceae genera *Litsea*, *Machilus* and *Phoebe*, were healthy despite the occurrence of both mating types of *P. cinnamomi* and a range of five other *Phytophthora* species, including *P. castaneae*, *P. heveae*, *P. parvispora*, *P. citricola* IX and *P.* sp. attenuata-like 3, in the soil. The only exception was a small patch dieback of *Dysoxylum juglans* with presence of *P.* sp. attenuata-like 3 in the rhizosphere (Figure 3g).

In the five *Phytophthora*-infested tropical lowland rainforest stands in Cuc Phuong (F18, F20, F21), Bù Gia Mập (F23), and Côn Đảo (F25) National Parks, no symptoms suggestive of *Phytophthora* diseases were observed.

4. Discussion

Vietnam harbours an extremely diverse flora probably due to its heterogeneous geology, geomorphology and climates and its transitional position between the eastern Himalayas, Yunnan, and the Indomalaysian archipelago on the Asian continental shelf [39–41]. The latter enabled repeated immigrations of plant and most likely also fungal and oomycete species during various glacial periods followed by subsequent speciations and species radiations in the interglacials. A similar scenario was proposed earlier for Taiwan [10,36–38]. We have shown here that the floristic and environmental diversity of Vietnam is reflected by the high diversity of oomycete taxa. In this survey of 25 natural forests and 16 rivers 13 described species, five informally designated taxa and 21 previously unknown taxa of *Phytophthora*, together with *N. vietnamensis* and a range of seven described and ten undescribed species of *Elongisporium*, *Pythium* and *Phytopythium* were obtained. Considering the relatively limited number and diversity of the sampled sites and ecosystem types it may be assumed that the true *Phytophthora* diversity of Vietnam is markedly higher. The finding of 20 *Phytophthora* taxa in 98 soil and four leaf samples from the 25 forest stands and an additional 15 *Phytophthora* taxa in 11 forest streams in Vietnam indicates a much higher diversity of forest *Phytophthora*s exists in Vietnam than occurs in Europe, the eastern US or the western US. In the latter areas, 39, 7, and 21 *Phytophthora* species, respectively, were detected in numerous surveys involving many more samples collected over much larger areas and a wider range of ecosystems [5,14,23,44,57–59].

The remote location of most sampled forest stands and forest streams in Vietnam, absence of introduced crop or tree species in the catchment areas and, apart from *P. cinnamomi* A2 in higher altitudes, the lack of association of *Phytophthora* with obvious disease symptoms suggest that most of the 35 forest *Phytophthora* species obtained are native to Vietnam. In contrast, only nine of the 32 *Phytophthora* species from European forests are considered indigenous [5,23,60]. The forest *Phytophthora* populations in Vietnam and Europe share only five species, *P. chlamydospora*, *P. cinnamomi* A2, *P. pseudocryptogea*, *P. ramorum* and *P.* sp. kelmania, while Vietnamese and North American forests have only *P. chlamydospora*, *P. cinnamomi* A2 and *P. ramorum* in common. Recent surveys in Taiwan, where floristic diversity is comparable to Vietnam, revealed a comparable *Phytophthora* diversity, with ten described and 17 previously unknown species from 30 forest stands and 25 streams [10,13]. Further, the *Phytophthora* communities revealed in Vietnam and Taiwan shared 12 taxa: *P. attenuata*, *P. capensis*, *P. castaneae*, *P. chlamydospora*, *P. cinnamomi* A1 and A2, *P. citricola* VII, *P. heveae*, *P. parvispora*, *P.* ×*heterohybrida*, *P.* sp. ×insolita-like, *P.* sp. ×kunnunara-like and *P.* sp. ×virginiana-like *s.l.*. In three areas in northern Yunnan, a Chinese province adjacent to northern Vietnam, eight *Phytophthora* species were isolated from streams running through sclerophyllous oak forests but only two of them, *P. chlamydospora* and *P. plurivora*, were recovered from forest soil samples [16]. The only *Phytophthora* species common to Vietnam and northern Yunnan were *P. chlamydospora* and *P. gregata*. In montane forests of the tropical island Hainan, located in the South China Sea close to Vietnam, six *Phytophthora* species were found [12] of which three species, *P. castaneae*, *P. cinnamomi* and *P. heveae*, also occurred in Vietnam. The lower *Phytophthora* diversities in the north Yunnan and Hainan surveys compared to Vietnam were most likely due to the smaller number of sites and forest types sampled and the use of different isolation techniques.

In recent years, an impressive diversity of both known and previously unknown *Phytophthora* species has been revealed from stream surveys in several countries, including the eastern and western USA, Chile, Australia, South Africa and Taiwan [10,11,14,17,18,44,61], as discussed previously [10]. By comparison the riparian *Phytophthora* communities identified here in Vietnam are remarkably rich, with seven described species, five informally designated taxa and 12 previously unknown taxa. Several montane streams with small catchments in the forests around the Fansipan harboured an unprecedented diversity of up to ten *Phytophthora* species per stream.

Interestingly, the most common *Phytophthora* species in Vietnamese forest soils, *P. cinnamomi*, *P. castaneae*, *P. heveae* and the four species from the '*P. attenuata* complex', were never isolated from streams running through or originating from infested forests. Overall, the *Phytophthora* communities found in the forest soils (20 taxa) and in the streams (24 taxa) shared only four species, *P. chlamydospora*, *P. citricola* VII and IX and *P.* sp. multivesiculata-like 1. Similar differences between terrestrial and aquatic *Phytophthora* populations were observed in comparable studies in Europe, Chile, Taiwan, South Africa and the USA [10,11,18,44,58]. This is consistent with previous observations that most *Phytophthora* species are adapted either to a soilborne and root-infecting or aerial foliage-infecting lifestyle, or are aquatic saprotrophs that tend to be opportunistic pathogens [6,10,11,21,53,62]. Consequently, when sampling *Phytophthora* diversity in a diverse environment both soils and streams should be analysed using optimal baiting methods for each or metagenomic approaches based on high-throughput pyrosequencing of environmental DNA with *Phytophthora*-specific primers [19,63]. Ideally, because metagenomic analyses can sometimes result in false molecular operational taxonomic units (MOTUs), and because living isolates are needed for taxonomic descriptions and host range testing, baiting, and metagenomic approaches should be carried out in parallel.

Altitude had a strong influence on *Phytophthora* distribution. The '*P. attenuata* complex', *P. castaneae* and *P. cinnamomi* occurred only in soils of submontane and montane forests above 700 m a.s.l. while *P. heveae* and most taxa from Clade 2a were restricted to forests below 1100 m altitude. The altitudinal influence on aquatic *Phytophthora*s was even more pronounced. While the 11 montane streams above 1000 m altitude with a subtropical to temperate climate contained mainly species belonging to Clades 2c, 2e, 6b, 7a, and 8c, the *Phytophthora* communities in the five lowland rivers with subtropical to tropical climate were dominated by species and hybrids from the high-temperature tolerant Clades 9a1 and 9a3 [28]. Clade 9 species and hybrids were also most common in lowland streams in Taiwan and South Africa [10,18].

The results of this survey offer new insights into the origin of several invasive *Phytophthora* pathogens and of clades and subclades of *Phytophthora*. Most notably, the finding of the highly invasive, wide-host range pathogen *P. ramorum* in eight forest streams around the Fansipan and Sau Chua mountains with both A1 and A2 mating types present, together with an apparent absence of overtly visible disease symptoms on potentially susceptible Ericaceae, Fagaceae, or Lauraceae, suceptible genera where *P. ramorum* is damaging and introduced in Europe and North America [6,7,64], suggests an equilibrium between the pathogen and the north Vietnamese vegetation as a consequence of long term endemism and co-evolution. This is supported by variability in the ITS and *cox1* sequences of the Vietnamese isolates and by *cox1* sequence differences between the Vietnamese isolates and the North American NA1 and NA2 and the European EU1 and EU2 lineages. Because of the implications both for the origin of the pathogen and for international biosecurity a detailed comparative phenotypic and molecular analysis of the Vietnamese *P. ramorum* isolates and the known EU1, EU2, NA1, and NA2 lineages [65] is currently ongoing to further characterise the Vietnamese population and its relationship to the known lineages. Since southern Yunnan, northern Laos, and the eastern Himalayas belong to the same biogeographic area as the Fansipan region mountain forests in these regions may also harbour endemic *P. ramorum* populations. Further surveys are needed to confirm this hypothesis.

Phytophthora cinnamomi was the most common soilborne *Phytophthora* species above 700 m. The A2 mating type of *P. cinnamomi* was more widespread, occurring in 11 forest stands between 713 and 2337 m, whereas the A1 occurred only in four forest stands located between 1108 and 2636 m a.s.l. In Taiwan and Papua New Guinea also the A1 mating type occurrs at higher altitudes than the A2 indicating higher tolerance to low temperatures [10,66]. However, in both of these locations the altitudinal differences between the mating types are larger than in Vietnam, the A2 being confined to the lowland forests. In each one stand in Hoàng Liên NP and Ba Vì NP both mating types co-occurred. In Hoàng Liên NP the A1 was present in the upper montane *Rhododendron* forest at 2636 m and in two lower montane stands at 1900 m. However, it was not detected in the eight forest stands between 2337 and 2022 m in which not only the A2 type was present but severe dieback of Fagaceae and Lauraceae was observed (notably *C. acuminatissima*, *Q. glauca*, *N. poilanei* and *N. polycarpa*). Pathogenicity trials are required to fulfill Koch' s postulates for these host-pathogen associations and confirm that *P. cinnamomi* A2 is causing the dieback of these native Fagaceae and Lauraceae species. In contrast, no dieback was observed in the three forest stands in Ba Vì NP between 713 and 1100 m, despite the presence of *P. cinnamomi* A2. In the two stands with the co-occurrence of both mating types the A1:A2 ratio of the 44 isolates was 59.1:40.9, whereas the overall mating type ratio of the 151 isolates from 13 *P. cinnamomi* infested stands was 30.5:69.5. Collectively, these results suggest that, as a consequence of current climatic warming, the more thermophilic but frost sensitive A2 mating type may be spreading into higher altitudes in Vietnam. Such a progression of *P. cinnamomi* A2 into higher latitudes and altitudes with climate change was predicted by CLIMEX modelling [67–69]. In the newly A2 invaded high-altitude forests in Vietnam the A2 may be outcompeting and replacing the native co-evolved A1, causing dieback in the susceptible non-coevolved hosts. The widespread distribution of *P. cinnamomi* in northern Vietnam, the co-occurrence of both mating types in several stands, and the absence of disease symptoms in lower altitudes also indicates that Vietnam lies within the origin of both mating types. *Phytophthora cinnamomi* is the most invasive member of the genus with a host range of almost 5000 woody plant species [2,70,71]. Two genotypes of the A2 mating type have reached a panglobal distribution causing epidemics in numerous natural and managed ecosystems while the A1 mating type has a limited distribution outside of Asia and has never been associated with epidemic disease [2,5,6,9–11,60,66,72–76].

Phytophthora attenuata, recently described from montane forests in Taiwan [22], and three closely related but previously unknown species, were found in the submontane and montane forests of northern Vietnam. *Phytophthora attenuata*, *P.* sp. attenuata-like 1 and *P.* sp. attenuata-like 2 were detected in the temperate, montane cloud forests around the Fansipan. However, *P.* sp. attenuata-like 3 was found only in the subtropical, humid submontane evergreen forests in Ba Vì NP. These four closely related species most likely result from sympatric species radiation, suggesting northern Indochina as the center of origin of the '*P. attenuata* complex'. A pathogenicity trial is required to confirm that *P.* sp. attenuata-like 3 is causing the dieback of *Dysoxylum juglans* in Ba Vi National Park. Another Clade 7a species that was first described from Taiwan, *P.* ×*heterohybrida*, was widespread in Vietnamese montane forest streams. This allopolyploid hybrid species has a functional but peculiar sexual system. In Taiwan, all isolates were selfsterile with both mating types being common and one isolate mating with both mating types [22]. In contrast, almost all Vietnamese isolates were self-sterile and belonged to the A1 mating type while one isolate was prolifically homothallic and stimulated oogonia formation in A2 tester strains.

The results indicate that the '*P. citricola* complex' from Clade 2c also underwent a species radiation process in Vietnam. Besides *P. capensis*, originally described from nursery plants in South Africa and also isolated from natural streams in Taiwan [10,77], and *P. citricola* VII, which was previously reported from a montane forest in Taiwan [10], four previously unknown taxa were found in this survey. Most common was *P. citricola* VII which occurred in mountain streams inside and outside of Hoàng Liên NP and in the rhizosphere of a montane *Alnus* forest, whereas the new taxa *P. citricola* VIII to XI had only cryptic distributions. It is noteable that *P. citricola* X and XI were only found in tropical lowland rainforests and that they differ from all other species of the '*P. citricola* complex' by producing mainly papillate instead of semipapillate sporangia. The occurrence of *P. citricola* VII to X within only ca 300 km in northern Vietnam and the co-occurrence of *P. citricola* VII, VIII and IX in individual streams suggest sympatric species radiation from a common ancestor. In contrast, the exclusive finding of *P. citricola* XI and also *P.* sp. botryosa-like 2 from Clade 2a on Côn Lôn island, situated on the Asian shelf 50 km off the southern Vietnamese coast, are more consistent with allopatric island speciation. The invasive wide-host range pathogen *P. plurivora* occurs in undisturbed, healthy, often deciduous temperate mountain forests in Taiwan, Nepal and Yunnan [10,15,16,43]. However, it was not found here in the subtropical and tropical forests of Vietnam. This suggests that *P. plurivora* is native to temperate mountainous regions of South and East Asia.

The detection of five new species and of *P. capensis* from the '*P. citricola* complex' in Clade 2c, three new species from Clade 2a and *P.* sp. tropicalis-like 2 and *P.* sp. multivesiculata-like 1 from Clades 2b and 2e in this survey, the findings of *P. bisheria*, *P. capensis*, *P. plurivora*, *P. citrophthora*, *P. tropicalis* and the three new Clade 2a species *P.* sp. ×botryosa-like, *P.* sp. ×meadii-like and *P.* sp. occultans-like from natural ecosystems in Taiwan [10,13] together with the widespread occurrence of *P. botryosa*, *P. citricola*, *P. colocasiae* and *P. meadii* across Southeast Asia [2,12,33–35] suggest South, Southeast and East Asia as the center of origin of *Phytophthora* major Clade 2.

Interestingly, all known isolates from the new Clade 2a taxa *P.* sp. meadii-like 1 and 2 and *P.* sp. botryosa-like 2 from Vietnam, as well as *P.* sp. ×botryosa-like and *P.* sp. ×meadii-like from Taiwan [10] are of A1 mating type and are characterised by oospore abortion rates exceeding 95% in mating tests with tester strains of *P. botryosa* and *P. meadii*. It appears that in this complex of aerial *Phytophthora* species the A1 is better adapted to and, hence, more common in natural forests than the A2. It is even possible that these self-sterile taxa, like many aquatic Clade 6 species [21,53], lack the A2 mating type and have abandoned sexual reproduction in favour of exclusive asexual reproduction, spreading via their caducous sporangia from infected to non-infected above-ground tissues. This possibility is supported by the extremely high oospore abortion rates in mating tests with tester strains of *P. botryosa* and *P. meadii*. More field surveys and laboratory tests are needed to verify this hypothesis.

Phytophthora castaneae and *P. heveae* from Clade 5 are also considered being native to Taiwan and Hainan [2,10,12,31]. Their widespread occurrence in Vietnamese forests and the lack of association with disease symptoms in the native vegetation indicate that Indochina also lies within the origin of both species.

As previously demonstrated in Australia, Chile, South Africa and Taiwan putative interspecific hybrids, indicated by multiple heterozygous sites in their ITS sequences, are common in watercourses and can also be found in forest soils [10,11,17,18,78,79]. As with predominantly aquatic species and hybrids from Clade 6, all Clade 9 hybrids from Vietnamese streams, with the exception of *P.* sp. ×insolita-like which produces oogonia without antheridia, are sterile and apparently adapted to rapid and continuous asexual proliferation via zoospores. Also, like many Clade 6 taxa, this may reflect adaptation to a mostly saprotrophic lifestyle as decomposers of naturally fallen leaves [21,53]. As with the Clade 6 hybrid *P. thermophila* × *P. amnicola* in the Valdivia River in Chile [11], no putative parents of the Clade 9 hybrids *P.* sp. ×Grenada 3-like, *P.* sp. ×insolita-like, *P.* sp. ×kunnunara-like, *P.* sp. ×Peru 4-like and *P.* sp. ×virginiana-like were detected in the Vietnamese rivers. Possibly the hybridisation events occurred in the Vietnamese streams and the parents were outcompeted by the better adapted hybrids. Alternatively, the hybrids could be introduced from elsewhere. Since the multicopy ITS locus is of limited use for hybrid studies sequencing of appropriate mitochondrial and nuclear genes are needed to confirm the hybrid status and elucidate the parents of the putative hybrid taxa.

Panglobally distributed pathogens from *Phytophthora* Clades 1 (*P. cactorum*, *P. infestans*, *P. nicotianae*) and 4 (*P. palmivora*) commonly cause diseases of horticultural crops and ornamental plants in mainland China, Hainan and Taiwan [12,33,35]. However, in this survey, as in previous surveys in Taiwan, species from Clades 1 and 4 (exception for one isolate of *P. palmivora* in Taiwan) were not detected in natural forests and streams [10,13], indicating that these two clades are not native to Taiwan and Southeast Asia. The same probably applies to Clades 3, 11 and 12 [23,28].

Although the natural hosts of the putatively endemic Vietnamese forest Phytophthoras obtained in this study are still unknown, it is evident that many native Asian forest Phytophthoras have co-evolved with a variety of tree genera also present in Europe and North America, including Fagaceae, Lauraceae, Aceraceae, Oleaceae, and Pinaceae. In this case high susceptibility of many non-coevolved European and North American trees to these Asian *Phytophthora* species is possible, as already well demonstrated for *P. cinnamomi*, *P. plurivora*, *P.* ×*cambivora* and, more recently, for six new Clade 7a species from Taiwan [22]. An extensive host range study with *Phytophthora* species from Asia, South and Central America has been initiated and will be published separately. In one part of this study, the pathogenicity of five Asian species (*P. castaneae*, *P. heveae* and the three new Vietnamese species *P. citricola* X, *P.* sp. multivesiculata-like 1 and *P.* sp. tropicalis-like 2) to *Castanea sativa*, *Quercus suber* and *Quercus robur* has been investigated and all five caused significant rot and loss of fine roots and suberised lateral roots in all three hosts, *C. sativa* being most susceptible [80].

Against this background, the annual importation of over three billion plants-for-planting into Europe [81], the large numbers of previously unknown *Phytophthora* species in natural and horticultural ecosystems being identified in Asia, South and Central America ([10,11,16,22,82], this study) and the occurrence of at least 47 exotic *Phytophthora* species in European nurseries and associated outplantings [60] represents a significant biosecurity risk for forestry, horticulture, and natural ecosystems in Europe and North America.

Many recent epidemics of trees and horticultural crops have been caused by introduced pathogens that were previously unknown to science, probably due to the organisms being co-evolved and benign in their centres of origin [6,10,83]. Although often introduced via the plants-for-planting pathway, none of them has ever been intercepted pre-emptively during routine phytosanitary controls at the ports of entry [60,81,83,84]. Despite overwhelming scientific evidence, current sanitary and phytosanitary (SPS) protocols largely ignore the risks from unknown, benign, co-evolved and unescaped organisms [6,60,83–85]. However, preventing further introductions of potentially harmful invasive Phytophthoras is a key issue for international forest biosecurity. A series of international research projects and organisations (listed in [10]) have come to similar conclusions. The current, outdated and scientifically flawed species-by-species regulation approach based on random visual inspections for symptoms of described pests and pathogens needs to be replaced by a sophisticated pathway regulation approach using pathway risk analyses, risk-based inspection regimes and molecular high-throughput detection tools [6,60,81,83,84,86,87].

To further define areas of *Phytophthora* diversity, including high-risk areas for the origin of potentially harmful pathogens, more *Phytophthora* surveys are needed in natural ecosystems in unsurveyed areas of Asia, Africa, and South and Central America, followed by host range testing of new taxa on naive tree hosts in Europe and elsewhere. Such surveys should also contribute to a better understanding of the global diversity of *Phytophthora*, the ancient biogeographic radiation of the *Phytophthora* species and Clades, and the influence of local environmental and host factors on breeding strategies and adaptation in the genus.

5. Conclusions

A remarkable diversity of 13 described species, five informally designated taxa and 21 previously unknown taxa of *Phytophthora* were obtained from 25 natural and semi-natural forest stands and 16 rivers in temperate and subtropical montane and tropical lowland regions of Vietnam. It is concluded that Vietnam is within the center of origin of most *Phytophthora* taxa found, including *P. cinnamomi* and *P. ramorum*, and that *Phytophthora* clades 2, 5, 6, 7, 8, 9, and 10 are native to Indochina.

Author Contributions: Conceptualization: T.J., M.H.J., C.M.B.; data curation: T.J., M.H.J., I.M.; formal analysis: T.J., M.H.J., B.S., M.T., M.P., J.B., H.R., A.P.-S., A.B., M.R., C.M.; investigation: T.J., M.H.J., C.M.B., J.W., T.C., B.S.; methodology: T.J., M.H.J., C.M.B.; writing—original draft: T.J.; writing—review and editing: T.J., C.M.B., B.S., T.C., I.M., M.T., J.B., M.H.J., A.P.-S. All authors have read and agree to the published version of the manuscript.

Acknowledgments: The Vietnamese Academy of Forest Sciences in Hanoi is gratefully acknowledged for providing lab space and technical support. We thank Henrieta Ďatková and Josef Janoušek (PRC, Brno, Czech Republic) for much appreciated technical support.

References

1. Brasier, C.M.; Robredo, F.; Ferraz, J.F.P. Evidence for *Phytophthora cinnamomi* involvement in Iberian oak decline. *Plant Pathol.* **1993**, *42*, 140–145. [CrossRef]
2. Erwin, D.C.; Ribeiro, O.K. *Phytophthora Diseases Worldwide*; APS Press: Saint Paul, MN, USA, 1996.
3. Hansen, E.M.; Goheen, D.J.; Jules, E.S.; Ullian, B. Managing Port–Orford–Cedar and the introduced pathogen *Phytophthora* lateralis. *Plant Dis.* **2000**, *84*, 4–14. [CrossRef] [PubMed]
4. Jung, T.; Blaschke, H.; Osswald, W. Involvement of soilborne *Phytophthora* species in Central European oak decline and the effect of site factors on the disease. *Plant Pathol.* **2000**, *49*, 706–718. [CrossRef]
5. Jung, T.; Vettraino, A.M.; Cech, T.L.; Vannini, A. *The impact of invasive Phytophthora species on European forests. Phytophthora: A Global Perspective*; Lamour, K., Ed.; CABI: Wallingford, UK, 2013; pp. 146–158, ISBN 978-1-78064-093-8.
6. Jung, T.; Pérez–Sierra, A.; Durán, A.; Horta Jung, M.; Balci, Y.; Scanu, B. Canker and decline diseases caused by soil- and airborne *Phytophthora* species in forests and woodlands. *Persoonia* **2018**, *40*, 182–220. [CrossRef]
7. Rizzo, D.M.; Garbelotto, M.; Davidson, J.M.; Slaugter, G.W. *Phytophthora ramorum* as the cause of extensive mortality of *Quercus* spp. and *Lithocarpus densiflorus* in California. *Plant Dis.* **2002**, *86*, 205–214. [CrossRef]
8. Hardham, A.R. *Phytophthora cinnamomi. Mol. Plant Pathol.* **2005**, *6*, 589–604. [CrossRef]

9. Cahill, D.M.; Rookes, J.E.; Wilson, B.A.; Gibson, L.; Mcdougall, K.L. Turner Review No. 17. *Phytophthora cinnamomi* and Australia's biodiversity: Impacts, predictions and progress towards control. *Aust. J. Bot.* **2008**, *56*, 279–310. [CrossRef]
10. Jung, T.; Chang, T.T.; Bakonyi, J.; Seress, D.; Pérez-Sierra, A.; Yang, X.; Hong, C.; Scanu, B.; Fu, C.H.; Hsueh, K.-L.; et al. Diversity of *Phytophthora* species in natural ecosystems of Taiwan and association with disease symptoms. *Plant Pathol.* **2017**, *66*, 194–211. [CrossRef]
11. Jung, T.; Durán, A.; Sanfuentes von Stowasser, E.; Schena, L.; Mosca, S.; Fajardo, S.; González, M.; Navarro Ortega, A.D.; Bakonyi, J.; Seress, D.; et al. Diversity of *Phytophthora* species in Valdivian rainforests and association with severe dieback symptoms. *Forest Pathol.* **2018**, *48*, e12443. [CrossRef]
12. Zeng, H.-C.; Ho, H.-H.; Zheng, F.-C. A survey of *Phytophthora* species on Hainan Island of South China. *J. Phytopathol.* **2009**, *157*, 33–39. [CrossRef]
13. Brasier, C.M.; Vettraino, A.M.; Chang, T.T.; Vannini, A. *Phytophthora lateralis* discovered in an old growth *Chamaecyparis* forest in Taiwan. *Plant Pathol.* **2010**, *59*, 595–603. [CrossRef]
14. Reeser, P.W.; Sutton, W.; Hansen, E.M.; Remigi, P.; Adams, G.C. *Phytophthora* species in forest streams in Oregon and Alaska. *Mycologia* **2011**, *103*, 22–35. [CrossRef] [PubMed]
15. Vettraino, A.M.; Brasier, C.M.; Brown, A.V.; Vannini, A. *Phytophthora himalsilva* sp. nov. an unusually phenotypically variable species from a remote forest in Nepal. *Fungal Biol.* **2011**, *115*, 275–287. [CrossRef] [PubMed]
16. Huai, W.X.; Tian, G.; Hansen, E.M.; Zhao, W.X.; Goheen, E.M.; Grünwald, N.J.; Cheng, C. Identification of *Phytophthora* species baited and isolated from forest soil and streams in northwestern Yunnan province, China. *Forest Pathol.* **2013**, *43*, 87–103. [CrossRef]
17. Hüberli, D.; Hardy, G.E.S.T.J.; White, D.; Williams, N.; Burgess, T.I. Fishing for *Phytophthora* from Western Australia' s waterways: A distribution and diversity survey. *Australas. Plant Pathol.* **2013**, *42*, 251–260. [CrossRef]
18. Oh, E.; Gryzenhout, M.; Wingfield, B.D.; Wingfield, M.J.; Burgess, T.I. Surveys of soil and water reveal a goldmine of *Phytophthora* diversity in South African natural ecosystems. *IMA Fungus* **2013**, *4*, 123–131. [CrossRef] [PubMed]
19. Burgess, T.I.; White, D.; McDougall, K.M.; Garnas, J.; Dunstan, W.A.; Català, S.; Carnegie, A.J.; Worboys, S.; Cahill, D.; Vettraino, A.M.; et al. Distribution and diversity of *Phytophthora* across Australia. *Pac. Conserv. Biol.* **2017**, *23*, 1–13. [CrossRef]
20. Jung, T.; Hansen, E.M.; Winton, L.; Oßwald, W.; Delatour, C. Three new species of *Phytophthora* from European oak forests. *Mycol. Res.* **2002**, *106*, 397–411. [CrossRef]
21. Jung, T.; Stukely, M.J.C.; Hardy, G.E.S.J.; White, D.; Paap, T.; Dunstan, W.A.; Burgess, T.I. Multiple new *Phytophthora* species from ITS Clade 6 associated with natural ecosystems in Australia: Evolutionary and ecological implications. *Persoonia* **2011**, *26*, 13–39. [CrossRef]
22. Jung, T.; Horta Jung, M.; Scanu, B.; Seress, D.; Kovács, D.M.; Maia, C.; Pérez-Sierra, A.; Chang, T.-T.; Chandelier, A.; Heungens, A.; et al. Six new *Phytophthora* species from ITS Clade 7a including two sexually functional heterothallic hybrid species detected in natural ecosystems in Taiwan. *Persoonia* **2017**, *38*, 100–135. [CrossRef]
23. Jung, T.; Horta Jung, M.; Cacciola, S.O.; Cech, T.; Bakonyi, J.; Seress, D.; Mosca, S.; Schena, L.; Seddaiu, S.; Pane, A.; et al. Multiple new cryptic pathogenic *Phytophthora* species from Fagaceae forests in Austria, Italy and Portugal. *IMA Fungus* **2017**, *8*, 219–244. [CrossRef] [PubMed]
24. Burgess, T.I.; Webster, J.L.; Ciampini, J.A.; White, D.; Hardy, G.E.S.J.; Stukely, M.J.C. Re-evaluation of *Phytophthora* species isolated during 30 years of vegetation health surveys in Western Australia using molecular techniques. *Plant Dis.* **2009**, *93*, 215–223. [CrossRef] [PubMed]
25. Burgess, T.I.; Simamora, A.V.; White, D.; Wiliams, B.; Schwager, M.; Stukely, M.J.C.; Hardy, G.E.S.J. New species from *Phytophthora* Clade 6a: Evidence for recent radiation. *Persoonia* **2018**, *41*, 1–7. [CrossRef] [PubMed]
26. Brasier, C. Phytophthora biodiversity: How many Phytophthora species are there? In *Phytophthoras in Forests and Natural Ecosystems: Fourth Meeting of the International Union of Forest Research Organizations (IUFRO) Working Party S07.02.09, USDA Forest Service*; Goheen, E.M., Frankel, S.J., Eds.; Pacific Southwest Research Station: Albany, NY, USA, 2009.
27. Kroon, L.P.; Brouwer, H.; de Cock, A.W.; Govers, F. The genus *Phytophthora* anno 2012. *Phytopathology* **2012**, *102*, 348–364. [CrossRef] [PubMed]

28. Yang, X.; Tyler, B.M.; Hong, C. An expanded phylogeny for the genus *Phytophthora*. *IMA Fungus* **2017**, *8*, 355–384. [CrossRef]
29. Cooke, D.E.L.; Drenth, A.; Duncan, J.M.; Wagels, G.; Brasier, C.M. A molecular phylogeny of *Phytophthora* and related Oomycetes. *Fungal Genet. Biol.* **2000**, *30*, 17–32. [CrossRef]
30. Ko, W.H.; Chang, H.S.; Su, H.J. Isolates from *Phytophthora cinnamomi* from Taiwan as evidence for an Asian origin of the species. *Trans. Br. Mycol. Soc.* **1978**, *71*, 496–499. [CrossRef]
31. Ko, W.H.; Wang, S.Y.; Ann, P.J. The possible origin and relation of *Phytophthora katsurae* and *P. heveae*, discovered in a protected natural forest in Taiwan. *Bot. Stud.* **2006**, *47*, 273–277.
32. Chang, T.T.; Wang, W.W.; Wang, W.Y. Use of random amplified polymorphic DNA markers for the detection of genetic variation in *Phytophthora cinnamomi* in Taiwan. *Bot. Bull. Acad. Sin.* **1996**, *37*, 165–171. Available online: https://ejournal.sinica.edu.tw/bbas/content/1996/3/bot373-01.html (accessed on 16 December 2019).
33. Ho, H.H.; Lu, J.Y. A synopsis of the occurrence and pathogenicity of *Phytophthora* species in mainland China. *Mycopathologia* **1997**, *138*, 143–161. [CrossRef]
34. Drenth, A.; Guest, D.I. *Diversity and management of Phytophthora in Southeast Asia*; ACIAR Monograph 114; Australian Centre for International Agricultural Research: Canberra, Australia, 2004; p. 238, ISBN 1-86320-405-9.
35. Ann, P.J.; Wong, I.T.; Tsai, J.N. New records of *Phytophthora* diseases of aromatic crops in Taiwan. *Plant Pathol. Bull.* **2010**, *19*, 53–68.
36. Chang-Fu, H.; Chung-Fu, S. Introduction to the flora of Taiwan, 1: Geography, Geology, Climate, and Soils. *Flora of Taiwan Second Edition*. Editorial Committee of the Flora of Taiwan Second Edition. 1994. Available online: http://tai2.ntu.edu.tw/ebook/ebookpage.php?volume=1&book=Fl.%20Taiwan%202nd%20edit.&page=1 (accessed on 11 January 2020).
37. Chang-Fu, H.; Chung-Fu, S.; Kuoh-Cheng, Y. Introduction to the Flora of Taiwan, 3: Floristics, Phytogeography, and Vegetation. *Flora of Taiwan Second Edition*. Editorial Committee of the Flora of Taiwan Second Edition. 1994. Available online: http://tai2.ntu.edu.tw/ebook/ebookpage.php?book=Fl.%20Taiwan%202nd%20edit.&volume=1&page=7 (accessed on 11 January 2020).
38. Chung-Fu, S. Introduction to the flora of Taiwan, 2: Geotectonic Evolution, Paleogeography, and the Origin of the Flora. *Flora of Taiwan Second Edition*. Editorial Committee of the Flora of Taiwan Second Edition. 1994. Available online: http://tai2.ntu.edu.tw/ebook/ebookpage.php?book=Fl.%20Taiwan%202nd%20edit.&volume=1&page=3 (accessed on 11 January 2020).
39. Gower, D.J.; Johnson, K.G.; Richardson, J.E.; Rosen, B.R.; Rüber, L.; Williams, S.T. *Biotic Evolution and Environmental Change in Southeast Asia; The Systematics Association Special Volume 82*; Cambridge University Press: New York, NY, USA, 2012; ISBN 13- 978-1107001305.
40. Averyanov, L.V.; Loc, P.K.; Hiep, N.T.; Harder, D.K. Phytogeographic review of Vietnam and adjacent areas of Eastern Indochina. *Komarovia* **2003**, *3*, 1–83.
41. Wurster, C.M.; Bird, M.I.; Bull, I.D.; Creed, F.; Bryant, C.; Dungait, J.A.J.; Paz, V. Forest contraction in north equatorial Southeast Asia during the last glacial maximum. *Proc. Natl. Acad. Sci. USA* **2010**, *107*, 15508–15511. [CrossRef] [PubMed]
42. Jung, T. Beech decline in Central Europe driven by the interaction between *Phytophthora* infections and climatic extremes. *Forest Pathol.* **2009**, *39*, 73–94. [CrossRef]
43. Jung, T.; Burgess, T.I. Re-evaluation of *Phytophthora citricola* isolates from multiple woody hosts in Europe and North America reveals a new species, *Phytophthora plurivora* sp. nov. *Persoonia* **2009**, *22*, 95–110. [CrossRef]
44. Hansen, E.M.; Reeser, P.W.; Sutton, W. *Phytophthora* beyond agriculture. *Annu. Rev. Phytopathol.* **2012**, *50*, 359–378. [CrossRef]
45. Scanu, B.; Hunter, G.C.; Linaldeddu, B.T.; Franceschini, A.; Maddau, L.; Jung, T.; Denman, S. A taxonomic re-evaluation reveals that *Phytophthora cinnamomi* and *P. cinnamomi* var. *parvispora* are separate species. *Forest Pathol.* **2014**, *44*, 1–20. [CrossRef]
46. White, T.J.; Bruns, T.; Lee, S.; Taylor, J. Amplification and direct sequencing of fungal ribosomal RNA genes for phylogenetics. In *PCR Protocols: A Guide to Methods and Applications*; Innis, M.A., Gelfand, D.H., Sninsky, J.J., White, T.J., Eds.; Academic Press: San Diego, CA, USA, 1990; pp. 315–322, ISBN 0123721806.
47. Kroon, L.P.N.M.; Bakker, F.T.; van den Bosch, G.B.M.; Bonants, P.J.M.; Flier, W.G. Phylogenetic analysis of

Phytophthora species based on mitochondrial and nuclear DNA sequences. *Fungal Genet. Biol.* **2004**, *41*, 766–782. [CrossRef]

48. Martin, F.N.; Tooley, P.W. Phylogenetic relationships among *Phytophthora* species inferred from sequence analysis of mitochondrially encoded cytochrome oxidase I and II genes. *Mycologia* **2003**, *95*, 269–284. [CrossRef]
49. Aragaki, M.; Uchida, J.Y. Morphological distinctions between *Phytophthora capsici* and *P. tropicalis* sp. nov. *Mycologia* **2001**, *93*, 137–145. [CrossRef]
50. Ilieva, E.; Man In't Veld, W.A.; Veenbaas-Rijks, W.; Pieters, R. *Phytophthora multivesiculata*, a new species causing rot in *Cymbidium*. *Eur. J. Plant Pathol.* **1998**, *104*, 677–684. [CrossRef]
51. Jung, T.; Scanu, B.; Bakonyi, J.; Seress, D.; Kovács, G.M.; Durán, A.; Sanfuentes von Stowasser, E.; Schena, L.; Mosca, S.; Thu, P.Q.; et al. *Nothophytophthora* gen. nov., a new sister genus of *Phytophthora* from natural and semi-natural ecosystems. *Persoonia* **2017**, *39*, 143–174. [CrossRef]
52. Werres, S.; Marwitz, R.; Man In't Veld, W.A.M.; Bonants, P.J.M.; De Weerd, M.; Themann, K.; Ilieva, E.; Baayen, R.P. *Phytophthora ramorum* sp. nov., a new pathogen on *Rhododendron* and *Viburnum*. *Mycol. Res.* **2001**, *105*, 1155–1165. [CrossRef]
53. Brasier, C.M.; Cooke, D.E.L.; Duncan, J.M.; Hansen, E.M. Multiple new phenotypic taxa from trees and riparian ecosystems in *Phytophthora gonapodyides–P. megasperma* ITS Clade 6, which tend to be high-temperature tolerant and either inbreeding or sterile. *Mycol. Res.* **2003**, *107*, 277–290. [CrossRef]
54. Safaiefarahani, B.; Mostowfizadeh-Ghalamfarsa, R.; Hardy, G.E.S.J.; Burgess, T.I. Re-evaluation of the *Phytophthora cryptogea* species complex and the description of a new species, *Phytophthora pseudocryptogea* sp. nov. *Mycol. Prog.* **2015**, *14*, 108. [CrossRef]
55. Jung, T.; Nechwatal, J. *Phytophthora gallica* sp. nov., a new species from rhizosphere soil of declining oak and reed stands in France and Germany. *Mycol. Res.* **2008**, *112*, 1195–1205. [CrossRef]
56. Yang, X.; Hong, C. *Phytophthora virginiana* sp. nov., a high-temperature tolerant species from irrigation water in Virginia. *Mycotaxon* **2013**, *126*, 167–176. [CrossRef]
57. Balci, Y.; Balci, S.; Eggers, J.; MacDonald, W.L.; Juzwik, J.; Long, R.P.; Gottschalk, K.W. *Phytophthora* spp. associated with forest soils in eastern and north-central U.S. oak ecosystems. *Plant Dis.* **2007**, *91*, 705–710. [CrossRef]
58. Jung, T.; La Spada, F.; Pane, A.; Aloi, F.; Evoli, M.; Horta Jung, M.; Scanu, B.; Faedda, R.; Rizza, C.; Puglisi, I.; et al. Diversity and distribution of *Phytophthora* species in protected natural areas in Sicily. *Forests* **2019**, *10*, 259. [CrossRef]
59. Milenković, I.; Keča, N.; Karadžić, D.; Radulović, Z.; Nowakowska, J.A.; Oszako, T.; Sikora, K.; Corcobado, T.; Jung, T. Isolation and pathogenicity of *Phytophthora* species from poplar plantations in Serbia. *Forests* **2018**, *9*, 330. [CrossRef]
60. Jung, T.; Orlikowski, L.; Henricot, B.; Abad-Campos, P.; Aday, A.G.; Aguín Casal, O.; Bakonyi, J.; Cacciola, S.O.; Cech, T.; Chavarriaga, D.; et al. Widespread *Phytophthora* infestations in European nurseries put forest, semi-natural and horticultural ecosystems at high risk of *Phytophthora* diseases. *For. Pathol.* **2016**, *46*, 134–163. [CrossRef]
61. Shrestha, S.K.; Zhou, Y.; Lamour, K. Oomycetes baited from streams in Tennessee 2010–2012. *Mycologia* **2013**, *105*, 1516–1523. [CrossRef]
62. Brasier, C.M. Evolutionary Biology of *Phytophthora*. I. Genetic system, sexuality and variation. *Annu. Rev. Phytopathol.* **1992**, *30*, 153–171. [CrossRef]
63. Català, S.; Peréz-Sierra, A.; Abad-Campos, P. The use of genus-specific amplicon pyrosequencing to assess *Phytophthora* species diversity using eDNA from soil and water in Northern Spain. *PLoS ONE* **2015**, *10*, e0119311. [CrossRef]
64. Grünwald, N.J.; Garbelotto, M.; Goss, E.M.; Heungens, K.; Prospero, S. Emergence of the sudden oak death pathogen *Phytophthora ramorum*. *Trends Microbiol.* **2012**, *20*, 131–138. [CrossRef]
65. Van Poucke, K.; Franceschini, S.; Webber, J.F.; Vercauteren, A.; Turner, J.A.; McCracken, A.R.; Heungens, K.; Brasier, C.M. Discovery of a fourth evolutionary lineage of *Phytophthora ramorum*: EU2. *Fungal Biol.* **2012**, *116*, 1178–1191. [CrossRef]
66. Arentz, F.; Simpson, J.A. Distribution of *Phytophthora cinnamomi* in Papua New Guinea and notes on its origin. *Trans. Br. Mycol. Soc.* **1986** *87*, 289–295. [CrossRef]

67. Brasier, C.M.; Scott, J.K. European oak declines and global warming: A theoretical assessment with special reference to the activity of *Phytophthora cinnamomi*. *Bull. OEPP EPPO Bull.* **1994**, *24*, 221–234. [CrossRef]
68. Brasier, C.M. *Phytophthora cinnamomi* and oak decline in southern Europe. Environmental constraints including climate change. *Ann. Des. Sci. For.* **1996**, *53*, 347–358. [CrossRef]
69. Burgess, T.I.; Scott, J.K.; McDougall, K.L.; Stukely, M.J.C.; Crane, C.; Dunstan, W.A.; Brigg, F.; Andjic, V.; White, D.; Rudman, T.; et al. Current and projected global distribution of *Phytophthora cinnamomi*, one of the world's worst plant pathogens. *Glob. Chang. Biol.* **2017**, *23*, 1661–1674. [CrossRef]
70. Shearer, B.L.; Crane, C.E.; Cochrane, A. Quantification of the susceptibility of the native flora of the South-West Botanical Province, Western Australia, to *Phytophthora cinnamomi*. *Aust. J. Bot.* **2004**, *52*, 435–443. [CrossRef]
71. Hardham, A.R.; Blackman, L.M. *Phytophthora cinnamomi*. *Mol. Plant Pathol.* **2018**, *19*, 260–285. [CrossRef] [PubMed]
72. Zentmyer, G.A. *Phytophthora Cinnamomi and the Diseases it Causes*; Monograph No. 10; The American Phytopathological Society: Saint Paul, MN, USA, 1980; ISBN 0890540306.
73. Linde, C.; Drenth, A.; Kemp, G.H.J.; Wingfield, M.J.; Von Broembsen, S.L. Population structure of *Phytophthora cinnamomi* in South Africa. *Phytopathology* **1997**, *87*, 822–827. [CrossRef] [PubMed]
74. Dobrowolski, M.P.; Tommerup, I.C.; Blakeman, H.D.; O'Brien, P.A. Non-Mendelian inheritance revealed in a genetic analysis of sexual progeny of *Phytophthora cinnamomi* with microsatellite markers. *Fungal Genet. Biol.* **2002**, *35*, 197–212. [CrossRef] [PubMed]
75. Dobrowolski, M.P.; Tommerup, I.C.; Shearer, B.L.; O'Brien, P.A. Three clonal lineages of *Phytophthora cinnamomi* in Australia revealed by microsatellites. *Phytopathology* **2003**, *93*, 695–704. [CrossRef] [PubMed]
76. Jung, T.; Colquhoun, I.J.; Hardy, G.E.S.J. New insights into the survival strategy of the invasive soilborne pathogen *Phytophthora cinnamomi* in different natural ecosystems in Western Australia. *Forest Pathol.* **2013**, *43*, 266–288. [CrossRef]
77. Bezuidenhout, C.M.; Denman, S.; Kirk, S.A.; Botha, W.J.; Mostert, L.; McLeod, A. *Phytophthora* taxa associated with cultivated *Agathosma*, with emphasis on the *P. citricola* complex and *P. capensis* sp. nov. *Persoonia* **2010**, *25*, 32–49. [CrossRef]
78. Nagel, J.H.; Gryzenhout, M.; Slippers, B.; Wingfield, M.J.; Hardy, G.E.S.J.; Stukely, M.; Burgess, T.I. Characterization of *Phytophthora* hybrids from ITS clade 6 associated with riparian ecosystems in South Africa and Australia. *Fungal Biol.* **2013**, *117*, 329–347. [CrossRef]
79. Burgess, T.I. Molecular characterization of natural hybrids formed between five related indigenous Clade 6 *Phytophthora* species. *PLoS ONE* **2015**, *10*, e0134225. [CrossRef]
80. Jung, T.; Maia, C.; Horta Jung, M. Host range testing of known and novel Phytophthora species from Asia, Central and South America among major forest tree species from Europe. Unpublished work. 2020.
81. Eschen, R.; Douma, J.C.; Grégoire, J.-C.; Mayer, F.; Rigaux, L.; Potting, R.P.J. A risk categorisation and analysis of the geographic and temporal dynamics of the European import of plants for planting. *Biol. Invasions* **2017**, *19*, 3243–3257. [CrossRef]
82. Rahman, M.Z.; Uematsu, S.; Takeuchi, T.; Shirai, K.; Ishiguro, Y.; Suga, H.; Kageyama, K. Two new species, *Phytophthora nagaii* sp. nov. and *P. fragariaefolia* sp. nov., causing serious diseases on rose and strawberry plants, respectively, in Japan. *J. Gen. Plant Pathol.* **2014**, *80*, 348–365. [CrossRef]
83. Brasier, C.M. The biosecurity threat to the UK and global environment from international trade in plants. *Plant Pathol.* **2008**, *57*, 792–808. [CrossRef]
84. Santini, A.; Ghelardini, L.; De Pace, C.; Desprez-Loustau, M.L.; Capretti, P.; Chandelier, A.; Cech, T.; Chira, D.; Diamandis, S.; Gaitniekis, T.; et al. Biogeographic patterns and determinants of invasion by alien forest pathogens in Europe. *New Phytol.* **2013**, *197*, 238–250. [CrossRef] [PubMed]
85. Liebhold, A.M.; Brockerhoff, E.G.; Garrett, L.J.; Parke, J.L.; Britton, K.O. Live plant imports: The major pathway for forest insect and pathogen invasions of the US. *Front. Ecol. Environ.* **2012**, *10*, 135–143. [CrossRef]
86. Eschen, R.; Rigaux, L.; Sukovata, L.; Vettraino, A.M.; Marzano, M.; Grégoire, J.-C. Phytosanitary inspection of woody plants for planting at European Union entry points: A practical enquiry. *Biol. Invasions* **2015**, *17*, 2403–2413. [CrossRef]
87. Eschen, R.; Britton, K.; Brockerhoff, E.; Burgess, T.; Dalley, V.; Epanchin-Niell, R.; Gupta, K.; Hardy, G.; Huang, Y.; Kenis, M.; et al. International variation in phytosanitary legislation and regulations governing importation of plants for planting. *Environ. Sci. Policy* **2015**, *51*, 228–237. [CrossRef]

13

Differences in the Response to Acute Drought and *Phytophthora cinnamomi* Rands Infection in *Quercus ilex* L. Seedlings

Francisco J. Ruiz Gómez [1,*], Alejandro Pérez-de-Luque [2], Rafael Sánchez-Cuesta [1], José L. Quero [1] and Rafael M. Navarro Cerrillo [1]

[1] Departamento de Ingeniería Forestal, Laboratorio de Ecofisiología de Sistemas Forestales ECSIFOR—ERSAF, Universidad de Córdoba, Campus de Rabanales, Crta, IV, km. 396, E-14071 Córdoba, Spain; rscuesta@uco.es (R.S.-C.); jose.quero@uco.es (J.L.Q.); rmnavarro@uco.es (R.M.N.C.)

[2] Área de Genómica y Biotecnología, IFAPA, Centro Alameda del Obispo, Avda. Menéndez Pidal s/n, Apdo 3092, 14080 Córdoba, Spain; alejandro.perez.luque@juntadeandalucia.es

* Correspondence: g72rugof@uco.es.

Abstract: The sustainability of "dehesas" is threatened by the Holm oak decline. It is thought that the effects of root rot on plant physiology vary depending on external stress factors. Plant growth and biomass allocation are useful tools to characterize differences in the response to drought and infection. The study of physiological responses together with growth patterns will clarify how and to what extent root rot is able to damage the plant. A fully factorial experiment, including drought and *Phytophtora cinnamomi* Rands infection as factors, was carried out with *Quercus ilex* L. seedlings. Photosynthesis, biomass allocation and root traits were assessed. Photosynthetic variables responded differently to drought and infection over time. The root mass fraction showed a significant reduction due to infection. *P. cinnamomi* root rot altered the growth patterns. Plants could not recover from the physiological effects of infection only when the root rot coincided with water stress. Without additional stressors, the strategy of our seedlings in the face of root rot was to reduce the biomass increment and reallocate resources. Underlying mechanisms involved in plant-pathogen interactions should be considered in the study of holm oak decline, beyond the consideration of water stress as the primary cause of tree mortality.

Keywords: biomass allocation; dehesas; drought; montados; oak decline; plant traits; root rot

---

1. Introduction

Holm oak (*Quercus ilex* L.) is a native species, widely distributed in the central-western part of the Mediterranean basin. This tree can be considered a key species due to its ecological and socioeconomic importance, and it shows a high rate of phenotypic plasticity in relation to conditions such as temperature, elevation, and soil composition [1]. It is considered a drought-tolerant species and thus can play an important role against desertification [2]. The most representative examples of its socioeconomic relevance are the "dehesas" and "montados" agroforestry systems, which are Mediterranean Savannah-like ecosystems, present mainly in Spain and Portugal [3,4].

Since the 1990s, oak decline has been recorded in Europe [5], holm oak being the species most affected in the Mediterranean area. In south-western Europe (Spain and Portugal), Holm oak ecosystems are threatened by management practices, climatic change, and biotic agents. This makes them some of the most vulnerable ecosystems in the Mediterranean area [6,7].

Habitat projection models for the south-west Iberian peninsula indicate that an increase in extreme rain events alongside extended drought periods, and rising mean temperatures are likely to influence the future decline of holm oak in the region [2,7].

The severity of root rot and decline symptoms in *Q. ilex*, including mortality, has been related to water stress, both drought and waterlogging, which, depending on microsite conditions, can occur sequentially [1,8]. Regarding biotic factors, root rot oomycetes, mainly *Phytophthora cinnamomi* Rands, are considered a causal agent or triggering factor in holm oak decline [9–11]. *Quercus ilex* is considered the most susceptible species of the genus to the root rot [12].

Although some unspecific responses were detected in previous histological studies of infected roots, like cell wall thickening, phenolic compound accumulation on the middle lamella and mucilage secretion of parenchymatous cells [13,14], the main changes in plant status caused by *P. cinnamomi* infection are related not to defensive responses (e.g., pathogen associated molecular pattern (PAMP)-triggered immunity, effectors-triggered immunity, hypersensitive response, hormonal signaling ...), but to physiological ones (e.g., stomatal closure, photosynthesis rate and water imbalance on plants) [11,15–17]. Root rot and water stress both impact tree health although some aspects of their relationship are unclear, such as the increase of water use efficiency on inoculated trees with strongly reduced water potential [16]. Not only drought, but also waterlogging treatments have been demonstrated to have significant effects on the mortality of infected plants [8,15,18]. Little has been published on physiological changes in holm oak plants as a response to *P. cinnamomi* infection, with high variability of results depending on the experimental conditions. Some reports have shown strong differences in photosynthesis and water potentials [15,17], while others have shown no significant effects [19–21], depending on whether the plants were subjected to continuous and acute waterlogging, drought stress, or only slight stress. All of the above highlight the relevance of studying the possible differences in the responses of holm oak to water stress and pathogen infection.

Variables related to plant morphology and biomass have been demonstrated to be powerful tools in the analysis of ecological, ontological, or physiological differences among species, ecotypes, or stress conditions [22,23]. The changes in plant growth patterns are related to physiological responses to stress conditions; in particular, biomass allocation is a key factor [24]. Root traits were found to be related to different key processes that vary in changing environmental conditions (for example, photosynthesis and photosynthates allocation, drought tolerance strategies, water potentials, transpiration rates ...) [25–28].

The main objective of this study was to characterize the effects of acute water stress and *P. cinnamomi* infection in seedlings of *Quercus ilex* L. subsp. *ballota*, both individually and in combination, focusing on physiological, growth, and biomass allocation changes. We hypothesized that the effects and plant responses would differ according to whether the seedlings were water stressed, inoculated, or both, and that there would be significant variation in the magnitude of the responses, depending on the type of stress. To reach our main objective, we defined three specific goals: (i) to describe changes in physiological plant status due to either *P. cinnamomi* infection or drought; (ii) to evaluate plant growth changes related to drought stress and *P. cinnamomi* infection, analyzing biomass allocation, and (iii) to determine and characterize differences in the effects of water stress (drought) and *P. cinnamomi* infection on *Q. ilex* roots. This will clarify how and to what extent root rot is able to severely damage the plant, thereby helping the development of more accurate strategies to palliate the consequences of holm oak decline.

2. Materials and Methods

2.1. Plant Material

Seedlings of *Quercus ilex* subsp. *ballota* were used in this experiment. Acorns were collected in a warm-temperate holm oak forest in Arenas del Rey (Granada, Spain, ETRS89, UTM 30N: 417 586, 4 095 930, elevation 490 m.a.s.l.). The site is characterized by dry summers and wet, mild winters: the mean temperatures over the last 30 years are 24.7 °C (warmest month) and 11.5 °C (coldest month), and the average annual rainfall is 489.3 mm. This parental tree was chosen because it was considered drought tolerant in previous studies [29].

The acorns were sown in a peat-perlite-vermiculite growth medium (4-2-1 by vol.), in black plastic containers (2.5 L). These were placed in a growth chamber (25 °C, 60% RH, photoperiod 14/10 light/dark) until acorn germination. Then, the seedlings were placed in a greenhouse at the University of Córdoba-Campus Rabanales (Córdoba, Spain; ETRS89, UTM 30N: 348 360, 4 198 200), in semi-controlled conditions (25 $\pm$ 7 °C, 60 $\pm$ 10% RH), and were watered twice a week to saturate the growth medium. For three months, artificial light (HPS lamps, 400 W, 48,000 lumen, $\geq$600 μmol (photons) m^{-2} s^{-1} at 50 cm above the table surface) was provided, to extend the photoperiod to 12 h and ensure that the photosynthetic photon flux density (PPFD) exceeded 1000 μmol (photons) m^{-2} s^{-1}. This value is considered to be enough to light-saturate *Q. ilex* leaves [30]. Before the experiment, seedlings were selected that were homogeneous in morphology (height, H = 31.79 $\pm$ 0.92 cm; diameter at the root collar, Øi = 6.98 $\pm$ 0.16 mm; total leaf area, LAt = 202.6 $\pm$ 9.3 cm^2; mean values with standard error), instantaneous chlorophyll fluorescence, and photosynthetic efficiency of photosystem II (Ft = 3205 $\pm$ 409; QY = 66.4 $\pm$ 5.2%; FluorPen FP100 fluorescence analyzer, Photon Systems, spol. s.r.o., Drásov, Czech Republic).

2.2. Experimental Design and Inoculation

The experiment had a completely randomized design in which inoculation (2 levels: with and without) and watering (2 levels: with and without) were the main factors, resulting in four different treatments: "Control" (pots watered and mock-inoculated), "Inoculation" (pots watered and inoculated with *P. cinnamomi*), "Drought" (Pots non-watered and mock-inoculated), and "I $\times$ D" (Pots non-watered and inoculated with *P. cinnamomi*). Each of the four treatments had 10 replicates, providing a total of 40 experimental units. The pots were placed in black plastic trays, five pots of the same treatment in each tray, and the trays were distributed randomly in the greenhouse.

Inoculation was carried out with carrot agar (CA) liquid inoculum at a concentration of >30 Infective Units (IU)/μL of *P. cinnamomi* chlamydospores [31]. The *P. cinnamomi* strain was isolated from *Q. ilex* roots in a previous survey in Puebla de Guzman (Huelva, Spain). The pathogen was grown in 9-cm-diameter Petri dishes containing CA medium for 15 days. Prior to inoculum preparation, the surface of the CA containing the pathogen mycelium was well rinsed and the CA was mixed with demineralized water. The concentration of IU (chlamydospores) was evaluated using a Neubauer chamber. A false inoculum was made by mixing the same number of CA plates, but without *P. cinnamomi*, with water. The methodology used for the inoculation treatment was adapted from the work of Turco et al. [21]. Three holes were made in the substrate of each pot, using a 10-mL syringe with a trimmed end. Then, approximately 15 mL of inoculum or false inoculum were placed in each hole (45 mL in each pot) and subsequently covered with the substrate extracted previously with the trimmed syringe.

Before inoculation, the pots were watered to substrate saturation, the plants were inoculated and the pots were subsequently weighed to obtain their weight at the field capacity of the substrate. The watering regime of the "Control" and "Inoculation" treatments consisted of a first manual watering 72 h after inoculation, and subsequent watering every 48 h with 100 mL of water. To boost the induction of water stress in the treatments that did not include watering, the environmental conditions of the greenhouse were altered to increase the evapotranspiration rate of the plants (T = 28 $\pm$ 3 °C; RH = 40 $\pm$ 10%).

2.3. Parameters Measured

The physiological status of the plants was evaluated through the stomatal conductance and the rate and efficiency of photosynthesis, at the beginning of the experiment (0 days post-inoculation—dpi) and at 7, 15, 22, and 30 dpi. Thirty days after inoculation, the plants were harvested and their water status, aboveground and belowground fractions, and root distribution parameters were characterized. The description and units of each measured or calculated variable are shown in Table 1.

Table 1. Description of studied variables.

Variable	Abv.	Units	Description
Water status			
Volumetric Water content	θ	$cm^3 (cm)^{-3}$	Relative humidity of pot substrate
Midday Water potential	Ψ_m	MPa	Water potential of leaves at midday (at solar noon)
Dry Matter of root	*DMr*	$g (100 g)^{-1}$	Dry matter of root relative to 100 g of fresh weight
Plant growth and biomass allocation			
Net Photosynthesis	*A*	$\mu mol\ CO_2\ m^{-2}\ s^{-1}$	Mean estimate of photosynthetic rate of leaves
Stomatal Conductance	*Gs*	$mmol\ H_2O\ m^{-2}\ s^{-1}$	Mean estimate leaf stomatal conductance
Photosynthetic Efficiency	*QY*	...	Max. quantum efficiency of photosystem II (Fv/Fm)
Root Dry Weight	*RDW*	g	Weight of root system after oven drying (root biomass)
Stem Dry Weight	*SDW*	g	Weight of stem after oven drying (stem biomass)
Leaf Dry Weight	*LDW*	g	Weight of leaves after oven drying (leaves biomass)
Root Mass Fraction	*RMF*	$g (g)^{-1}$	Root biomass relative to total plant biomass
Stem Mass Fraction	*SMF*	$g (g)^{-1}$	Stem biomass relative to total plant biomass
Leaf Mass Fraction	*LMF*	$g (g)^{-1}$	Leaves biomass relative to total plant biomass

The volumetric water content (θ) of the pot substrate was measured during the experiment, in five pots per treatment, using time domain reflectometry probes (Decagon ECH$_2$O Ec-5-Decagon Devices, Inc., Washington, WA, USA), previously calibrated for the specific pot substrate composition (accuracy $\pm$ 1%, resolution 0.1% of θ). Prior to harvest, the midday stem water potential (Ψ_m) of three leaves per pot was measured using an SKPM 1400 pressure chamber (Skye Instruments, Ltd.; Llandrindod Wells, Powys, UK) [31]. The minimum Ψ_m considered was -6 MPa, due to the technical limitations of the pressure chamber; in *Q. ilex* the minimum values of Ψ_m before sap flux failure range between -4 and -6 MPa [32].

The maximum quantum efficiency of photosystem II (*QY*) was measured using a Hansatech PEA portable chlorophyll fluorimeter (Hansatech Instrument, Ltd.; King's Lynn, Norfolk, UK), for dark-adapted leaves after covering them for 20 min with portable leaf clips (Hansatech Instrument, Ltd., Narborough, UK). Five fully-expanded leaves per plant were measured at 0, 7, 15, 22, and 30 dpi.

The net photosynthesis rate (*A*) and stomatal conductance (*Gs*) were measured in three fully-expanded leaves of each plant, using a portable infrared CO_2 gas analyzer (LiCor Li6400XT, Li-Cor, Inc.; Lincoln, NE, USA) fitted with a 6-cm^2 leaf cuvette. The measurements were taken using a CO_2 concentration of 390 $\pm$ 1.7 ppm, a flow of 300 $\pm$ 1.2 cm^3 min^{-1}, and PPFD >1000 µmol (photons) m^{-2} s^{-1}. When a leaf did not fit completely in the leaf cuvette, a photograph of it was taken using a digital camera (HP Photosmart R827, Hewlett Packard Inc., Palo Alto, CA, USA). In order to correct the measurements according to the actual leaf area, the photographs taken at the moment of each measurement were analyzed using ImageJ image analysis software [33]. All the measurements of physiological variables were taken at 11:20–13:20 h UTC (Universal Time Coordinates), considering a 2-h window around the solar noon (13:20–15:20 h CET—Central European Time).

After the physiological measurements, the stems and leaves were excised from the root collar and the root ball was extracted from each pot. A subsample of fine roots was collected for pathogen isolation and, subsequently, the root ball was carefully washed with tap water on a 0.5-mm sieve, avoiding the loss of fine roots [34]. The fine and very fine roots lost in this process were recovered from the detached substrate using tweezers; subsequently, the root fraction was scanned in a Regent LA1600+ densitometer (Regent Instruments Inc., Quebec, QC, Canada).

The plant biomass was estimated for the stem, leaves, and roots to assess biomass allocation changes, using gravimetric methodology. The stems and leaves were dried immediately after excision (85 °C for 48 h; JP Selecta Conterm, Barcelona, Spain), in paper bags of known weight. The root fraction was dried after scanning, following the same procedure conducted for the aboveground biomass. All the dried samples were cooled in a desiccator at room temperature for 30 min prior to weighing. All the biomass fractions were expressed as dry biomass (stem dry weight, *SDW*, leaves dry weight, *LDW*, and root dry weight, *RDW*, in g).

2.4. Pathogen Isolation

To confirm the presence or absence of *P. cinnamomi* in the root system, a representative subsample of fine roots was collected for each plant, prior to root-ball detachment. Fifty pieces of fine and very fine roots (Ø < 2 mm), approximately 1 cm in length, were excised randomly from different regions of the root-ball. They were surface-disinfected by immersion in 70% ethanol for 10 s, washed in sterilized-deionized water, trimmed, and placed in 9-cm Petri dishes containing the PARPBH selective medium [35]. They were stored at room temperature, in darkness, for 14 days and were assessed every 48 h.

The colonies that grew were sub-cultured and sown in selective PARPBH medium and Carrot Agar (CA) to obtain axenic cultures. The colonies were observed under a microscope to identify the genus or species according to their morphology, following the indications of Erwin and Ribeiro [36]. The pathogen *P. cinnamomi* was recovered and isolated from fine roots of all the samples of the inoculation treatments and was not isolated from any of the studied roots from mock-inoculated plants. No other oomycete species were isolated from the samples.

2.5. Data Analysis

The effects of the factors on the root variables were calculated, according to Olmo et al. [27], as the ratio of the mean value of a root trait under the treatment to its mean value under control conditions. A ratio greater than 1 means that the treatment increased the value of this trait and thus was considered (+), and (−) for values between 0 and 1.

The scanned images of roots were analyzed with WinRHIZO Pro 2004a software (Regent Instruments Inc., Quebec, QC, Canada) to estimate root traits, for plasticity index (Pi) calculation. The Pi of each variable was calculated as described by Valladares and Sánchez-Gómez [37], through the expression (1):

$$\mathrm{Pi} = \left[\sum_{i=1}^{n} (X_{i-max} - X_{i-min}) / X_{i-max} \right] / \mathrm{n} \tag{1}$$

where $X_{i\text{-}max}$ represents the maximum value of the variables avoiding statistically extreme outliers, $X_{i\text{-}min}$ represents the minimum value of the variables avoiding statistically extreme outliers, and n represents the number of the variables used to calculate the Pi.

The Pi varies between 0 and 1. The mean root Pi was calculated as the average plasticity of all the root variables, to compare treatments.

To represent the differences in stem biomass avoiding the influence of aboveground biomass differences, *SDW* was divided by total aboveground biomass to calculate relative *SDW* proportion. This operation resulted in a better understanding of stem biomass changes.

The normality and homoscedasticity of the variables were assessed using the Shapiro-Wilk and Levene tests, respectively. Variables that did not fit a normal distribution were transformed by applying log(x) or 1/x [38], and the normality of the transformed variables was re-analyzed. After the normality test, two-way Analysis of Variance (ANOVA) was carried out, considering inoculation and watering as the independent factors. When no interaction between factors was detected in the two-way ANOVA, the Student's t test was used to study the effects of inoculation and watering on biomass variables. Repeated measures ANOVA (RMANOVA) was carried out for the physiological variables and θ, using dpi as the repeated measure and considering the watering and inoculation treatments as between-subject factors. To avoid test failure due to noncompliance with the ANOVA assumptions, Greisser-Greenhouse correction of the degrees of freedom was used for univariate within-subjects' analysis. The post-hoc comparison of treatments, in all other cases, was carried out using Tukey's High Significant Difference (HSD) test. Null hypotheses were rejected at the $p < 0.05$ level. All the statistical analyses were performed using IBM SPSS Statistics 19 (IBM Corp., Armonk, NY, USA).

3. Results

3.1. Plant and Root Symptoms

Plants of the "Drought" and combined ("I × D") treatments presented strong chlorosis and wilting of the leaves at the end of the assay. In the "Inoculation" treatment (watered), the symptoms were more variable—with chlorosis and partial wilting in several plants, while others had only slight symptoms in the upper part of the stem. No plants from the "Inoculation" (watered) treatment were dead after 30 days. No chlorosis or wilting was seen in plants from the "Control" treatment (Figure 1).

Figure 1. Plant symptoms after 30 dpi. (**a**) Leaf symptoms for the different treatments (**b**) Detail of roots showing root rot symptoms of inoculated plants in comparison to control. "I × D" = Pots non-watered and inoculated with *P. cinnamomi* Rands.

The root system of plants from the "Drought" and combined ("I × D") treatments exhibited taproot necrosis and root frailty. The plants from the "Inoculation" treatment (watered) showed root rot symptoms—consisting of dark-brown coloration, necrotic lesions in tips, and root softness—when compared with control plants. Another important symptom was the lack of fine lateral roots on the coarse roots (Ø > 2 mm).

3.2. Water Status

The RMANOVA showed significant differences in the volumetric water content of the substrate (θ) regarding watering (F = 91.7; $p < 0.001$) and inoculation (F = 6.5; $p < 0.05$). Time (dpi) strongly influenced θ (F = 373.8; $p < 0.01$) (Table 2, Figure 2a), with a significant effect of the interaction between dpi and the watering treatment (dpi × D; F = 60.9; $p < 0.001$).

Table 2. Repeated-measures ANOVA (RMANOVA) results for the effect of inoculation and water stress on physiological variables and volumetric water content. The degrees of freedom (df), type III sum of squares are shown for each variable. Geisser-Greenhouse adjusted probabilities were used for the within-subject analysis. Values of sum of squares are highlighted in bold type when significant (* $p < 0.05$; ** $p < 0.01$; *** $p < 0.001$).

Factors	df	Experimental Variable			
		A	*Gs*	*QY*	*θ*
Between-subjects source					
Inoculation (I)	1	31.531 ***	0.016 **	0.183 **	0.047 *
Drought (D)	1	86.041 ***	0.010 *	0.001	0.660 **
I × D	1	0.394	0.000	0.351 ***	0.002
Error	16	31.698	0.023	0.304	0.058
Within-subjects source					
Days (dpi)	4	22.821	0.007	0.262 ***	0.901 ***
dpi × I	4	3.254	0.003	0.015	0.002
dpi × D	4	77.014 ***	0.007	0.162 **	0.147 ***
dpi × I × D	4	8.279	0.001	0.051	0.001
Error	64	139.518	0.051	0.472	0.02

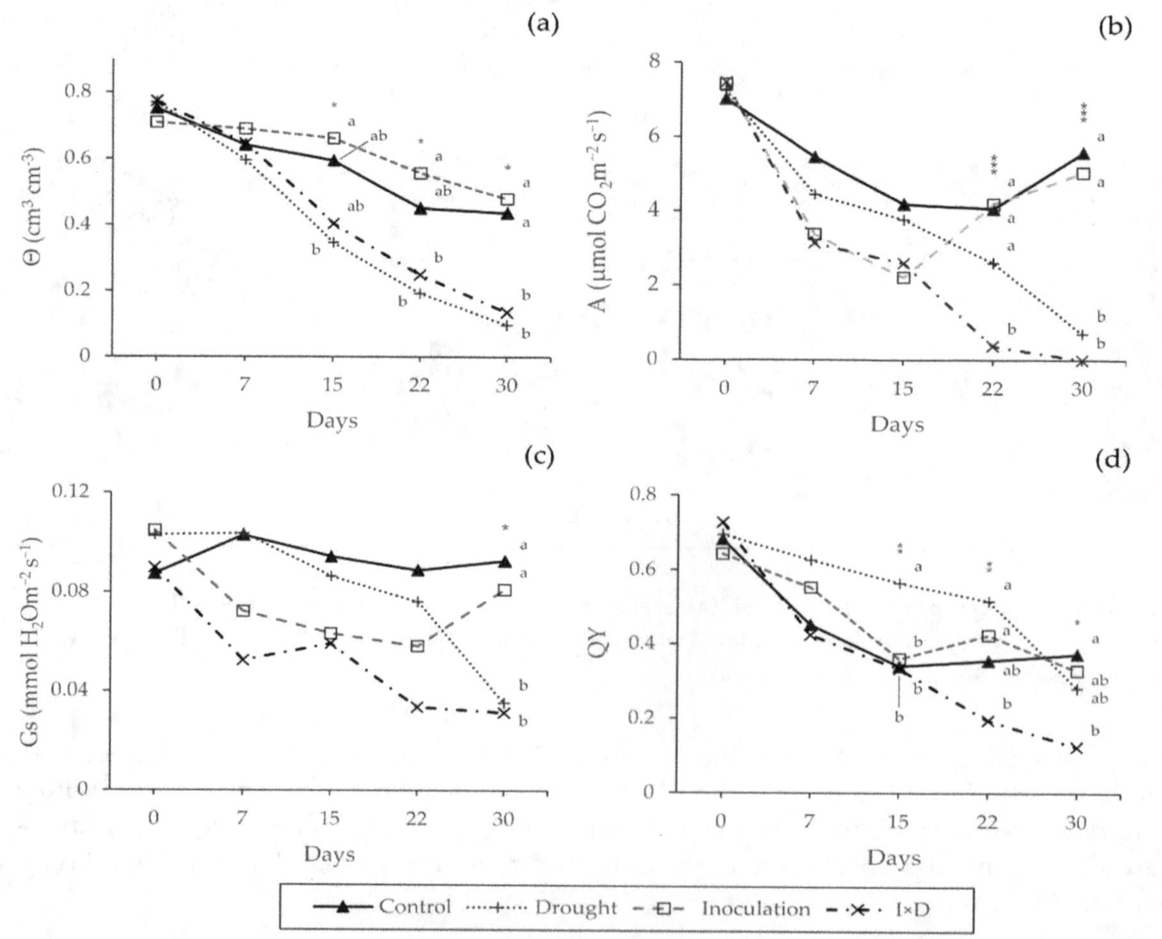

Figure 2. Physiological characterization of *Quercus ilex* L. seedlings during the experiment. (**a**) Volumetric water content (*θ*). (**b**) Net photosynthesis (*A*). (**c**) Stomatal conductance (*Gs*). (**d**) Photosynthetic efficiency of Photosystem II (*QY*). Points with the same letter are not significantly different for the analyzed dpi ($p > 0.05$). In sets of points where there were no significant differences for dpi according to the ANOVA test, the letters are not presented. Differences between values of the same days are represented only when significant at $p < 0.05$ (*), $p < 0.01$ (**) and $p < 0.001$ (***).

The differences due to inoculation were not influenced by dpi. At the end of the experiment, water stressed plants ("Drought" and "I × D" treatments) presented values of θ around 0.1 cm^3 cm^{-3}, and midday water potential (Ψ_m) values equal or below −6 MPa, with differences only between watered and non-watered plants (Supplementary Material, Table S1).

The root dry matter content (*DMr*) was influenced by a significant interaction between the watering and inoculation treatments (Table 3). Although no significant differences in *DMr* due to the inoculation factor were found at 30 dpi, plants of the "Inoculation" (watered) treatment had a significantly lower value than "Control" plants, without differences between the "I × D" and "Drought" treatments (Figure 3).

Table 3. Two-way ANOVA results for variables measured 30 dpi. Values of F statistic are highlighted in bold type when significant (* $p < 0.05$; ** $p < 0.01$; *** $p < 0.001$; n/s, not significant). R^2 represents corrected ANOVA model adjustment. (+) or (−) mean that the factor has a positive or negative effect, respectively.

Variable	Factors			
	Inoculation (I)	**Drought (D)**	**I × D**	R^2
Ψ_m	2.4	6763.7 *** (−)	2.4	0.99 ***
DMr	0.0	25.1 *** (+)	8.0 **	0.61 ***
RDW	0.7	4.3 * (+)	0.0	0.24 **
SDW	5.0 * (−)	3.3	5.5 *	0.46 ***
RMF	10 ** (−)	41.7 *** (+)	0.0 n/s	0.76 ***
SMF	5.0 * (−)	12.4 ** (−)	7.6 **	0.61 ***
LMF	18.8 *** (+)	8.8 ** (−)	3.9 n/s	0.66 ***

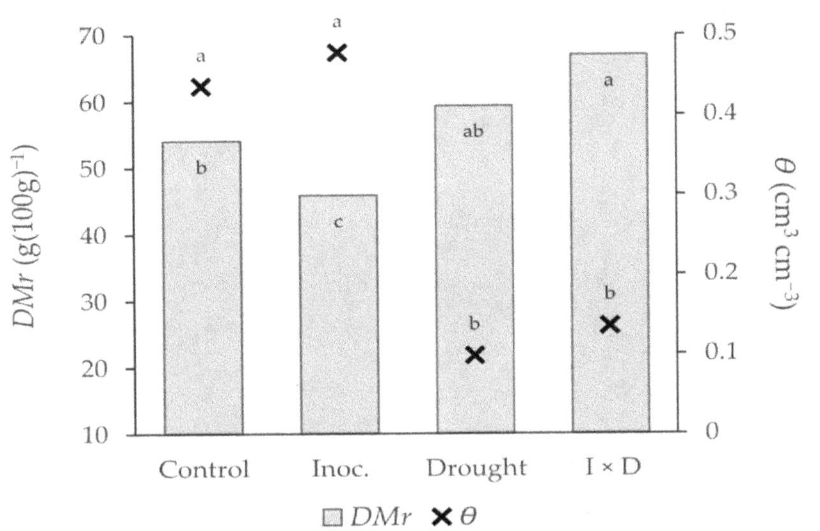

Figure 3. Mean volumetric water content (θ) of soil (points) and mean dry matter of roots (columns) at 30 dpi. Treatments with the same letter for each variable are not significantly different ($p > 0.05$).

3.3. Photosynthesis

At 30 dpi, only the watering treatment had a significant effect on the net photosynthesis rate (*A*) and *Gs* (Figure 2b,c; Supplementary Material, Table S1), while *QY* differed significantly only between the "Control" and "I × D" treatments (Figure 2d). However, significant effects of both inoculation and watering on *A* and *Gs* were found when the time-trend was analyzed (RMANOVA, Table 2). The time after inoculation did not influence *A* or *Gs*, but *A* varied significantly among the dpi depending on the water stress level, as shown by the significant interaction between dpi and the watering factor for this variable. No significant differences in *A* were found between treatments until day 22, when the value was lower for the "I × D" plants, coinciding with the lower values of *Gs*. At this time, a significant change in the trend was found in all physiological variables for the "Inoculation" treatment (inoculated and watered plants) (Figure 2). The QY was not affected significantly by the watering treatment, but it was by the inoculation and the interaction between factors, being significantly influenced by dpi and, as in the case of A as well, by the interaction between dpi and drought. The QY values for the "I × D" treatment at 15 and 22 dpi were significantly lower than those of the "Drought" treatment, for which QY and Gs did not decrease until the end of the experiment, when the substrate showed values of θ around 0.1 $cm^3\ cm^{-3}$.

3.4. Growth and Biomass Allocation

The root biomass (RDW) was higher in water-stressed plants ($t = 2.2$, df = 28, $p < 0.05$) but did not differ significantly when inoculation was considered (Figure 4a; Table 3). However, despite a lack of statistical significance, inoculation gave the lowest values of RDW, for both watering treatments (Figure 4a; Supplementary Material, Table S1).

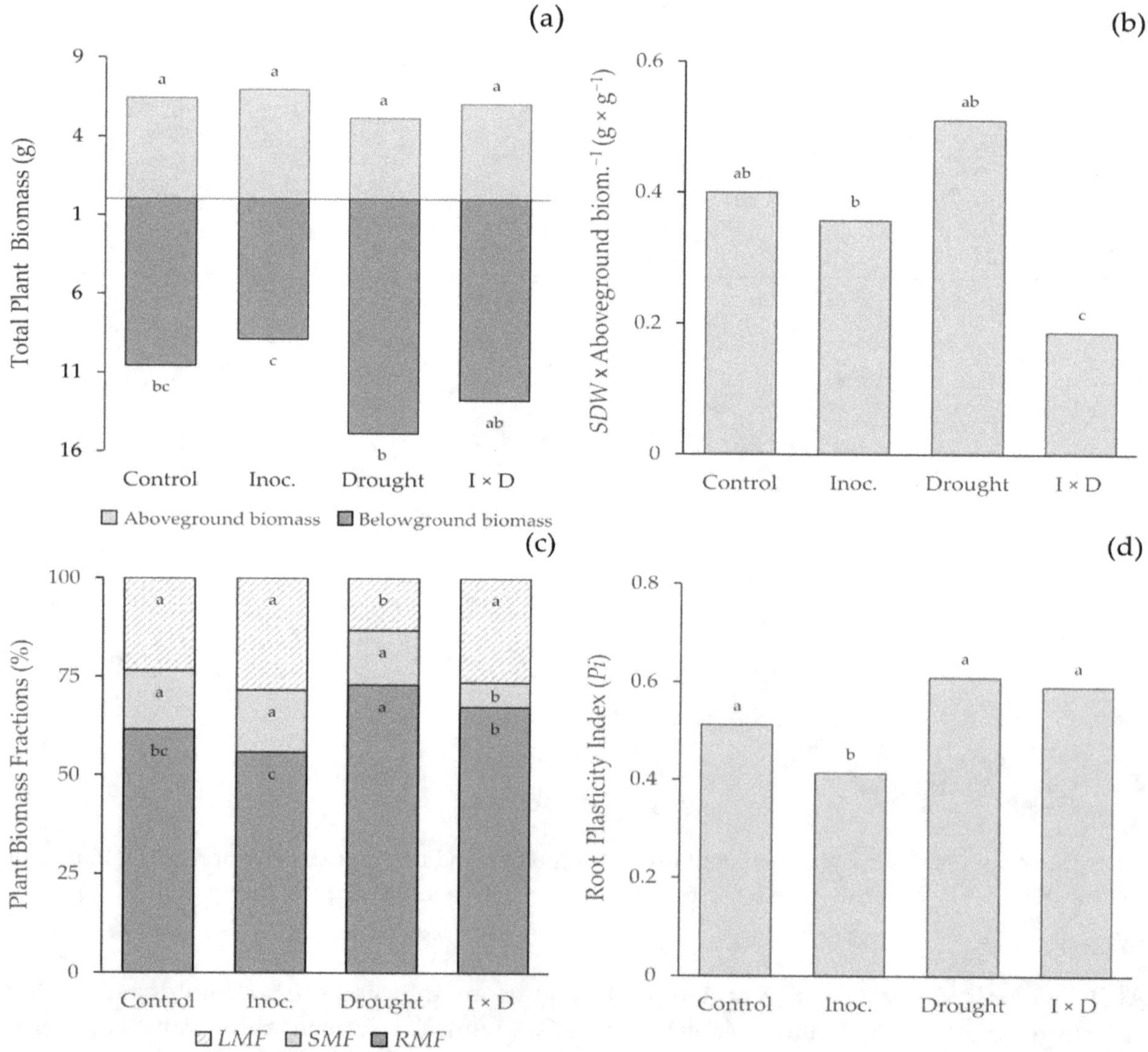

Figure 4. Plant biomass analysis (**a**) Biomass distribution per treatment. Upside mean aboveground (Stem and leaves) biomass. Downside mean belowground (root) biomass. (**b**) Relative stem biomass with respect to total aboveground biomass, per treatment. (**c**) Plant fractions for each treatment. LMF = Leaf mass fraction. SMF = Stem mass fraction. RMF = Root mass fraction. (**d**) Average plasticity index for each treatment. Error bars show ±SE. Bars with the same letter are not significantly different ($p > 0.05$).

Aboveground biomass (leaves plus stem) did not differ significantly among treatments, but the factor inoculation had a negative influence on *SDW*, with a significant interaction between the two experimental factors (Table 3). The relative proportion of *SDW*, regarding aboveground biomass (Figure 4b), was lower for the "I × D" treatment, *SDW* representing, on average, 45.6% of the total aboveground biomass in mock-inoculated plants ("Control" and "Drought") and 27.2% for the inoculated ones. For the plants of the "I × D" treatment, on average, both *SDW* ($F = 4.6$; $p < 0.01$) and relative *SDW* ($F = 3.6$; $p < 0.01$) were significantly lower than for the "Drought" treatment plants.

The biomass fractions (*RMF*, *SMF*, and *LMF*) were significantly influenced by both inoculation and watering, but the interaction between them was only significant for *SMF* (Table 3). The *RMF* decreased due to the effect of inoculation ($t = -2.7$, df = 38, $p < 0.05$) and increased due to water stress ($t = 5.4$, df = 38, $p < 0.001$), while *LMF* increased significantly due to inoculation ($t = 3.4$, df = 38, $p < 0.01$). The graphical comparison of fractions and treatments (Figure 4c) shows different trends in the aboveground (*SMF*, *LMF*) and root (*RMF*) fractions. All the biomass allocation parameters differed between the "Drought" and "I × D" treatments. The "Drought" treatment gave the maximum value of *RMF* (0.73 ± 0.02) and the minimum value of *LMF* (0.13 ± 0.01).

The minimum value of *RMF* occurred in the "Inoculation" treatment (0.56 ± 0.01), but without differences from the "Control" treatment. The values of *RMF* and *SMF* for the "I × D" treatment were lower than those of the "Drought" treatment, the plants of the "I × D" treatment showing a clear decrease in *SMF* ($F = 8.3$; $p < 0.001$). The average Pi for all the studied variables shows that the roots of watered plants had lower plasticity ($t = 4.5$, df = 110, $p < 0.001$). When the means of the Pi were compared for each of the four treatments in a one-way ANOVA, significant differences appeared ($F = 7.87$; $p < 0.001$), the mean Pi of the "Inoculation" treatment having the lowest value (Figure 4d).

4. Discussion

Although previous authors have proposed this idea as a hypothesis [16,18,21], to the best of our knowledge this is the first work to evidence differences in the physiological response of *Quercus ilex* to the stress of combining *P. cinnamomi* and drought, in comparison with each stress applied separately. All the studied variables responded to the experimental conditions, showing differences between both factors. The main changes on physiology presented different trends due to water stress induction in inoculated plants. The seedlings inoculated with *P. cinnamomi* recovered if no additional stress was induced. The differential responses of physiological parameters, growth, and biomass allocation were influenced by inoculation and water supply, without an interaction between them in most cases.

General Symptoms and Stress Indicators

The plant symptoms as a result of *P. cinnamomi* inoculation described here were similar to those reported for leaves [11,17] and roots [14,16] of holm oak seedlings in previous work. The root rot symptoms were differentiated from water stress damage in roots mainly by the color and general aspect of the root ball, the aboveground symptoms being less specific (i.e., chlorosis and wilting) (Figure 1).

Regarding water balance, θ was increased in inoculated pots where the infected (damaged) roots were less able to take up water. Time-trend analysis showed this significant effect, but only non-watered plants underwent water stress. Previous works have reported the early accumulation of pectidic and mucilaginous materials in xylem vessels of fine *Q. ilex* roots as a result of *P. cinnamomi* infection, and also the invasion of pathogenic structures in xylem cells in advanced stages of infection [14], this obstruction of conductive vessels causing the reduction of xylem conductance. In small plants, when water availability alternates between low and high, blocked vessels display the two stages characteristic of active vessels, alternating between embolism due to evaporative stress and refill by reverse osmosis [39]. In well-irrigated plants, the refill of blocked vessels in coarse roots is related to higher Ψ, and could be the cause of a high moisture content in root tissue [40], the blockage of vessels causing different effects in water stressed plants, in which reverse osmosis did not occur. However, the high *DMr* of the "I × D" treatment plants seems to be in conflict with this hypothesis (Figure 3), which might be a consequence of the combined effect of acute drought, inducing root growth.

Leaf water potential (Ψ) is often used as a water stress indicator [41]. This variable was only affected significantly by drought in our study (Tables 2 and 3), agreeing with Turco et al. [21]—who found that Ψ in *Q. ilex* seedlings was not changed by *P. cinnamomi* inoculation. Holm oak is considered an isohydric species [42], the detection of the first drought stress symptoms resulting in quick stomatal closure. However, in this work, inoculated seedlings responded to the root rot with early partial stomatal closure and a reduction in their photosynthetic activity at 7 dpi, at which time the pathogen would not have invaded to a significant extent the xylem vessels [14]. At this time, growth cessation and loss of root uptake ability due to root rot occurred [14,16]. This imbalance might be related to alterations in plant osmoregulation, and thus to the increase in root tissue water content in the "Inoculation" treatment plants, agreeing with previous studies in other species [43]. Oßwald et al. [16] indicated that the infection of woody plants by *Phytophthora* spp. could trigger a generalized dysfunction in plant water status related to hormonal changes, with an alteration in the balance between abscisic acid (ABA) and other plant hormones involved in stomatal regulation.

Physiological Changes

Water stress produced significant reductions in *A*, *Gs*, and *QY* at the end of the experiment, agreeing with the expected response of *Q. ilex* to water stress [42], but analyzing time-trend, inoculated plants responded in a different way, without influence of inoculation in photosynthetic efficiency (*QY*), when the influence of time (dpi) was eliminated (Table 2). The early reduction in photosynthetic activity as a result of inoculation, supported by the results of RMANOVA, might be related to the lower values of root biomass and root proportion in inoculated plants and agrees with other results which evidenced an early cessation of root growth, only 24 h after inoculation [14]. Also, it must be considered that *A* represents net CO_2 assimilation. A high *QY* accompanied by low rates of *A* could result from the increase of secondary metabolism activity, which increased intracellular leaf CO_2 concentration [44].

At 15 dpi, Inoculated plants had recovered their photosynthetic activity, their *Gs* and *QY* being equal to those of "Control" plants at 30 dpi, but plants of the "I × D" treatment did not exhibit photosynthetic activity after 22 dpi. Other works obtained similar results when inoculated plants were not subjected to acute water stress, without important changes in physiology, nor mortality of seedlings [15,19,45]. Turco et al. [21] did not find changes in the final physiological status of *Q. ilex* seedlings due to *P. cinnamomi* infection under slight drought conditions. In their work, inoculated plants (both watered and water-stressed) recovered their water status after 14 dpi, only being water-stressed after 42 days.

Root rot caused by *P. cinnamomi* intensifies the damage resulting from physiological stress [15]. The plant responses detected in previous works as a result of *P. cinnamomi* inoculation [14], should be linked to changes in the secondary metabolism, agreeing with the suggestions of other authors about the involvement of secondary metabolites in stomatal closure or osmotic imbalance [16,46]. These changes might be related with the reduction of carbon compounds storage in roots under water stress [47]. However, when the normal metabolism of plants was only altered by pathogen infection, the responses detected in roots [14,48] could be considered as evidence of a set of physiological and morphological changes which led to a recovery of our seedlings after 15 days, time in which the first infection cycle is completed [14].

The effects of *P. cinnamomi* root rot are not homogeneous throughout the root system; they depend on root diameter distribution [49] and other factors such as availability of inoculum. In infected plants, some functional roots could still be active, or new fine roots could grow as a response to root uptake reduction [45], taking up water and preventing total stomatal closure in well-irrigated plants. Previous work with *Q. ilex* described the maintenance of root growth independently of soil moisture and the ability to increase the growth of smaller-diameter fractions, leading to a tendency towards a thinning of roots, as a response to water deficiency in the plant [50]. All this evidence might explain the recovery of plants receiving the "Inoculation" treatment in our experiment at 15 dpi. According to our physiological data (*A*, *Gs*, and *QY*), the alterations of plant physiology produced by the "Inoculation" treatment were uncoupled from water stress; this indicates that root rot by itself was not enough to seriously disturb the functionality of the vascular system in our infected and well-watered seedlings, at least up to 30 dpi.

Biomass allocation

The root volume and biomass increments of *Q. ilex* seedlings in response to water stress are consistent with other results [51,52], increasing total root biomass and root mass fraction. These changes in response to drought are considered an adaptative response of drought tolerant plants [22,24,27], such as holm oak [21]. Hence, if we consider that root rot due to *P. cinnamomi* infection reduces water uptake as a consequence of fine roots loss [11,16], a response to water stress similar to the one triggered by drought could be expected in inoculated plants—increasing fine root turnover and root biomass in secondary roots. But, if no water stress signaling was triggered, the fine root "rot/growth" rate would have altered this response in inoculated plants.

Detailed evaluation of the aboveground biomass showed a decrease in the proportion of stem, relative to the overall aboveground biomass, and a lack of differences in leaf biomass, confirmed by the effects of the combined treatment ("I × D") (Figure 4b,c). Evaluating the plant fractions, it might be considered that the main effects of the treatments were observed in roots—the aboveground biomass not being sensitive to the effects of either treatment, except in the case of stem growth decline, which agreed with the lower *SMF* in "I × D" plants. Jönsson [53] reached a similar conclusion for *Quercus robur* L. infected with *Phytophthora quercina* T.Jung & T.I. Burgess and *Phytophthora cactorum* (Lebert & Cohn) J. Schröt. since the aboveground biomass showed no significant response to the fine root loss produced.

One of the main symptoms associated with the root rot caused by *P. cinnamomi* is the lack of lateral fine roots [13,14,49,54–57], as described in this work. However, our data do not show a significant reduction of fine roots or effects on related traits in the inoculation treatments ("Inoculation" and "I × D"). One of the main objectives of our work was to assess biomass allocation differences in plants, since we recovered all the fine roots found in the substrate. We considered that 30 days is too short a time for the significant disappearance of excised or rotted roots in the pot substrate, neither with controlled watering (without flooding) nor without watering. Nevertheless, root rot symptoms and lateral root excisions were clearly identified in the "Inoculation" treatment, which might explain the significant increment in the number of tips in coarse roots (roots of >2 mm Ø, data not shown).

The Pi is frequently calculated to assess differences in the tolerance of stress factors between species or phenotypes, and to indicate the variability of different root traits in limiting conditions [27,37]. In our case, Pi might explain some of the differences in the root changes, agreeing with the idea of the reduction of the plant's ability to explore the soil due to *P. cinnamomi* infection. The "Inoculation" treatment gave lower Pi values, statistically different from the rest, with the maximum value corresponding to "Drought" (Figure 4d). High plasticity levels are correlated with tolerance of stress factors; Bongers et al. [22] found, for functional traits correlated with a drier climate and stress conditions, that greater phenotypic plasticity was related to traits associated with rapid recovery and growth after drought. Stress caused by *P. cinnamomi* inoculation provoked lower plasticity of root traits than water stress, agreeing with the high susceptibility of *Q. ilex* to *P. cinnamomi* [12] and the high tolerance of this species to hydric stress [42]. Thus, it can be hypothesized that, without additional stressors, the strategy of our seedlings in the face of root rot was to reduce the biomass increment and reallocate resources to the equilibration of osmotic and hormonal imbalances, enabling them to recover their physiological status after the first infection cycle. On the other hand, changes in physiology and decreases in root plasticity caused by root infection could reduce the drought tolerance of *Q. ilex* plants, this being another possible cause of tree decline.

5. Conclusions

This work shows that the responses of *Q. ilex* to *P. cinnamomi* infection and water stress are different. *P. cinnamomi* root rot altered mainly the growth patterns of plants, while the plants could not recover from the physiological effects of infection only when the root rot coincided with water stress. It must be considered that the experiment showed the response of plants growing in ideal conditions, subjected to acute stress, being possible that plants adapted to drought in field conditions respond in a different way. In addition, no long-term impact in plant survival could be deduced from our results. However, we demonstrated the existence of underlying mechanisms of plant responses different to the one that they show against water stress, which could drive the plant recovery after one cycle of infection.

The differing responses of the roots to drought and infection were reflected in the early reduction of photosynthetic activity and relative changes in biomass allocation under water deprivation; the effects of the pathogen, without additional stress, focused on the reduction of plant growth and of the ability of the root system to explore the substrate, confirmed by the low Pi.

The infection of *Q. ilex* seedlings by *P. cinnamomi* was not enough, in this case, to kill the plants or to cause permanent damage to their physiological status. The differing degrees of susceptibility among provenances [29] should be considered as one of the causes of this observation, but, doubtless, the pathogen aggravated the consequences of hydric stress, since the results provide no evidence to support the induction of acute water stress by root rot.

Author Contributions: F.J.R.G., R.M.N.C. and A.P.-d.-L. designed the study; F.J.R.G., J.L.Q. and R.S.-C. performed the experiment and laboratory analysis; F.J.R.G., R.M.N.C. and A.P.L. conducted the statistical analysis; R.M.N.C. and J.L.Q. obtained the funds for the research; F.J.R.G., A.P.L., R.M.N.C., R.S.-C. and J.L.Q. participated in the writing and editing of the manuscript.

Acknowledgments: The authors thank the "Consejería de Medio Ambiente, Junta de Andalucía" for its support in acorn collection, and the "Campus de Excelencia Internacional en Agroalimentaión-CeiA3" through the "Servicio Centralizado de Apoyo a la Investigación-SCAI" (University of Córdoba) for their services.

References

1. De Rigo, D.; Cadullo, G. Quercus ilex in Europe: Distribution, habitat, usage and threats. In *European Atlas of Forest Tree Species*; Publication Office of the European Union: Luxembourg, 2016; pp. 152–153.
2. Quercus, L.; Gil-Pelegrín, E.; Peguero-Pina, J.J.; Sancho-Knapik, D. *Oaks Physiological Ecology. Exploring the Functional Diversity of Genus*; Tree Physiology; Springer International Publishing: Cham, Switzerland, 2017; Volume 7, ISBN 978-3-319-69098-8.
3. Guzmán Álvarez, J.R. The image of a tamed landscape: Dehesa through history in Spain. *Cult. Hist. Digit. J.* **2016**, *5*, 003. [CrossRef]
4. Pinto-Correia, T.; Azeda, C. Public policies creating tensions in Montado management models: Insights from farmers' representations. *Land Use Policy* **2017**, *64*, 76–82. [CrossRef]
5. Thomas, F.M.; Blank, R.; Hartmann, G. Abiotic and biotic factors and their interactions as causes of oak decline in Central Europe. *For. Pathol.* **2002**, *32*, 277–307. [CrossRef]
6. Consejería de Agricultura y Pesca. *Caracterización Socioeconómica de la Dehesa en Andalucía*; Secretaría General Técnica, Servicio de Publicaciones y Divulgación; Junta de Andalucía Consejería de Agricultura y Pesca: Sevilla, Spain, 2008.
7. Gea-Izquierdo, G.; Fernández-de-Uña, L.; Cañellas, I. Growth projections reveal local vulnerability of Mediterranean oaks with rising temperatures. *For. Ecol. Manag.* **2013**, *305*, 282–293. [CrossRef]
8. Swiecki, T.J.; Berndhart, E.A. Testing and implementing methods for managing *Phytophthora* root diseases in California native habitats and restoration sites. In Proceedings of the Sudden Oak Death Sixth Science Symposium, San Francisco, CA, USA, 20–23 June 2016; p. 53.

9. Brasier, C.M.; Robredo, F.; Ferraz, J.F.P. Evidence for *Phytophthora cinnamomi* involvement in Iberian oak decline. *Plant Pathol.* **1993**, *42*, 140–145. [CrossRef]
10. Brasier, C.M. *Phytophthora cinnamomi* and oak decline in southern Europe. Environmental constraints including climate change. *Ann. Sci. For.* **1996**, *53*, 347–358. [CrossRef]
11. Hardham, A.R.; Blackman, L.M. *Phytophthora cinnamomi*. *Mol. Plant. Pathol.* **2018**, *19*, 260–285. [CrossRef] [PubMed]
12. Moralejo, E.; García-Muñoz, J.A.; Descals, E. Susceptibility of Iberian trees to *Phytophthora ramorum* and *P. cinnamomi*. *Plant Pathol.* **2009**, *58*, 271–283. [CrossRef]
13. Ruiz Gómez, F.J.; Sanchez-Cuesta, R.; Navarro-Cerrillo, R.M.; Perez-de-Luque, A. A method to quantify infection and colonization of holm oak (*Quercus ilex*) roots by *Phytophthora cinnamomi*. *Plant Methods* **2012**, *8*, 39. [CrossRef] [PubMed]
14. Ruiz Gómez, F.J.; Navarro-Cerrillo, R.M.; Sánchez-Cuesta, R.; Pérez-de-Luque, A. Histopathology of infection and colonization of *Quercus ilex* fine roots by *Phytophthora cinnamomi*. *Plant Pathol.* **2015**, *64*, 605–616. [CrossRef]
15. Corcobado, T.; Cubera, E.; Juárez, E.; Moreno, G.; Solla, A. Drought events determine performance of *Quercus ilex* seedlings and increase their susceptibility to *Phytophthora cinnamomi*. *Agric. For. Meteorol.* **2014**, *192–193*, 1–8. [CrossRef]
16. Oßwald, W.; Fleischmann, F.; Rigling, D.; Coelho, A.C.; Cravador, A.; Diez, J.; Dalio, R.J.; Horta Jung, M.; Pfanz, H.; Robin, C.; et al. Strategies of attack and defence in woody plant–*Phytophthora* interactions. *For. Pathol.* **2014**, *44*, 169–190. [CrossRef]
17. Sghaier-Hammami, B.; Valero-Galvàn, J.; Romero-Rodríguez, M.C.; Navarro-Cerrillo, R.M.; Abdelly, C.; Jorrín-Novo, J. Physiological and proteomics analyses of Holm oak (*Quercus ilex* subsp. ballota [Desf.] Samp.) responses to Phytophthora cinnamomi. *Plant Physiol. Biochem.* **2013**, *71*, 191–202. [CrossRef] [PubMed]
18. Corcobado, T.; Cubera, E.; Moreno, G.; Solla, A. *Quercus ilex* forests are influenced by annual variations in water table, soil water deficit and fine root loss caused by *Phytophthora cinnamomi*. *Agric. For. Meteorol.* **2013**, *169*, 92–99. [CrossRef]
19. León, I.; García, J.; Fernández, M.; Vázquez-Piqué, J.; Tapias, R. Differences in root growth of *Quercus ilex* and *Quercus suber* seedlings infected with *Phytophthora cinnamomi*. *Silva Fenn.* **2017**, *51*. [CrossRef]
20. Maurel, M.; Robin, C.; Capron, G.; Desprez-Loustau, M.-L. Effects of root damage associated with *Phytophthora cinnamomi* on water relations, biomass accumulation, mineral nutrition and vulnerability to water deficit of five oak and chestnut species. *For. Pathol.* **2001**, *31*, 353–369. [CrossRef]
21. Turco, E.; Close, T.J.; Fenton, R.D.; Ragazzi, A. Synthesis of dehydrin-like proteins in *Quercus ilex* L. and *Quercus cerris* L. seedlings subjected to water stress and infection with *Phytophthora cinnamomi*. *Physiol. Mol. Plant Pathol.* **2004**, *65*, 137–144. [CrossRef]
22. Bongers, F.J.; Olmo, M.; Lopez-Iglesias, B.; Anten, N.P.R.; Villar, R. Drought responses, phenotypic plasticity and survival of Mediterranean species in two different microclimatic sites. *Plant Biol.* **2017**, *19*, 386–395. [CrossRef] [PubMed]
23. Poorter, H.; Jagodzinski, A.M.; Ruiz-Peinado, R.; Kuyah, S.; Luo, Y.; Oleksyn, J.; Usoltsev, V.A.; Buckley, T.N.; Reich, P.B.; Sack, L. How does biomass distribution change with size and differ among species? An analysis for 1200 plant species from five continents. *New Phytol.* **2015**, *208*, 736–749. [CrossRef] [PubMed]
24. Poorter, H.; Niklas, K.J.; Reich, P.B.; Oleksyn, J.; Poot, P.; Mommer, L. Biomass allocation to leaves, stems and roots: Meta-analyses of interspecific variation and environmental control: Tansley review. *New Phytol.* **2012**, *193*, 30–50. [CrossRef] [PubMed]
25. Chen, H.; Dong, Y.; Xu, T.; Wang, Y.; Wang, H.; Duan, B. Root order-dependent seasonal dynamics in the carbon and nitrogen chemistry of poplar fine roots. *New For.* **2017**, *48*, 587–607. [CrossRef]
26. Lopez-Iglesias, B.; Villar, R.; Poorter, L. Functional traits predict drought performance and distribution of Mediterranean woody species. *Acta Oecol.* **2014**, *56*, 10–18. [CrossRef]
27. Olmo, M.; Lopez-Iglesias, B.; Villar, R. Drought changes the structure and elemental composition of very fine roots in seedlings of ten woody tree species. Implications for a drier climate. *Plant Soil* **2014**, *384*, 113–129. [CrossRef]
28. Roumet, C.; Birouste, M.; Picon-Cochard, C.; Ghestem, M.; Osman, N.; Vrignon-Brenas, S.; Cao, K.; Stokes, A. Root structure–function relationships in 74 species: Evidence of a root economics spectrum related to carbon economy. *New Phytol.* **2016**, *210*, 815–826. [CrossRef] [PubMed]

29. Navarro Cerrillo, R.M.; Ruiz Gómez, F.J.; Cabrera-Puerto, R.J.; Sánchez-Cuesta, R.; Palacios Rodriguez, G.; Quero Pérez, J.L. Growth and physiological sapling responses of eleven *Quercus ilex* ecotypes under identical environmental conditions. *For. Ecol. Manag.* **2018**, *415*, 58–69. [CrossRef]
30. Quero, J.L.; Villar, R.; Marañón, T.; Zamora, R. Interactions of drought and shade effects on seedlings of four *Quercus* species: Physiological and structural leaf responses. *New Phytol.* **2006**, *170*, 819–834. [CrossRef] [PubMed]
31. Scholander, P.F.; Bradstreet, E.D.; Hemmingsen, E.A.; Hammel, H.T. Sap pressure in vascular plants: Negative hydrostatic pressure can be measured in plants. *Science* **1965**, *148*, 339–346. [CrossRef] [PubMed]
32. Tognetti, R.; Longobucco, A.; Miglietta, F.; Raschi, A. Transpiration and stomatal behaviour of *Quercus ilex* plants during the summer in a Mediterranean carbon dioxide spring. *Plant Cell Environ.* **1998**, *21*, 613–622. [CrossRef]
33. Schneider, C.A.; Rasband, W.S.; Eliceiri, K.W. NIH Image to ImageJ: 25 Years of Image Analysis. Available online: https://www.nature.com/articles/nmeth.2089 (accessed on 14 March 2018).
34. Phillips, J.M.; Hayman, D.S. Improved procedures for clearing roots and staining parasitic and vesicular-arbuscular mycorrhizal fungi for rapid assessment of infection. *Trans. Br. Mycol. Soc.* **1970**, *55*, 158–161. [CrossRef]
35. Jeffers, S.N. Comparison of two media selective for *Phytophthora* and *Pythium* species. *Plant Dis.* **1986**, *70*, 1038–1043. [CrossRef]
36. Erwin, D.C.; Ribeiro, O.K. *Phytophthora Diseases Worldwide*; APS Press: St. Paul, NY, USA, 1996; ISBN 978-0-89054-212-5.
37. Valladares, F.; Sánchez-Gómez, D. Ecophysiological traits associated with drought in Mediterranean tree seedlings: Individual responses versus interspecific trends in eleven species. *Plant Biol.* **2006**, *8*, 688–697. [CrossRef] [PubMed]
38. Milton, J.S.; Tsokos, J.O. *Statistical Methods in the Biological and Health Sciences*; McGraw-Hill: New York, NY, USA, 1983; ISBN 978-0-07-042359-6.
39. Canny, M. Tyloses and the maintenance of transpiration. *Ann. Bot.* **1997**, *80*, 565–570. [CrossRef]
40. Borghetti, M.; Grace, J.; Raschi, A. Water transport in plants under climatic stress. In Proceedings of the International Workshop, Firenze, Italy, 29–31 May 1990.
41. Urban, L.; Aarrouf, J.; Bidel, L.P.R. Assessing the effects of water deficit on photosynthesis using parameters derived from measurements of leaf gas exchange and of chlorophyll a fluorescence. *Front. Plant Sci.* **2017**, *8*, 2068. [CrossRef] [PubMed]
42. Quero, J.L.; Sterck, F.J.; Martínez-Vilalta, J.; Villar, R. Water-use strategies of six co-existing Mediterranean woody species during a summer drought. *Oecologia* **2011**, *166*, 45–57. [CrossRef] [PubMed]
43. Aigbe, S.O.; Remison, S.U. The Influence of root rot on dry matter partition of three cassava cultivars planted in different agro-ecological environments. *Asian J. Plant Pathol.* **2010**, *4*, 82–89. [CrossRef]
44. Else, M.A.; Janowiak, F.; Atkinson, C.J.; Jackson, M.B. Root signals and stomatal closure in relation to photosynthesis, chlorophyll a fluorescence and adventitious rooting of flooded tomato plants. *Ann. Bot.* **2009**, *103*, 313–323. [CrossRef] [PubMed]
45. Tuset, J.J.; Hinarejos, C.; Mira, J.L.; Cobos, J.M. Implicación de *Phytophthora cinnamomi* Rands en la enfermedad de la «seca» de encinas y alcornoques. *Bol. Sanid. Veg. Plagas* **1996**, *22*, 491–499.
46. Robin, C.; Capron, G.; Desprez-Loustau, M.L. Root infection by *Phytophthora cinnamomi* in seedlings of three oak species. *Plant Pathol.* **2001**, *50*, 708–716. [CrossRef]
47. Simova-Stoilova, L.P.; Romero-Rodríguez, M.C.; Sánchez-Lucas, R.; Navarro-Cerrillo, R.M.; Medina-Aunon, J.A.; Jorrín-Novo, J.V. 2-DE proteomics analysis of drought treated seedlings of *Quercus ilex* supports a root active strategy for metabolic adaptation in response to water shortage. *Front. Plant Sci.* **2015**, *6*, 627. [CrossRef] [PubMed]
48. Redondo, M.A.; Perez-Sierra, A.; Abad-Campos, P.; Torres, L.; Solla, A.; Reig-Arminana, J.; Garcia-Breijo, F. Histology of *Quercus ilex* roots during infection by *Phytophthora cinnamomi*. *Trees-Struct. Funct.* **2015**, *29*, 1943–1957. [CrossRef]
49. Blaschke, H. Decline symptoms on roots of Quercus robur. *For. Pathol.* **1994**, *24*, 386–398. [CrossRef]
50. Manes, F.; Vitale, M.; Donato, E.; Giannini, M.; Puppi, G. Different ability of three Mediterranean oak species to tolerate progressive water stress. *Photosynthetica* **2006**, *44*, 387–393. [CrossRef]

51. Leiva, M.J.; Fernández-Alés, R. Variability in seedling water status during drought within a *Quercus ilex* subsp. *ballota population, and its relation to seedling morphology. For. Ecol. Manag.* **1998**, *111*, 147–156. [CrossRef]
52. Villar-Salvador, P.; Planelles, R.; Enriquez, E.; Rubira, J.P. Nursery cultivation regimes, plant functional attributes, and field performance relationships in the Mediterranean oak *Quercus ilex* L. *For. Ecol. Manag.* **2004**, *196*, 257–266. [CrossRef]
53. Jönsson, U. *Phytophthora* species and oak decline—Can a weak competitor cause significant root damage in a nonsterilized acidic forest soil? *New Phytol.* **2004**, *162*, 211–222. [CrossRef]
54. Cahill, D. Cellular and histological changes induced by *Phytophthora cinnamomi* in a group of plant species ranging from fully susceptible to fully resistant. *Phytopathology* **1989**, *79*, 417–424. [CrossRef]
55. Sánchez, M.; Caetano, P.; Ferraz, J.; Trapero, A. *Phytophthora* disease of *Quercus ilex* in south-western Spain. *For. Pathol.* **2002**, *32*, 5–18. [CrossRef]
56. Sánchez, M.E.; Andicoberry, S.; Trapero, A. Pathogenicity of three *Phytophthora* spp. causing late seedling rot of *Quercus ilex* ssp. *ballota*. *For. Pathol.* **2005**, *35*, 115–125. [CrossRef]
57. Vettraino, A.M.; Belisario, A.; Maccaroni, M.; Vannini, A. Evaluation of root damage to English walnut caused by five *Phytophthora* species. *Plant Pathol.* **2003**, *52*, 491–495. [CrossRef]

PERMISSIONS

All chapters in this book were first published in MDPI; hereby published with permission under the Creative Commons Attribution License or equivalent. Every chapter published in this book has been scrutinized by our experts. Their significance has been extensively debated. The topics covered herein carry significant findings which will fuel the growth of the discipline. They may even be implemented as practical applications or may be referred to as a beginning point for another development.

The contributors of this book come from diverse backgrounds, making this book a truly international effort. This book will bring forth new frontiers with its revolutionizing research information and detailed analysis of the nascent developments around the world.

We would like to thank all the contributing authors for lending their expertise to make the book truly unique. They have played a crucial role in the development of this book. Without their invaluable contributions this book wouldn't have been possible. They have made vital efforts to compile up to date information on the varied aspects of this subject to make this book a valuable addition to the collection of many professionals and students.

This book was conceptualized with the vision of imparting up-to-date information and advanced data in this field. To ensure the same, a matchless editorial board was set up. Every individual on the board went through rigorous rounds of assessment to prove their worth. After which they invested a large part of their time researching and compiling the most relevant data for our readers.

The editorial board has been involved in producing this book since its inception. They have spent rigorous hours researching and exploring the diverse topics which have resulted in the successful publishing of this book. They have passed on their knowledge of decades through this book. To expedite this challenging task, the publisher supported the team at every step. A small team of assistant editors was also appointed to further simplify the editing procedure and attain best results for the readers.

Apart from the editorial board, the designing team has also invested a significant amount of their time in understanding the subject and creating the most relevant covers. They scrutinized every image to scout for the most suitable representation of the subject and create an appropriate cover for the book.

The publishing team has been an ardent support to the editorial, designing and production team. Their endless efforts to recruit the best for this project, has resulted in the accomplishment of this book. They are a veteran in the field of academics and their pool of knowledge is as vast as their experience in printing. Their expertise and guidance has proved useful at every step. Their uncompromising quality standards have made this book an exceptional effort. Their encouragement from time to time has been an inspiration for everyone.

The publisher and the editorial board hope that this book will prove to be a valuable piece of knowledge for researchers, students, practitioners and scholars across the globe.

LIST OF CONTRIBUTORS

Marco Pautasso
Animal and Plant Health Unit, European Food Safety Authority (EFSA), 43126 Parma, Italy
Forest Pathology & Dendrology, Institute of Integrative Biology (IBZ), ETH Zurich, 8092 Zurich, Switzerland

Linda Rigerte, Kathrin Blumenstein and Eeva Terhonen
Forest Pathology Research Group, Büsgen-Institute, Department of Forest Botany and Tree Physiology, Faculty of Forest Sciences and Forest Ecology, University of Göttingen, Büsgenweg 2, 37077 Göttingen, Germany

Richard C. Cobb
Department of Plant Pathology, University of California, Davis, One Shields Ave, Davis, CA 95616, USA

Margaret R. Metz
Department of Biology, Lewis & Clark College, 0615 S.W. Palatine Hill Road MSC 53, Portland, OR 97219, USA

Beatriz Mora-Sala, Paloma Abad-Campos and Mónica Berbegal
Instituto Agroforestal Mediterráneo, Universitat Politècnica de València, Camino de Vera s/n, 46022 Valencia, Spain

David Gramaje
Instituto de Ciencias de la Vid y del Vino, Consejo Superior de Investigaciones Científicas-Universidad de la Rioja-Gobierno de La Rioja, Ctra. de Burgos km. 6, 26007 Logroño, Spain

Federico La Spada, Antonella Pane, Maria Evoli, Roberto Faedda, Cinzia Rizza, Ivana Puglisi and Santa Olga Cacciola
Department of Agriculture, Food and Environment (Di3A), University of Catania, Via Santa Sofia 100, 95123 Catania, Italy

Francesco Aloi
Department of Agriculture, Food and Environment (Di3A), University of Catania, Via Santa Sofia 100, 95123 Catania, Italy
Department of Agricultural, Food and Forest Sciences, University of Palermo, Viale delle Scienze Ed., 4, 90128 Palermo, Italy

Gaetano Magnano di San Lio and Leonardo Schena
Dipartimento di Agraria, Mediterranean University of Reggio Calabria, località Feo di Vito, 89122 Reggio Calabria, Italy

Tomáš Kudláček
Phytophthora Research Centre, Mendel University in Brno, Zemědělská 1, 61300 Brno, Czech Republic

Tamara Corcobado
Phytophthora Research Centre, Mendel University in Brno, Zemědělská 1, 61300 Brno, Czech Republic
Federal Research and Training Centre for Forests, Natural Hazards and Landscape, Unit of Phytopathology, Department of Forest Protection, Seckendorff-Gudent-Weg 8, 1131 Vienna, Austria

Thomas L. Cech, Martin Brandstetter, Andreas Daxer and Christine Hüttler
Federal Research and Training Centre for Forests, Natural Hazards and Landscape, Unit of Phytopathology, Department of Forest Protection, Seckendorff-Gudent-Weg 8, 1131 Vienna, Austria

Marília Horta Jung and Thomas Jung
Phytophthora Research Centre, Mendel University in Brno, Zemědělská 1, 61300 Brno, Czech Republic
Phytophthora Research and Consultancy, Am Rain 9, 83131 Nussdorf, Germany

Kamyar Aram and David M. Rizzo
Department of Plant Pathology, University of California, One Shields Drive, Davis, CA 95616, USA

Xiaoxue Xu, Wenxia Huai and Wenxia Zhao
The Key Laboratory of National Forestry and Grassland Administration on Forest Protection, Research Institute of Forest Ecology Environment and Protection, Chinese Academy of Forestry, Beijing 100091, China

Hamiti
Forest Bureau of Emin County, Xinjiang Uygur Autonomous Region, Emin 834600, China

Xuechao Zhang
Institute of Agricultural Sciences of Ili Kazakh Autonomous Prefecture, Yining 835000, China

Clizia Sechi and Salvatore Seddaiu
Settore Difesa Delle Piante Forestali, Agris Sardegna, Via Limbara 9, 07029 Tempio Pausania (SS), Italy

Andrea Brandano
Dipartimento di Agraria, Sezione di Patologia Vegetale ed Entomologia, Università degli Studi di Sassari, Viale Italia 39, 07100 Sassari, Italy

Pino Angelo Ruiu
Settore Sughericoltura e Selvicoltura, Agris Sardegna, Via Limbara 9, 07029 Tempio Pausania (SS), Italy

Bruno Scanu
Nucleo Ricerca Desertificazione, Università degli Studi di Sassari, Viale Italia 39, 07100 Sassari, Italy
Dipartimento di Agraria, Sezione di Patologia vegetale ed Entomologia (SPaVE), Università degli Studi di Sassari, Viale Italia 39, 07100 Sassari, Italy

Rafael Sánchez-Cuesta, Rafael M. Navarro-Cerrillo, José L. Quero and Francisco J. Ruiz-Gómez
Department of Forest Engineering, Laboratory of Dendrochronology, Silviculture and Global Change-DendrodatLab-ERSAF, University of Córdoba, Campus de Rabanales, Ctra. N-IV, km. 396, E-14071 Córdoba, Spain

Ivan Milenković
Phytophthora Research Centre, Mendel University in Brno, Zemědělská 1, 61300 Brno, Czech Republic
Faculty of Forestry, University of Belgrade, Kneza Višeslava 1, 11030 Belgrade, Serbia

Nenad Keča and Dragan Karadžić
Faculty of Forestry, University of Belgrade, Kneza Višeslava 1, 11030 Belgrade, Serbia

Zlatan Radulović
Institute of Forestry, Belgrade, Kneza Višeslava 3, 11030 Belgrade, Serbia

Justyna A. Nowakowska
Faculty of Biology and Environmental Sciences, Cardinal Stefan Wyszynski University in Warsaw, Wóycickiego 1/3 Street, 01-938 Warsaw, Poland

Tomasz Oszako
Faculty of Forestry, Białystok University of Technology, Marszałka J. Piłsudskiego 1A, 17-200 Hajnówka, Poland

Katarzyna Sikora
Forest Research Institute-IBL, Braci Leśnej 3, 05-090 Raszyn, Poland

Michal Tomšovský, Aneta Bačová and Milica Raco
Phytophthora Research Centre, Mendel University in Brno, 61300 Brno, Czech Republic

Clive M. Brasier, Ana Pérez-Sierra and Joan Webber
Forest Research, Alice Holt Lodge, Farnham, Surrey GU10 4LH, UK

Matěj Pánek
Phytophthora Research Centre, Mendel University in Brno, 61300 Brno, Czech Republic
Crop Research Institute, Drnovská 507/73, 16106 Prague 6, Czech Republic

József Bakonyi
Plant Protection Institute, Centre for Agricultural Research, Herman Ottó út 15, 1022 Budapest, Hungary

Cristiana Maia
Centre of Marine Sciences (CCMAR), University of Algarve, 8005-139 Faro, Portugal

Helen Rees
Forest Research, Alice Holt Lodge, Farnham, Surrey GU10 4LH, UK
School of Life Sciences, University of Bristol, 24 Tyndall Avenue, Bristol BS8 1TQ, UK

Francisco J. Ruiz Gómez and Rafael M. Navarro Cerrillo
Departamento de Ingeniería Forestal, Laboratorio de Ecofisiología de Sistemas Forestales ECSIFOR – ERSAF, Universidad de Córdoba, Campus de Rabanales, Crta, IV, km. 396, E-14071 Córdoba, Spain

Alejandro Pérez-de-Luque
Área de Genómica y Biotecnología, IFAPA, Centro Alameda del Obispo, Avda. Menéndez Pidal s/n, Apdo 3092, 14080 Córdoba, Spain

Index

www.ingramcontent.com/pod-product-compliance
Lightning Source LLC
LaVergne TN
LVHW080948210926
848485LV00004B/804